FERMENTATION PROCESS DEVELOPMENT IN
INDUSTRIAL ORGANISMS
(Bioprocess Technology Series/4)

Edited by: Justin O. Neway

Published: 1989
344 pages, bound, illustrated

marcel dekker, inc.
270 MADISON AVENUE • NEW YORK, NY 10016 • 212-696-9000

Fermentation Process Development of Industrial Organisms

Bioprocess Technology

Series Editor

W. Courtney McGregor

Xoma Corporation
Berkeley, California

Volume 1 Membrane Separations in Biotechnology, *edited by W. Courtney McGregor*

Volume 2 Commercial Production of Monoclonal Antibodies: A Guide for Scale-Up, *edited by Sally S. Seaver*

Volume 3 Handbook on Anaerobic Fermentations, *edited by Larry E. Erickson and Daniel Yee-Chak Fung*

Volume 4 Fermentation Process Development of Industrial Organisms, *edited by Justin O. Neway*

Additional Volumes in Preparation

Fermentation Process Development of Industrial Organisms

edited by

JUSTIN O. NEWAY
Cetus Corporation
Emeryville, California

MARCEL DEKKER, INC. New York and Basel

Library of Congress Cataloging-in-Publication Data

Fermentation process development of industrial organisms
/ edited by Justin O. Neway
p. cm. -- (Bioprocess technology ; v. 4)
Includes bibliographies and index.
ISBN 0-8247-7917-7 (alk. paper)
1. Microbial biotechnology. 2. Animal cell biotechnology.
3. Fermentation. I. Neway, Justin O. II. Series.
TP248.27.M53F47 1989
660'.62--dc20 89-34741
CIP

This book is printed on acid-free paper.

MARCEL DEKKER, INC.
270 Madison Avenue, New York, New York 10016

Current printing (last digit):
10 9 8 7 6 5 4 3 2 1

PRINTED IN THE UNITED STATES OF AMERICA

Series Introduction

The revolutionary developments in recombinant DNA and hybridoma technologies that began in the mid-1970s have helped to spawn several hundred new business enterprises. Not all these companies are aimed at producing gene products or cell products, as such. Many are supportive in nature; that is, they provide contract research, processing equipment, and various other services in support of companies that actually produce cell products. With time, some small companies will probably drop out or be absorbed by larger, more established firms. Others will mature and manufacture their own product lines. As this evolution takes place, an explosive synergism among the various industries and the universities will result in the conversion of laboratory science into industrial processing. Such a movement, necessarily profit driven, will result in many benefits to humanity.

New bioprocessing techniques will be developed and more conventional ones will be revised because of the influence of the new biotechnology. As bioprocess technology evolves, there will be a need to provide substantive documentation of the developments for those who follow the field. It is expected that the technologies will continue to develop rapidly, just as the life sciences have developed rapidly over the past 10-15 years. No single book will cover all of these developments adequately. Indeed, any single book will be in need of replacement or revision every few years. Therefore, our continuing series in this rapidly moving field will document the growth of bioprocess technology as it happens.

The numerous cell products already in the marketplace, and the others expected to arrive, in most cases come from three types of bioreactors: (a) classical fermentation; (b) cell culture technology; and (c) enzyme bioreactors. Common to the production of all cell products or cell product analogs will be bioprocess control, downstream processing (separation and purification), and bioproduct finishing and formulation. These major branches of bioprocess technology will be represented by cornerstone books, even though they may not appear first. Other subbranches will appear, and over time, the bioprocess technology "tree" will take shape and continue growing by natural selection.

W. Courtney McGregor

Preface

Fermentation processes have been operating at the industrial scale for many years. Once the bench-scale description is available, it has been the traditional role of the engineer to scale the process up to the industrial level. Much has been written about the quantitative aspects of this task, and there is a consequent paucity of information, at least in the literature, about the biological aspects of the particular organism involved. This volume presents material useful to those interested in manipulating the biology of organisms to produce economic levels of industrial products of the appropriate quality for downstream processing. This book is directed at those rare individuals who are neither all engineer nor all biologist, but who work in the middle ground between the geneticists and expression laboratories, on the one hand, and the production facilities, on the other.

In the past, fermentation process economics have dictated the use of low-cost, but nutritionally complex, waste streams as feedstocks. This, along with our relatively poor understanding of the physiology and regulation of industrially important metabolic pathways, has created a significant "magic" content in the physiological control of fermentation processes. The advent of "recombinant" technologies is changing this situation for the better. It has now become possible to genetically alter industrial organisms in ways not thought possible just a few years ago. Modern fermentation development thus includes an increasingly higher degree of physiological control, due to these improvements in our understanding and ability to control the metabolic pathways involved.

When properly folded pharmaceutical proteins with modified amino acids or glycosylation, or both, are needed, or if monoclonal antibodies are the product, then the use of mammalian cells is indicated. This has turned mammalian cells into industrial organisms, although fermentation development in this area is in its infancy in comparison to most other fermentation processes. The media used often contain a considerable undefined component, and the fermentation physiology of secretion, posttranslational modifications, and glycosylation are poorly understood. Because many important pharmaceutical proteins can be made using mammalian cells, development of these fermentation processes is a significant challenge receiving considerable attention.

The chapters presented here are arranged by organism rather than by process. The authors have gathered material that illustrates both the diversity and the similarity of industrial processes based on a particular organism. In addition to the more traditional fermentation processes, such as those for antibiotics and chemicals, there are sections dealing with more modern industrial processes based on the application of contemporary genetic techniques. In each case, the intent is to present material that shows the interactive nature of the art and science of process development research and the role of the individuals who mediate between the abilities of the geneticists and the needs of the production engineers. The areas of medium development, control of growth and expression, control of product quality and harvestability, and definition of biological process parameters for scaleup, all require considerable biological instinct and engineering support. They are the subject of the chapters in this book.

Justin O. Neway

Contents

Contributors

DAVID R. BERRY Department of Bioscience and Biotechnology, University of Strathclyde, Glasgow, Scotland

JANET P. DeWITT Fermentation Development, Abbott Laboratories, North Chicago, Illinois

RICHARD P. ELANDER Fermentation Research and Development, Bristol-Myers Company, Industrial Division, Syracuse, New York

JAY V. JACKSON Fermentation Development, Abbott Laboratories, North Chicago, Illinois

ANTHONY S. LUBINIECKI* Cell Culture Research and Development, Genentech, Inc., South San Francisco, California

SHIGERU NAKAMORI Central Research Laboratories, Ajinomoto Co., Inc., Yokohama, Japan

THOMAS J. PAULUS Fermentation Development, Abbott Laboratories, North Chicago, Illinois

FERGUS G. PRIEST Department of Brewing and Biological Sciences, Heriot-Watt University, Edinburgh, Scotland

---

**Present affiliation:* Biopharmaceutical Manufacturing and Development, Smith, Kline & French Laboratories, King of Prussia, Pennsylvania.

C. L. PROPST Affiliated Scientific, Inc., Ingleside, Illinois

RICHARD J. SHARP Microbiology and Fermentation Research Group, Division of Biotechnology, Public Health Laboratory Service, Centre for Applied Microbiology and Research, Porton Down, Salisbury, England

ISAMU SHIIO Central Research Laboratories, Ajinomoto Co., Inc., Yokohama, Japan

RANDALL J. VON WEDEL CytoCulture International, Inc., San Francisco, California

Fermentation Process Development of Industrial Organisms

1

Actinomycetes

JANET P. DeWITT, JAY V. JACKSON, and THOMAS J. PAULUS
Abbott Laboratories, North Chicago, Illinois

I. INTRODUCTION

The actinomycetes are a diverse group of filamentous bacteria, many of which display extraordinary biosynthetic capabilities. An example of this biosynthetic diversity can be seen in their ability to produce antibiotics. Actinomycetes produce two-thirds of the known antibiotics (Berdy, 1974). *Streptomyces* spp. alone produce 2769 of the 4973 naturally occurring antibiotics reported as of 1978 (Berdy, 1980). It is because of these unique biosynthetic properties that actinomycetes have been of industrial interest for several decades.

Although the primary use of actinomyces has been the production of antibiotics, other economically important uses include steroid transformations (Miller, 1985) and glucose isomerase production (Chen, 1980). Other actinomycete metabolites are being developed to be used as insecticides, herbicides, and anticancer agents. The use of *Streptomyces* as a host for production of recombinant DNA products is being explored.

In this chapter we try to present an overview of the various aspects of process development for actinomycete fermentation products. We have usually relied on antibiotic-related studies because of their important industrial applications and because in most cases they are the only ones available. In-depth discussions of any one aspect of process development have been avoided, and only a brief summary of the subject and current literature has been reviewed. References for more complete discussions on a particular subject are provided in each section.

II. STRAIN IMPROVEMENT

Strain improvement, in combination with medium development, has been one of the primary means by which the yields of industrial fermentation products have been increased. An example of these types of improvements can be seen in Fig. 1. In this example, yields of antimycin A were increased from 620 to 7800 μg/ml by a combination of mutation, natural selections, and media improvement. In addition to increasing yields, strain improvement programs have been used to adapt strains to more economical or desirable fermentation substrates or to eliminate the production of undesirable products in the fermentation.

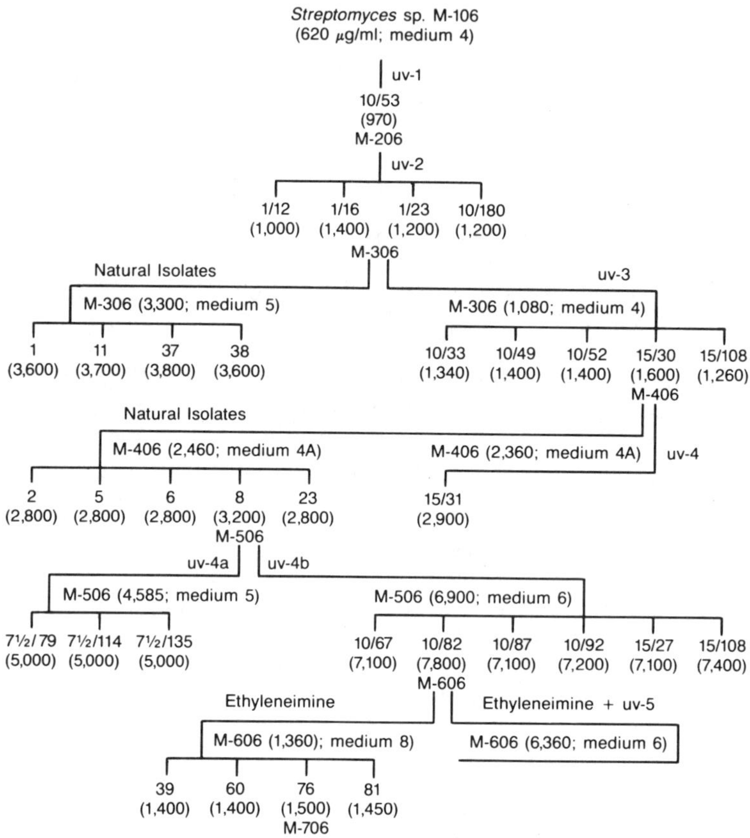

Figure 1 Family of antimycin A-producing strains of *Streptomyces* sp. M-106. Yields in milligrams per liter are given in parentheses. (From Vezina et al., 1976; reprinted from the *Journal of Antibiotics.*)

In this section, various aspects of the strain improvement process are discussed. Strain improvement programs have been carried out for over 30 years and involve three basic steps: (1) generation of genetic variation in the microbial population; (2) selection of individuals from the population, and (3) testing of individuals for increased yields or other desired properties. A number of general reviews on strain improvement methods have been written (see Elander and Chang, 1979; Rowlands, 1984a,b), and only an overview of the strain improvement process as it applies to actinomycetes is presented here. Reviews on the genetics of *Streptomyces*, the best characterized actinomycete, may also be useful in the development of a strain improvement program (Hopwood, 1983; Hopwood and Chater, 1984; Hopwood and Merrick, 1977; Hopwood et al., 1973).

A. Genetic Variation

1. Mutation

Improvement of most industrial strains has been accomplished by mutation (Holt and Sanders, 1986). Recent advances in the molecular biology of actinomycetes have increased the number of methods by which genetic variation can be generated, but mutation is still the most important means of quickly and efficiently creating genetic variation in a population. Some information about the mechanisms of mutagenesis in actinomycetes is known. Clark and Hopwood (1976) observed that mutation in *Streptomyces coelicolor* follows two-hit kinetics, suggesting the existence of an inducible, error-prone repair system. More recent work on mutagenesis in *Streptomyces fradiae* has shown mutagenesis to occur via an error-prone DNA repair system similar to that of *Escherichia coli* (Stonesifer and Baltz, 1985). Very little induced mutation was found to be caused by direct mispairing, suggesting that streptomycetes may be more evolved than *E. coli* and may more closely resemble *Saccharomyces cerevisiae*, a lower eukaryote (Stonesifer and Baltz, 1985). Isolation of repair-deficient mutants of this organism would also suggest that streptomycetes may have two error-prone DNA repair pathways or a more complex error-prone repair pathway than is found in *E. coli* (Baltz and Stonesifer, 1985). See Baltz (1986b) for a review of mutagenic mechanisms in streptomycetes.

An understanding of the mechanisms of mutagenesis is not needed, however, to select mutagens to modify a given strain. Since it is difficult to predict what type of mutation at the molecular level is needed to improve a strain, a series of mutagens of complementary types should be tested (Rowlands, 1984a). The importance of mutagen selection was recognized early in the history of strain improvement (Alikhanian, 1962). The optimum dose forming other types of mutants or for causing optimum kill is often not the optimum for generating improved strains. This emphasizes that mutagens must be optimized not only for kill, but also for their effect on the traits being selected.

Many physical and chemical agents useful in mutating other microorganisms have been successfully used in actinomycetes. These include shortwave ultraviolet light (Hopwood and Sermonti, 1962), x-rays (Mracek et al., 1969), *N*'-methyl-*N*'-nitro-*N*'-notrosoguanidine (NTG) (Delic et al., 1970), 8-methoxypsoralen and near ultraviolet light (Townsend et al., 1971), ethyl methanesulfonate, methyl methanesulfonate, and 4-nitroquinoline 1-oxide (Stonesifer and Baltz, 1985). Conditions of mutagensis, however, may be specific for the actinomycetes. For example, mutagenesis with NTG is usually most effective at pH 9.0, in contrast to the pH ranges in the acidic region that are usually used with other bacteria (Delic et al., 1970). Hopwood et al. (1973) and Baltz (1986a) give general procedures for mutagenesis of both sporulating and nonsporulating actinomycetes.

2. Recombination

In addition to mutation, genetic variation can also be generated by recombination of existing genetic material from two or more pre-existing strains. Genetic recombination can be used to combine characteristics from divergent strains derived from the same parent by several rounds of mutagenesis and selection or from strains from related species. For example, a high-yielding strain can be recombined with an ancestral strain to eliminate negative mutations introduced by repeated rounds of mutation. Several mechanisms for recombination are known among the actinomycetes. Recombination via conjugation is widespread (Hopwood, 1976; see Hopwood and Merrick, 1977) and, when studied, has been found to be mediated by plasmid sex factors (Bibb et al., 1977; Friend et al., 1978; Hopwood et al., 1973; Puglia et al., 1973; DeWitt, 1985). Recombination via phage transduction has been found to occur in a few streptomycetes, such as *Streptomyces venezuelae* (Studdard, 1979), *S. hygroscopcus* (Suss and Klaus, 1981) and *S. fradiae* (Chung and Molvar, 1982). Classic transformation of competent mycelia has only been reported for one actinomycete, *Thermoactinomyces vulgaris* (Hopwood and Wright, 1972).

There are several major drawbacks to using classic recombination systems to generate genetic variation. Transduction is limited to a few species and usually involves transfer of only small regions of the chromosome. Conjugation, although transferring larger regions of the chromosome, usually occurs at frequencies of 10^{-5}-10^{-6} (Hopwood et al., 1977). At these low frequencies strains must carry genetic markers, such as auxotrophy or drug resistances for selection following matings. Adding genetic markers is both time consuming and

Table 1 Protoplast Fusion Systems in Actinomycetes

Organism	Reference
Streptomyces acrimycini	Hopwood et al., 1977
S. albus	Baltz and Matsushimi, 1981
S. ambofaciens	Baltz and Matsushimi, 1981
S. antibioticus	Ochi et al., 1979
S. azureus	Ogata et al., 1985
S. bikiniensis	Godfrey et al., 1978
S. chrysomallus	Keller et al., 1983
S. coelicolor	Hopwood et al., 1977
S. fradiae	Baltz, 1978; Godfrey et al., 1978; Ikeda et al., 1983
S. griseofuscus	Baltz and Matsushimi, 1981
S. griseus	Hopwood et al., 1977; Kuklin et al., 1983
S. griseus subsp. *cryophilus*	Kitano et al., 1985
S. hygroscopicus	Ogawa et al., 1983
S. incarnatus	Malina et al., 1985
S. lactamdurans	Wesseling and Lago, 1981
S. lavendulae	Nakano et al., 1982
S. lividans	Hopwood et al., 1977
S. narbonensis	Ikeda et al., 1983
S. parvulus	Hopwood et al., 1977; Ochi et al., 1979
S. rimosus	Hranueli et al., 1983
S. setonii	Petty and Crawford, 1984
S. tenjimariensis	Yamashita et al., 1985
S. viridosporus	Petty and Crawford, 1984
S. sp. AM4900	Ikeda et al., 1983
Streptosporangium viridogriseum	Oh et al., 1980
Micromonospora rosaria	Ryu et al., 1983
Micromonospora	Szvoboda et al., 1980
Actinoplanes brasiliensis	Palleroni, 1983

often detrimental to the organism. Also, genetic recombination does not occur naturally in all actinomycetes.

Protoplast fusion can be used to overcome some of these problems. This technique involves removing the bacterial cell wall with lysozyme, fusion of the resulting bacterial protoplasts using a fusogenic agent, such as polyethylene glycol, and then regeneration of the cell walls (Hopwood et al., 1977; Baltz, 1978). During the fusion process two or more protoplasts can form hybrids and recombination can occur (Hopwood and Wright, 1978). Table 1 lists some of the protoplast fusion systems that have been developed for actinomycetes. Recombination frequencies are increased 100-fold to greater than 10,000-fold over naturally occurring frequencies. Frequencies of up to 31% have been reported for some strains (Ochi et al., 1979). These high levels of recombination make feasible the use of unmarked strains. Recombination has also been reported to occur in strains for which there is no known naturally occurring mechanism of recombination (Baltz, 1978) or between species that do not normally recombine (Godfrey et al., 1978). Protoplast fusion offers an additional advantage, greater heterogeneity among recombinants. In *S. coelicolor*, chromosomal exchange was found to occur over wider regions of the chromosome and more double crossover events were observed when protoplast fusion was compared with conjugation (Hopwood et al., 1983a). Additional uses for protoplasts are in the transformation by plasmid and phage DNA or chromosomal DNA entrapped in liposomes (Bibb et al., 1978; Krugel et al., 1980; Suarez and Chater, 1980a; Makins and Holt, 1981).

General reviews on the applications of protoplast fusion to strain improvement can be found in Hopwood (1981) and Hopwood and Merrick (1977). Protoplast fusion has been applied successfully to improve several actinomycetes. Yields of cephamycin C were increased 10-15% by protoplast fusion of two recently diverged *Nocardia lactamdurans* strains (Wesseling and Lago, 1981). Lignin degradation in *Streptomyces* spp. was increased by fusing lignin degrading streptomycetes, *Streptomyces viridosporus* T7a and *Streptomyces setonii* 75Vi2 (Petty and Crawford, 1984). The *Streptomyces griseus* subsp. *cryophilus*, protoplast fusion between sulfate transport-negative mutants and a positive parent gave higher producers of sulfated carbapenam antibiotic C-19393 S2 than did reversion of the sulfate transport-negative mutant, suggesting that more than simple reintroduction of the sulfate transport system occurred (Kitano et al., 1985).

It appears that in some cases the simple regeneration of protoplasts can effect yields of an antibiotic. Protoplast formation followed by regeneration, without fusion of the protoplasts, resulted in a severalfold increase in sinefungin production in *Streptomyces incarnatus* NRRL 8089 (Malina et al., 1985). Improvement of a macrolide-producing streptomycete has also been reported after protoplast regeneration (Ikeda et al., 1983). This phenomenon may be due to changes in the chromosome, such as amplifications and deletions, or the loss of plasmids that has been reported to occur following protoplast regeneration in several streptomycetes (Aguilar and Hopwood, 1982; Robinson et al., 1981; Fishman and Hershberger, 1983).

The primary difficulty with protoplast fusion is in the regeneration of the protoplasts. A regeneration system must be optimized for the species of interest. Some even closely related strains do not regenerate efficiently under the same set of conditions, thus complicating fusion experiments and necessitating extensive systems development, which can impede the progress of fusion programs.

3. Recombinant DNA

The newest means of creating genetic variation in actinomycetes is by the method of gene cloning. Gene cloning was first described in streptomycetes in 1980 (Bibb et al., 1980; Suarez and Chater, 1980b; Thompson et al., 1980). Progress in this field has been rapid, and a laboratory manual is now available that describes a wide range of procedures for the isolation and analysis of DNA and RNA from *Streptomyces* spp. (Hopwood et al., 1985). Additional information on plasmid and phage vectors can also be found in Hopwood et al. (1987).

The potential industrial applications of gene cloning have been reviewed (Chater et al., 1982; Hopwood et al., 1983b) and include its use in the amplification of genes for rate-limiting steps, manipulation of cloned genes (e.g., by in vitro mutagenesis or alteration of regulatory sequences), and introduction of these genes into independently improved strains. Numerous examples of the application of gene cloning to antibiotic yield improvement with examples from tylosin biosynthesis can be found in Cox et al. (1987). Other examples include using insertion mutation cloning (Chater and Bruton, 1985) of vector KC575 into a putative repressor region of methylenomycin, causing overproduction of methylenomycin, and cloning of the gene for aminoglycoside 6'-*N*-acetyltransferase, an antibiotic resistance gene, on a high copy vector in *Streptomyces kanamyceticus*, a kanamycin producer, and *S. fradiae*, a neomycin producer, substantially increasing the production of kanamycin and neomycin (Cameri and Davies, 1986).

B. Random and Rational Selections

Once genetic variation has been created in the population, individuals must be selected from that population for screening. Selection can be random or can be based on a physiological, biochemical, or morphological characteristic. The latter of these selections has been referred to as rational or directed selection (Elander and Chang, 1979). Random selection is the most straightforward method of selection; no prior knowledge of the biochemistry of the organism or synthesis of the product is required. Individuals are selected at random and tested, usually in shake flasks, for product formation. This method, although effective, is time consuming but can be improved using miniaturization, automation, and microprocessor technology (see later discussion).

Rational selection is based on known or proposed biochemical and/or genetic pathways. Rational selection often selects individuals for traits other than increases in product yield. Selections are made to alter metabolic controls, relieve feedback inhibition, change permeability, or reduce toxicity to the product being formed. The advantage of these types of selections is that they can often be designed to screen large numbers of isolates rapidly, reducing the number of isolates to be screened by shake flask fermentations. Rational selections allow selection for desired properties (e.g., derepression by phosphate), and some knowledge about the type of mutation being made is known. This can aid the scaleup or exploitation of the strain. The primary disadvantage of most rational selections is that it is not known a priori whether a given mutation will necessarily increase yields, so in this respect rational selections are still indirect and can only be used as prescreening prior to testing for product yield. Most strain improvement programs are a combination of random selections and well-designed rational selections.

The best examples of rational selection have been reported by Chang and Elander (1979) for the improvement of cephalosporin C production by *Cephalosporium acremonium*, but several applications of these techniques have been reported for actinomycetes. Many antibiotic-producing actinomycetes are sensitive to the antibiotic they produce, especially during early stages of growth. Lincomycin production by *Micromonospora* spp. (Ivanitskoya et al., 1980), chlortetracycline production by *Streptomyces aureofaciens* (Valerianova et al., 1981; Kitagiri, 1954), and olivomycin production by *Streptomyces olivoreticuli* (L'vova and Lapchinskaya, 1981) have been increased by selecting mutants with increased levels of resistance to the antibiotic they produce. This technique would also presumably be effective for nonantibiotic products that can potentially build to toxic levels in the producing organism. The reversion of mutations causing the loss of antibiotic production has also resulted in increases in antibiotic production (Dulaney and Dulaney, 1967; Unowsky and Hoppe, 1978). Second-site mutations are thought to occur that not only restore antibiotic production but also affect expression of the gene or possibly affect the catalytic properties or stability of the gene product.

Mutations that affect precursor formation or regulation of primary and/or secondary metabolism are candidates for rational selections. Mutations causing overproduction of amino acids that serve as precursors of some antibiotics have resulted in increases in product formation. These mutations can be isolated by selecting resistance to amino acid analogs that are normally toxic to the organism (Godfrey, 1973). Mutants escape the toxic effect of the analog by overproducing the amino acids and diluting the analog. A second class of mutations that are isolated with this selection are transport mutants, which escape toxicity by not transporting the analog. These mutants may not be as useful and are often harmful since they limit nutrient uptake. Mutants overproducing amino acids, vitamins, or bases can also be obtained by reverting auxotrophic mutations, the mechanism being similar to that noted for the reversion of antibiotic nonproducers.

Relief of carbon, nitrogen, or phosphate repression of primary and/or secondary metabolism can impact on product formation directly or can allow the use of substrates that might normally cause a depression of product formation. This is particularly important in industrial fermentations in which raw materials are often complex agricultural products containing fluctuating amounts of carbon, nitrogen, or phosphate. Phosphate derepression can be relieved by direct selection for the production of product in the presence of high levels of phosphate. Phosphate repression of candicidin production was relieved using an agar-piece method that allowed the rapid screening of large numbers of isolates (Martín et al., 1979). Changes in phosphate regulation can also be obtained by selecting mutants resistant to phosphate analogs, such as arsenate or vanadate. In *E. coli* and fungi, resistance to these compounds results from changes in the phosphate uptake mechanisms. Increases in streptothricin yields were obtained in some strains using arsenate resistance as a selection (Friedrich et al., 1984). Relief of carbon repression can be obtained in some cases by the selection of mutants resistant to 2-deoxyglucose, a glucose analog. Derepression of the glycerol operon was obtained in *S. coelicolor* by the selection of mutants able to grow on glycerol in the presence of 2-deoxyglucose (Hodgson, 1982). These mutants were found to be deficient in glucokinase activity (Seno and Chater, 1983).

C. Screening

The final stage of strain development is the screening of strains for the desired property, usually an increase in product yield. Strains must be screened whether they were randomly selected or selected using rational selection techniques. Agar medium can be used as a preliminary screen if the product is diffusible and an assay can be developed for activity in a zone around the colony. Ichikawa et al. (1971) have used this technique, in combination with media development, to improve kasugamycin yields to 7 g/liter. The radically different environment of a semisolid surface versus a submerged culture can, however, lead to many artifacts. Better correlations to submerged culture can be obtained if agar is made of fermentation-type medium and corrections are made for colony biomass (Trilli et al., 1982).

Automated screens are an alternative for evaluating large numbers of isolates. These screens avoid the problems associated with solid surfaces and are applicable to products that cannot be easily assayed on an agar surface. Automated screens are reviewed by Queener and Lively (1986) and rely on miniaturization and mechanization. Conventional screens for strain improvement are done in shake flasks. Davies (1964), Dahl et al. (1972), and Calam (1970) present strategies for designing these screens. Generally, isolates are tested in single replication and isolates that appear superior in this test are retested with replication to identify truly superior strains.

A key element of the screening program, irrespective of the system being used, is the screening medium. Approaches to developing a screening medium are somewhat dependent on whether a large-scale fermentation process is available. If no large-scale tank process is available, a screening medium must be developed to reflect a potential, commercially viable process. Considerations are raw materials, temperature, and reasonable inoculum development that are applicable to production scale. If a fermentation process is already available, then it may be more difficult to develop the screening medium. The medium must reflect the current process. This is particularly difficult with a highly developed, mature process. Generally, the same or similar raw materials in ratios similar to those found in the production process are used. The medium is usually lower in concentration because of lower oxygen transfer. Many factors are difficult if not impossible to reproduce in flasks or smaller vessels (e.g., oxygen transfer, shear, hydrostatic head, which affects dissolved gas levels, and autoclaving conditions) or are difficult to simulate (e.g., pH control or feeds). Some, but not all, of these problems can be overcome by the use of small stirred vessels for primary screening, but this severely limits the number of isolates that can be tested. Because of the difficulty in simulating tank conditions on a small scale, it is important to routinely test putative improved strains under production conditions.

D. Genetic Instability

Genetic instability of actinomycetes used in industrial processes is a well-known phenomenon (Reusser, 1963). In many streptomycetes, aerial mycelium formation, pigment production, and antibiotic production are lost spontaneously (Gregory and Huang, 1964; Okanishi, 1979) or following treatment with plasmid-curing agents (see Hopwood, 1978). This latter result led to speculation that genes for these traits were carried on extrachromosomal elements. Rigorous evidence for this was not obtained however, and only the genes for methylenomycin production by *S. coelicolor* have been shown to be plasmid borne (Kirby and Hopwood, 1977; Kirby et al., 1975). Rather, it appears that the genomes of many streptomycetes are highly plastic. Changes in the structure of the chromosomes, many of which affect secondary metabolites and differentiated functions, have been found to occur spontaneously and after manipulation of the organism.

Robinson et al. (1981) were the first to note that DNA isolated from strains produced by protoplast fusion contained a tandemly repeated 6 kb pair sequence that constituted 15-18% of the total DNA. Since this initial observation, a number of streptomycetes exhibiting genetic instability have been found to form variants containing highly repeated sequences of DNA. In most cases these sequences were found in single or only a few copies in the parent strain (Hasegawa et al., 1985; Ono et al., 1982; Fishman and Hershberger, 1983; Altenbuchner and Cullum, 1985). The appearance of a reiterated sequence was observed in a strain of *S. fradiae* that had lost the ability to produce tylosin following protoplast fusion (Fishman and Hershberger, 1983). *Streptomyces glaucescens* exhibit a high degree of genetic instability, with mutants arising at high frequency that have lost one or several properties: melanin formation, hydroxystreptomycin production, resistance to streptomycin group antibiotics, laminarinase production, or chitinase excretion (Suter et al., 1978; Hutter et al., 1981). Analysis of DNA from spontaneous mutants and mutants derived from ethidium bromide treatment revealed that these strains contained tandemly reiterated DNA sequences within the chromosome (Ono et al., 1982; Hasegawa et al., 1985). Amplified sequences represented up to 45% of the total genomic DNA and repeated units were 2.9-35 kb in length, no two sequences being identical (Hasegawa et al., 1985). In *Streptomyces reticuli*, the loss of melanin production was also associated with amplified sequences and also with the loss of the structural gene for tyrosinase, the enzyme catalyzing melanin formation (Schrempf, 1983). Changes in the chromosome can affect properties other than the production of secondary metabolites. Spontaneous chloramphenicol-sensitive mutants of *Streptomyces lividans* 66, which occurred at a frequency of approximately 1%, gave rise to arginine auxotrophs at a frequency of up to 25% (Altenbuchner and Cullum, 1984). The Arg mutants contained an amplified 5.7 kb DNA sequence and also contained a deletion to one side of the amplified element (Altenbuchner and Cullum, 1985). Although not all genetic instabilities can be explained by the presence of amplified DNA, this chromosomal change may be an example of the types of chromosomal aberrations common in streptomycetes. For a consideration of the effects of genetic instability on scaleup, see Sec. IV.D.

III. ASPECTS OF MEDIUM DEVELOPMENT

A. Raw Materials

Generally, a suitable fermentation medium for shake flasks does not need to be designed from scratch because the existing medium (or media) used in discovery is usually available. However, a research microbiologist approaching industrial fermentations for the first time will be impressed immediately with the crudeness of the medium components used on a large scale. Reagent-grade chemicals purchased in small quantities and of high purity are seldom used; instead, bulk agricultural products of a relatively, but not completely, undefined nature are the rule. Cost and availability are the major factors. Medium components can contribute up to 80% of the total cost of a fermentation (Kalk and Langlykke, 1986). A complete review of fermentation raw materials is not warranted here. Miller and Churchill (1986) have written an excellent discussion of the substrates used in large-scale fermentations for a variety of products. Also of great value to the industrial microbiologist is the *Trader's Guide to Fermentation Media Formulation* (Zabriskie et al., 1980).

Naturally, the construction of the medium is dependent on the type of product or process. In many actinomycete antibiotic fermentations, a long production phase is highly desirable, making complex, insoluble carbon and nitrogen sources preferable to more refined, soluble components. Oils (vegetable, fish, and lard) are often used in conjunction with carbohydrates; carbohydrate provides readily available carbon and energy for quick and extensive production of biomass, whereas oil provides the long-term carbon and energy for product synthesis. Oils are relatively easy to handle, do not generally cause pH problems, and avoid the potential metabolic effects of glucose feeds. Oils have the added benefit of foam control.

The most difficult aspect of antibiotic fermentation medium development is balancing the carbon, nitrogen, and phosphorus ratios. Most antibiotic fermentations are either nitrogen or phosphorus limited during the production phase (see later), and there may be a critical level for one of these elements. The crude nature of the raw materials is a disadvantage in this situation. Nitrogen sources, such as soybean meal, corn gluten meal, or yeast, naming only a few, also contain substantial amounts of phosphate compounds. The total phosphorus content can be determined by chemical methods, but the chemical nature and availability of the phosphorus are not as readily measured. Moreover, because these nutrients are bulk agricultural products, the composition can change from one lot to the next or from one vendor to another in significant yet elusive ways.

Actinomycete fermentation for products other than antibiotics are less dependent on medium composition (Miller and Churchill, 1986). Extensive and reproducible growth are generally the main considerations. In steroid transformations, some adjustments to the medium can be made to accommodate the relative polarities of the precursors or products of the transformation. Steroid transformations can also require the addition of inducers (see Sec. III. B.4). Glucose isomerase production requires growth on the appropriate sugar-xylose (Chen, 1980). Other enzyme fermentations are similar and obvious; for example, protease production is greatly stimulated by protein as the carbon and/or nitrogen source. In most situations, a knowledge of the physiology of the organism and the metabolic regulation of the process is a valuable aid to medium development.

B. Physiology and Regulation

Actinomycetes used in antibiotic fermentations are primarily soil bacteria widely distributed in nature. Numerous reviews and entire books have been devoted to the biology of the actinomycetes. *The Actinomycetes*, by Waksman (1959), is still a useful resource providing background information on the basic phenomenology of growth and metabolism. Other excellent recent resources are the reviews, Hirsch and McCann-McCormick (1985), Lechevalier and Lechevalier (1985), and the books, *Nocardia and Streptomyces* (Modarski et al., Eds.), 1978, and *Actinomycetes* (Schaal and Pulverer, Eds.), 1981.

1. Actinomycete Biology

For the fermentation microbiologist, the most important aspects of actinomyces biology are the differentiated life cycle and the formation of mycelia in submerged culture. Germination of the small immotile spore occurs by hyphal outgrowth. Vegetative growth on solid media is characterized by elongation and branching of the hyphae to produce a compact colony with a smooth moist surface. This is the substrate mycelium. Sporulation occurs, presumably in response to nutrient limitation, with the formation of aerial mycelia. Aerial mycelia have an additional wall layer, called a sheath, that makes them slightly thicker than the substrate mycelia. During vegetative growth, the colony is largely coenocytic with few crosswalls formed. Spores form on the aerial mycelia by fragmentation of the hyphal tip. Not all actinomycetes produce aerial mycelia, however, and spore formation is variable. Taxonomic classification of actinomyces by morphology alone is difficult. Chemical composition has been found to be a useful taxonomic device (Lechevalier and Lechevalier, 1980). In submerged culture, streptomycetes form pellets with branched hyphae emanating from a growth center (point of germination). The size of the mycelial pellet is shear sensitive and medium dependent, but in general, they are small in comparison with the pellets formed by the filamentous fungi. Certain strains of *Streptoverticillium*, *Nocardia*, and *Micromonospora* have a propensity to fragment in liquid culture with hyphal tips sealing off to appear almost coccoid.

Streptomycetes generally do not sporulate in submerged culture. However, Kendrick and Ensign (1983) have been able to obtain sporulation of *Streptomyces viridiochromogenes* in liquid medium. Microcycle sporulation was demonstrated for the same organism by Koepsel and Ensign (1984). In both cases the culture was allowed to germinate in complex medium and then transferred to a minimal medium containing only arginine as the carbon and nitrogen source.

2. Actinomycete Physiology

Actinomycetes are, in general, not fastidious organisms. Most species can grow in minimal salts medium with glucose; however, the growth of some species can be significantly enhanced by the addition of thiamine. A variety of sugars can be utilized by actinomycetes, with acid production under some conditions (Waksman, 1959). The acids formed (acetic, pyruvic, and lactic) can be utilized via the tricarboxylic acid cycle (TCA cycle) (Waksman, 1959; Surowitz and Pfister, 1985). Some studies on the mechanism of sugar utilization have been performed. Sabater and Asensio (1973) were unable to detect a phosphoenolpyruvate phosphotransferase system (PTS) for active sugar transport in *Streptomyces violaceoruber*. Glucose, mannose, and fructose were each transported by specific, constitutive transport system, and the phosphorylation of these sugars occurred via specific kinases (Sabater et al., 1972a, b). Novotna and Hostalek (1985) also concluded that a PTS was not operative in *S. aureofaciens*.

Phosphoenolpyruvate-dependent phosphorylation of glucose and fructose was detected. However, the kinetics of glucose-6-phosphate formation suggested that the phosphorylation was indirect, and only fructose-6-phosphate was formed from fructose, suggesting the activity of a hexokinase, not PTS. *Nocardia* and *Mycobacterium* also appear to lack the PTS (Saier, 1977). Thus actinomycetes appear to resemble other strict aerobes that do not possess the PTS; carbon regulation in these organisms does not appear to involve cyclic nucleotides.

Hodgson (1982) has demonstrated that *S. coelicolor* possesses constitutive active transport systems for glucose, galactose, and fructose and inducible active transport systems for arabinose and glycerol. However, a process resembling carbon catabolite regulation appears to be operative. Glucose represses the synthesis of the arabinose and glycerol transport systems and inhibits the uptake of galactose, fructose, and glycerol. Mutants resistant to the glucose analog, 2-deoxyglucose, were depressed for glycerol and arabinose transport. These mutants also produced derepressed levels of extracellular agarase (Hodgson, 1982). Seno and Chater (1983) demonstrated that the enzymes of glycerol degradation are glucose repressed and that this repression was absent from deoxyglucose-resistant (DOG-r) strains. Although several types of DOG-r strains were isolated by Hodgson (1982), the characteristic of most of them is low glucokinase activity. The glucokinase gene and DNA from the glycerol operon from *S. coelicolor* have been cloned, and the molecular mechanisms behind carbon catabolite regulation should be forthcoming (Ikeda et al., 1984; Seno et al., 1984).

Nitrate and ammonia are readily assimilated by actinomycetes. Nitrate utilization is marked by a rapid alkalinization of the medium by the formation of ammonia. Rapid ammonia utilization is often accompanied by acidification of the medium. Consequently, a mixture of ammonia and nitrate is often used. Amino acid and protein utilization also lead to ammonia accumulation and high pH through the activity of deaminases.

The regulation of nitrogen metabolism in *Streptomyces* is only beginning to be understood. The glutamine synthetase of *Streptomyces cattleya* is regulated by adenylation and deadenylation in a manner similar to that in other bacteria (Streicher and Tyler, 1981). Glutamine synthetase activity declined following the addition of high levels of ammonia to *S. cattleya*; however, the major mode of regulation appears to be at the transcriptional level (Wax et al., 1982).

Not much is known regarding the regulation of phosphate metabolism in actinomycetes. Except for an apparent lack of a PTS, detailed investigations of the biochemistry of phosphate acquisition are not yet available. Martín (1977) suggests that phosphate and carbohydrate metabolism may be tightly associated. This assertion seems reasonable considering the important metabolic positions of such compounds as glucose-6-phosphate and ATP. Cyclic AMP and other highly phosphorylated nucleotides have been detected in *Streptomyces* (Gersch et al., 1979). The metabolic roles for these compounds are not well defined. Riesenberg et al. (1984) have presented evidence that (p)ppGpp is probably involved in a stringent response in *Streptomyces hydroscopicus* similar to that found in *E. coli.* The pool size of (p)ppGpp is 100-fold lower in *S. hygroscopicus* than it is in *E. coli*, however. An interesting report demonstrated that decoyinine, an inhibitor of GMP synthetase, decreased GTP levels and initiated aerial mycleia formation in a *Streptomyces* sp. MA406-a-1 (formycin producing) (Ochi, 1986). *Streptomyces* may be similar to *Bacillus* in the control of differentiation (Freese, 1981; Ochi, 1987).

Evidence is accumulating that sulfur metabolism in actinomycetes may be more similar to that in fungi than in other bacteria. These studies often surround the production of β-lactam antibiotics. The enzyme L-cystathionine-γ-lyase, which cleaves L-cystathionine to L-cysteine, α-ketobutyrate, and ammonia, was found to be widely distributed in actinomycetes but not detectable in other bacteria (Nagasawa et al., 1984). Thus, actinomycetes possess the reverse transsulfuration pathway commonly found in fungi (Delavier-Klutchko and Flavin, 1965). Enteric bacteria contain L-cystathionine-β-lyase (products: homocysteine, pyruvate, and ammonia). Cystathionine-γ-lyase provides cysteine for thienamycin production in *S. cattleya* (Williamson et al., 1985) and probably also for cephamycin C synthesis in *N. lactamdurans* (Kern and Inamine, 1981).

Actinomycetes are well known for the ability to degrade a large range of biopolymers through the activity of extracellular enzymes (Waksman, 1959). The polymers degraded include starch, protein, chitin, cellulose, hemicellulose, lignin, pectin, and peptidoglycan. The streptomycetes are probably second only to the genus *Bacillus* in protease production. Most of these enzymes are probably inducible (Iwamoto et al., 1983; Fairbairn et al., 1986). Extracellular amylase from *Streptomyces limosus* was inducible by maltose and starch substrates (Fairbairn et al., 1986). Amylase from *S. erythreus* appears to be regulated in a similar manner (Paulus and DeWitt, unpublished observations). *Nocardia* and other nonstreptomycete actinomycetes are well known for the ability to degrade a wide range of hydrocarbons (Cain, 1981; Lechevalier and Lechevalier, 1985). The ability to perform biotransformations of steroids may be part of this biodegradative ability.

3. Antibiotic Production

Secondary Metabolism. Secondary metabolism has been studied extensively in microorganisms (Campbell, 1984; Demain et al., 1983). Despite the amount of effort, the principles governing the control of secondary metabolism are not well understood. This is due, in part, to the difficulty in defining secondary metabolism and secondary metabolites. For the fermentation scientist, a *working* definition can be stated simply: A secondary metabolite is a compound produced by the organism that is not required for growth. Most compounds generally thought of as secondary metabolites, including all antibiotics, fall within this definition. Distinctions between production during the growth phase and production after growth is completed need not be made as, in practice, these are only operational problems.

Sporulation and its associated products may also be considered part of secondary metabolism. The concept of sporulation as a secondary metabolic process may be useful in the study of the control of antibiotic synthesis. Historically, correlations have been made between spore formation, protease production, and antibiotic synthesis (Hanlon and Hodges, 1981; Gibb and Strohl, 1987). The diversity of products and processes commonly considered part of secondary metabolism suggest that a single control mechanism does not exist. More likely, there are levels of control, some shared between processes and some unique to a given process, operating simultaneously.

An empirical view of antibiotic synthesis indicates that the primary control factor is nutrient limitation to some degree. Antibiotics are often produced after rapid growth has stopped. The organism enters a phase of slower or no growth in which production of the antibiotic occurs. This period, sometimes called the idiophase, has nutritional and environmental requirements that can be significantly different from the requirements in the growth phase (trophophase). A well-studied example is the requirement for trace metals during secondary metabolism (Weinberg, 1970). For actinomycetes, the key trace metals during secondary metabolism are iron and zinc, whereas in bacilli, the critical metals are iron and manganese; secondary metabolism in fungi is most affected by zinc only. Trace metal requirements are determined and optimized empirically. If an increased level of a metal, such as iron, is required for production, it can usually be added to the medium prior to sterilization. This excess iron will most likely not adversely affect the growth phase.

The induction of secondary metabolism and differentiation by low-molecular-weight molecules has been reported in several instances (reviewed by Khokhlov, 1982). The best studied example is the stimulation of streptomycin biosynthesis by A factor [2(*S*)-isocapryloyl-3-(*R*)-hydroxymethyl-γ-butyrolactone] (Kleiner et al., 1976; Khokhlov and Tovarova, 1979). A factor is produced by streptomycin-producing *S. griseus* and *Streptomyces bikiniensis*, but not by *Streptomyces albus* or *Streptomyces mashuenis*, organisms that also produce streptomycin. Mutants of *S. griseus* unable to produce A factor cannot form streptomycin. The stimulation only occurs when A factor is added early in the fermentation (before 48 hr) and need not be present in the culture for extended time periods. A brief exposure, 3-4 min, is sufficient to produce the effect. A factor also affects morphological development. The mechanism of action of A factor is unknown; however, the effect appears to be a derepression of the streptomycin biosynthetic enzymes (Hara and Beppu, 1982).

The occurrence of A factorlike molecules may be widespread. Eritt et al. (1984) reported recently that 35 different species produced inducing factor(s) (L factors) for leukaemycin production and aerial mycelium formation by *S. griseus*. Even species of fungi and nonactinomycete bacteria produced inducerlike effects. Kawaguchi et al. (1984) described the induction of rifamycin production in a *Nocardia* sp. by B factor. B factor was isolated from Difco yeast extract and found to be 3'-(1-butylphosphoryl)adenosine. B factor was an effective stimulant of rifamycin biosynthesis at 10 ng/ml in a nonproducing mutant.

Although it is clear that A factor does affect antibiotic synthesis and differentiation in some streptomycin-producing organisms, the isolation of substances from ecologically and taxonomically distinct organisms (e.g., B factor from yeast) suggests a lack of specificity unexpected in a highly controlled process. It is possible that, in some cases, these chemicals are affecting gene expression in a general way that is more related to nonspecific chemistry of the affected genes rather than by a highly evolved, endogenous process of metabolic regulation. The industrial significance of inducing factors is not known.

Carbon Regulation. Carbon catabolite regulation of antibiotic synthesis is the process by which certain carbon sources, most notably glucose, suppress or delay antibiotic formation. Carbon catabolite regulation has been reviewed by Drew and Demain (1977), Martín and Demain (1980), and Demain et al. (1983). Although glucose is the most often cited regulating carbon source, other compounds are capable of the same effect. Novobiocin production by *Streptomyces niveus* is suppressed by citrate but not glucose (Kominek, 1972). Cycloserine production by *Streptomyces garyphalus* is subject to strong regulation by glycerol and to a lesser extent by glucose (Svensson et al., 1983). Glutamate regulation of tryptophan oxygenase, required for NAD and actinomycin formation in *Streptomyces parvulus*, has also been reported (Foster and Katz, 1981).

The mechanism of carbon catabolite regulation is, in many instances, not well understood and may not be uniform. In some cases, antibiotic biosynthetic enzymes are repressed. For actinomycin synthesis, glucose represses phenoxazinone synthetase, an enzyme involved in ring formation (Gallo and Katz, 1972). Glucose also appears to repress *N*-acetylkanamycin amidohydrolase required for kanamycin synthesis in *S. kanamyceticus* (Satoh et al., 1976). These cases resemble the glucose repression of alternative carbon source uptake and of the enzymes of glycerol utilization described in a previous section. A common mechanism for regulation of alternative carbon source utilization and antibiotic biosynthetic enzymes may exist in certain organisms. In many bacteria, cAMP is a positive effector in the induction of catabolic enzymes (Pastan and Adhya, 1976). Cyclic AMP was reported to reverse the glucose repression of *N*-acetylkanamycin amidohydrolase in *S. kanamyceticus.* Chatterjee and Vining (1982) have, however, published convincing evidence *against* the involvement of cAMP in the regulation of chloramphenicol synthesis and the induction of β-galactosidase in *S. venezuelae.* The lack of a PTS for carbohydrate uptake in actinomycetes and other strictly aerobic organisms correlates with the noninvolvement of cAMP in carbon regulation.

Other mechanisms for carbon catabolite regulation of antibiotic synthesis besides antibiotic enzyme repression are possible. Glycerol suppression of cycloserine production was attributed to inhibition of cycloserine synthetase by accumulation of glycerol metabolites, 3-phosphoglyceric acid and glyceraldehyde-3-phosphate. Apparent carbon catabolite regulation may also be due to processes regulating or related to the rate of growth (Martín and Demain, 1980).

Industrially, carbon catabolite regulation can be circumvented by genetic modification of the producing organism or by media and process design. The latter method is probably more common: Nonrepressing carbon sources, such as the insoluble starches or oils found in agricultural materials, can be used; alternatively, the slow feeding of glucose stimulates the formation of polyene macrolides (Martín and McDaniel, 1974).

Nitrogen Regulation. Nitrogen metabolite regulation of antibiotic production has been reviewed by Aharonowitz (1980) and Gräfe (1982). Sections on nitrogen metabolite regulation appear in Demain et al. (1983), Martín and Demain (1980), and Drew and Demain (1977). When it has been examined thoroughly, the mechanism of nitrogen metabolite regulation appears to involve the repression of enzymes necessary for antibiotic synthesis. Recently, Braña et al. (1985) demonstrated that ammonia repressed the formation of isopenicillin *N* synthetase and deacetoxycephalosporin C synthetase; these two enzymes are part of the cephalosporin pathway in *Streptomyces clavuligerus.* Ammonium repression of these two enzymes and, additionally, repression of isopenicillin *N* isomerase, was shown to occur in *Streptomyces lactamdurans* (Castro et al., 1985). Glutamine synthetase is also repressed by ammonia, and an inverse relationship has been demonstrated between glutamine synthetase activity and antibiotic production in *S. clavuligerus* (Aharonowitz, 1979; Aharonowitz and Demain, 1979). The role of glutamine synthetase in nitrogen metabolite regulation is not known. Conceivably the intracellular regulating metabolite is glutamine.

Ōmura and coworkers have shown that ammonia suppresses the amino acid catabolism required for tylosin synthesis by *S. fradiae* (Ōmura et al., 1983, 1984a,b; Masuma et al., 1983). Specifically, incorporation of [^{14}C]amino acids into tylosin aglycone was decreased by high levels of ammonium ion (Ōmura et al., 1984a,b); valine dehydrogenase, the first enzyme of valine catabolism, was repressed and inhibited by ammonia (Ōmura et al., 1983). Vu-Trong and Gray (1987) have shown that high levels of ammonium ion added just before the onset of tylosin synthesis suppressed tylosin biosynthetic enzymes.

Ammonium suppression of tylosin synthesis could be reduced by the addition of agents (tribasic magnesium phosphate, tribasic calcium phosphate, and natural zeolites) that bind ammonium ion and release it slowly to the organism (Masuma et al., 1983). This phenomenon appears to be widespread; ammonia repression of several macrolide antibiotics was reversed by ammonium trapping agents Ōmura et al., 1980, 1981). Moreover, the addition of tribasic magnesium phosphate prevented the ammonia repression of isopenicillin *N* synthetase and stimulated cephalosporin synthesis in *S. clavuligerus* (Braña et al., 1985).

Nitrogen metabolite regulation can also be circumvented by empirical optimization of the level and source of nitrogen. Insoluble protein, such as soybean meal, is often effective because of the slow breakdown during the fermentation process. Use of protein does necessitate the production of extracellular proteases by the organism. Caution must be exercised when interpreting the results of the direct addition of ammonium ion to a fermentation. Rapid ammonia utilization can lead to a drastic drop in the pH of the medium, which may confuse the effect of nitrogen excess.

Phosphate Regulation. Suppression of antibiotic biosynthesis by inorganic phosphate is a pervasive observation in the literature. Numerous, structurally unrelated products of secondary metabolism are suppressed by levels of phosphate that are well below phosphate levels inhibitory to growth. Listings of secondary metabolites and suppressing levels of phosphate have been compiled by Weinberg (1970) and Martín (1977).

Phosphate regulation of antibiotic synthesis has been thoroughly reviewed by Martín (1977). The possible mechanisms by which phosphate suppresses antibiotic biosynthesis, proposed and discussed by Martín (1977), are listed in Table 2. Despite the importance of phosphate metabolism to the production of antibiotics, it is not yet possible to disregard any of these hypotheses. The lack of understanding may be due, in part, to the critical role that phosphate and phosphate acquisition occupies in the physiology of microorganisms. The complicated genetic control mechanisms for phosphate acquisition elucidated for *E. coli* (Wanner, 1983) and *Neurospora* (Metzenberg, 1979) confirm this observation.

Table 2 Postulated Mechanisms of the Phosphate Effect on Antibiotic Biosynthesis

1. Phosphate favors primary metabolism and growth at the expense of secondary metabolism; macromolecular synthesis is stimulated by phosphate
2. Phosphate shifts carbohydrate catabolic pathways; phosphate favors glycolysis over the hexose monophosphate pathway
3. Phosphate represses antibiotic pathway enzymes or limits the synthesis of inducers
4. Phosphate suppresses the formation of antibiotic precursors
5. Phosphatases associated with antibiotic biosynthesis are repressed by phosphate

Source: Abstracted from Martín (1977).

The biosynthesis of the macrolide antibiotic tylosin is a good example of the complexity of phosphate regulation. The effect of phosphate on tylosin biosynthesis has been investigated in detail by two groups of workers (Madry et al., 1979; Madry and Pape, 1982; Vu-Trong et al., 1980).

Madry and Pape (1982) demonstrated that phosphate represses the formation of three tylosin biosynthetic enzymes. However, the total effect of phosphate on tylosin production was found to be more complicated and interwoven with glucose metabolism (Madry et al., 1979). Glucose and phosphate suppressed tylosin production and the oxidation of methyloleate, the probable source of carbon for tylosin production. Excess phosphate (30 mM, defined medium) was found not to affect growth but did raise ATP levels and, to a greater extent, glucose-6-phosphate levels. It was suggested by these workers that glucose-6-phosphate was a regulatory intermediate that would respond to both glucose and phosphate variations. Supporting evidence was drawn from the effects of glucose analogs. 2-Deoxyglucose, which can be phosphorylated, suppressed tylosin production and methyloleate oxidation, whereas 3-O-methylglucose, which cannot be phosphorylated, had no effect (Madry et al., 1979).

Vu-Trong et al. (1981) demonstrated that phosphate suppressed the activities of methylmalonyl-coenzyme A carboxytransferase and propionyl-coenzyme A carboxylase; these enzymes supply tylosin precursors. The degree of suppression of these enzymes was not great enough to account for the reduction in tylosin production, however. In this study, excess phosphate (>30 mM, defined medium) was claimed to stimulate growth slightly. Phosphate was again found to inhibit methyloleate utilization and raise ATP levels.

Moreover, the total adenylate pool was elevated. Significantly, energy charge was clearly *not* involved (Vu-Trong et al., 1981; Vu-Trong and Gray, 1982). Instead, the total concentration of adenylates was found to be more clearly correlated to antibiotic production. Glucose added during the idiophase raised adenylate levels and suppressed tylosin production (Vu-Trong et al., 1980).

ATP has been implicated as a regulatory element in other antibiotic production systems. The ATP level of *S. griseus* doubles within 5 min after the addition of 10 mM phosphate to antibiotic-producing cells (Martín and Demain, 1976). An inverse correlation between ATP levels and antibiotic production in antibiotic-producing strains has been reported recently for nourseothricin production by *Streptomyces noursei* (Hañel et al., 1984). Hänel et al. (1984) presented evidence against the involvement of glucose-6-phosphate in antibiotic regulation; higher producing strains did not necessarily have lower levels of this metabolite.

The tylosin example and the postulation of ATP as the key metabolic signal illustrate the complexity of phosphate regulation and of the regulation of antibiotic biosynthesis in general. Clearly, ATP levels cannot be lowered infinitely without adversely affecting antibiotic production. In practice, phosphate suppression is controlled by limiting the amount of phosphate in the medium or by the addition of complexing agents (Payne and Wang, 1988; Müller et al., 1984). Phosphate is a significant component of many fermentation raw materials used for carbon and nitrogen sources, including cornflour and soybean meal. The use of yeast or yeast extract as a supplemental nitrogen or vitamin source can also introduce large amounts of phosphate. The form of the phosphate may be as important as the absolute level. Much of the phosphate in corn flour exists as phytic acid. Generally, inorganic phosphate is a more potent suppressor of antibiotic production than organic phosphate. The ability of metals, such as zinc, to reduce the suppression of antibiotic production by phosphate supports the hypothesis that free phosphate is the critical factor (Müller et al., 1984).

Attempts to circumvent phosphate suppression genetically have been performed primarily by the selection of arsenate-resistant strains or the direct selection of phosphate-deregulated antibiotic production on solid media. The former technique has been reported for nourseothricin production by *S. noursei* (Friedrich et al., 1984). The latter technique has been developed by Martin (1977) for candicidin production. Preliminary evidence suggests that phosphorylated glycolytic intermediates, such as fructose-1,6-diphosphate, may be important.

As a final note on phosphate regulation, an important area of phosphate metabolism that has been somewhat overlooked in actinomycetes and antibiotic production is the possible synthesis of polyphosphate as a phosphate storage mechanism. Numerous microorganisms synthesize polyphosphate and use it during times of phosphate starvation (Harold, 1966). The occurrence of polyphosphate has been reported in *Streptomyces*, but its significance has not been thoroughly investigated. The ability to store phosphate may explain why additions of phosphate early in an antibiotic fermentation can have long-lasting effects.

4. Steroid Transformations

Modification of steroids by actinomycetes is a well-known phenomenon exploited by industry (Miller, 1985; Schoemer and Martín, 1980). The ability of actinomycetes, especially *Nocardia*, to modify steroids is probably related to the ability to degrade and assimilate a large variety of aliphatic and aromatic hydrocarbons (reviewed by Cain, 1981). This catabolic diversity suggested to Cain that the enzymes involved are regulated by induction as in the case of *Pseudomonas*. In fact, enzyme induction has been shown to occur in a number of cases (Feller and Träger, 1985; Buckland et al., 1976). The transforming enzymes can sometimes be induced by several steroids; however, the substrate to be transformed is generally the best inducer (Miller, 1985). An interesting recent finding demonstrated the loss of steroid-transforming activities in aerial mycelia-negative and/or arginine auxotrophs of *Streptomyces hydrogenans* (Feller and Träger, 1985). These phenotypes are characteristic of "bald" mutants of streptomycetes that include the loss of other differentiated functions, including antibiotic biosynthesis. Whether the induction of steroid-transforming enzymes can be related to A factorlike induction of antibiotic pathways remains to be seen.

5. Glucose Isomerase

Glucose isomerase (GI) catalyzes the conversion of glucose of fructose and is used in the production of high-fructose corn syrup (Verhoff et al., 1985; Chen, 1980). GI can be obtained from a wide variety of bacteria, including many actinomycetes. Chen (1980) has compiled a list of GI-producing organisms, approximate medium compositions, and yields. Of the 70 microorganisms listed by Chen, approximately half are actinomycetes.

GI is actually xylose isomerase and catalyzes the aldose-ketose isomerization of a number of sugars. The enzyme is regulated in most organisms by xylose induction and glucose repression (reviewed by Verhoff et al., 1985). Bacterial mutants constitutive for GI have been isolated as resistant to 2-deoxyglucose. Verhoff et al. (1985) described a process for GI production by a deoxyglucose-resistant strain of *Streptomyces olivaceous* developed at Miles Laboratories.

C. Metabolic Studies in Shake Flasks

1. Shake Flasks Versus Fermentors

There are, of course, substantial differences between shake flasks and stirred tanks of any size. Oxygen transfer, shear forces, and sterilization are factors that are difficult to mimic in a shake flask. These factors, and others, are discussed in more detail in subsequent sections relating to scaleup. Despite the inherent problems with shake flasks, there are still distinct advantages to their use in fermentation development and analysis. Because of the large number of flasks that can be set, compared with tanks, the number of conditions and replicates that can be analyzed simultaneously is greatly increased. Samples, usually taken as individual flasks, can be processed immediately in a laboratory. The cost of materials and maintenance of equipment is lower for shake flasks than for tanks. Exotic and expensive compounds can be added in much smaller quantities to shake flasks. Radioisotopes used to analyze metabolic pathways can be added to shake flasks and adequately contained. The intent here is to generate information that may lead to strain or process improvements. A major disadvantage of flasks is that all analyses are run "off-line," which excludes measurements of some important parameters, such as dissolved oxygen. Nevertheless, useful measurements can be made with shake flasks and compared with values obtained from a limited number of tanks.

It is impractical to attempt to model a tank fermentation exactly with shake flasks. However, care should be taken to ensure that key parameters are matched as nearly as possible. Nutrient depletion profiles, growth rate, and yield should be comparable. This can generally be accomplished by using similar medium components. A medium used in the pilot plant can sometimes be "scaled down" for use in shake flasks. A reduced strength, say 60%, is recommended to accommodate the decrease in oxygen transfer in flasks. Product synthesis profiles and product intermediate levels in shake flasks should be similar to those in tanks. The disproportionate accumulation of an intermediate in shake flasks that does not occur in tanks suggests a different rate limitation.

2. Enzyme Measurements

Enzyme levels are measurements that can be made readily using shake flasks. Enzyme levels are dynamic quantities directly related to the metabolic state of the organism and provide useful information regarding the metabolic controls present in a fermentation at a particular point in time. In practice, enzymes can be subdivided into two categories, extracellular and intracellular.

Extracellular enzymes are generally hydrolases produced by the microbe to release transportable nutrients from medium components.

Examples are proteases, amylases, lipases, and phosphatases. Extracellular enzymes are often very stable in vitro and easily obtained by filtration or centrifugation of the medium. Dye-linked substrates, such as azocasein and azocollagen for proteases and starch azure for amylases, are convenient for assay. Phosphatases, lipases, and oligosaccharidases can be assayed using *p*-nitrophenyl-linked substrates. The type of information obtained is varied. Comparison of cephalosporin production with extracellular protease activity was made, demonstrating that increased antibiotic production was correlated with increased proteolytic activity (Ginther, 1979). In other instances, process comparisons are possible. Amylase, protease, and erythromycin production were measured for *S. erythreus* grown with two starch protein media in shake flasks. The complex medium used cornstarch and soy flour, whereas the soluble medium substituted soluble starch and soy hydrolyzate. It was found that in the complex medium both enzyme levels and antibiotic production closely paralleled the results obtained in the pilot plant (T. Paulus, unpublished observations).

The amount of information obtainable from extracellular enzymes does have limits. Whereas high amylase activity in vitro suggests rapid starch utilization, the loss of total medium carbohydrate is more indicative of the rate of starch utilization. The pattern or profile of extracellular enzymes is probably more informative than absolute levels and must be interpreted in conjunction with nutrient measurements.

Intracellular enzymes can provide information on metabolic pathways operating in a fermentation process. Intracellular enzymes are, however, more difficult to measure, largely because of a decreased stability in vitro. Moreover, the use of complex, insoluble medium components complicates the preparation of cell-free extracts. The presence of proteases can also be a problem; the inclusion of protease inhibitors, such as PMSF, is advised. Nevertheless, intracellular enzyme measurements are extremely important. Especially important are the antibiotic synthetases, which can shed light on the mechanisms of product synthesis and metabolic regulation. Reports of antibiotic synthetases were discussed in preceding sections. Other uses include the characterization of strains. Seno and Baltz (1982) have compared the activities from various strains of macrocin O-methyltransferase, the terminal enzyme in tylosin biosynthesis by *S. fradiae*. They were able to show that increased tylosin biosynthesis was correlated with increased activity of this enzyme. A similar study was performed using propionate kinase activity from erythromycin-producing *S. erythreus* (Raczyńska-Bojanowska et al., 1973).

3. Pool Measurements

Intricate theories have been developed to describe the control of flux in metabolic pathways (Kacser and Burns, 1973; Westerhoff et al., 1984). A determination of the rate-limiting step(s) in a fermentation is potentially of great use in the design of new strains and process improvements. Unfortunately, the information required to utilize the current models is often in excess of available information. In the case of antibiotic fermentations, cell-free systems are unavailable, many of the pathway intermediates are still unknown, and in vitro enzyme assays are just beginning to become commonplace. However, useful information is obtainable by studying precursor and intermediate metabolite pools that can shed light on process kinetics and control. The crossover theorem originally developed for mitochondrial electron transport can be applied to point to possible rate constrictions. An excellent description of the crossover theorem and application has been written by Rolleston (1972).

4. Tracer Studies

The use of radioactive and stable isotopes in fermentation research has historically been associated with establishing precursor-product relationships. The use of stable isotopes, such as ^{13}C, is rapidly supplanting the more traditional radioisotopes because of the positional information that can be obtained through the use of nuclear magnetic resonance (NMR) spectroscopy without chemical degradation of recovered products. The extensive work of Ōmura et al. (1975, 1976, 1977) on the origin of the lactone ring of 16-membered macrolide antibiotics demonstrates the power of ^{13}C-NMR in biosynthetic studies. Besides ^{13}C-labeled tracers, other stable isotopes can be used (reviewed by Rinehart, 1983). Oxygen 18 was used by Cane et al. (1981) to demonstrate that the oxygens attached to the lactone ring of erythromycin were derived from propionate, not water.

Radioactive isotopes are still frequently used because of the simplicity and sensitivity of the measurements. Recently, amino acids were determined to provide precursors for the lactone ring of tylosin by the use of ^{14}C-labeled materials (Dotzlaf et al., 1984). Often a combination of ^{13}C- and ^{14}C-labeled precursors is used. Byrne et al. (1985) investigated the biosynthesis of a new antitumor polyketide antibiotic, fredericamycin A, produced by *S. griseus*, using both stable and radioisotopes. Radioactive glucose and acetate were used to establish general pathways, whereas ^{13}C-labeled material was used to confirm specific carbon labeling and to establish the direction of polyketide chain formation.

An important use of ^{13}C-NMR in fermentation development may be the ability to determine the pathways needed to form precursors in addition to the identification of the precursors themselves.

Ōmura et al. (1976) demonstrated that the propionate units used for leukomycin biosynthesis were synthesized via the TCA cycle instead of the glyoxalate shunt. Other recent ^{13}C-NMR studies have been able to distinguish between these two competing pathways (Walsh and Koshland, 1984). This type of information should have a direct impact on strain and process improvements.

IV. SCALEUP

A. Background

Historically, reference to fermentor scaleup has involved the areas of oxygen transfer and mixing, usually in the context of the initial transfer of a culture from the laboratory to production. As the disciplines of biotechnology have become more sophisticated, so has the understanding of other parameters important in the control of the microbial environment in the transition from laboratory studies to production operations. [See Young (1979) for a discussion of the total environment approach to scaleup.] Table 3 summarizes the parameters that influence the microbial environment and the relevant controlling factors. Each scale has a different balance between the mechanical control mechanisms that are under the control of the fermentologist and the biological control mechanisms of the microorganisms. Understanding how the balance can change is the essence of scaleup. In order to successfully scale up a fermentation, one must not introduce a rate limitation on the large scale that is more severe than the rate-limiting step on the small scale. What can make scaleup so difficult is that it is not practical to evaluate all the *potential* factors that could introduce a lower rate limitation.

In the last 20 years the emphasis has shifted away from the initial scaleup of new products to the optimization of existing processes. Large-scale actinomycete fermentations seem to present almost unlimited possibilities for improvements. Wang et al. (1979) illustrated this type of productivity improvement for an industrial antibiotic fermentation. In 20 years, a 20-fold increase was obtained through the combined disciplines of engineering and microbiology. As one might expect, yield improvements do not follow a steady increase because of the inherent variability in fermentation processes. Initial scaleup work can show impressive yield increases. Hamilton et al. (1980) increased daunorubicin productivity of an unspecified streptomycete by a factor of 10 during 40 production runs at the 10,000 liter scale through optimization and mutant selection. Florent (1985) documented the increase in literature-reported values of streptomycin yields from 0.1 g/liter in 1944 to 10.5 g/liter in 1979.

Table 3 Scaleup Parameters

Parameter	Mechanical control mechanism	Biological control mechanism
Physical		
Temperature	Cooling or heating system	Heat of metabolism
Shear	Tank, impeller design	Rheology
Interphase mass transfer	rpm, agitation	Rheology
Chemical		
pH	Acid or base addition	Uptake rates
Dissolved gasses	Head pressure, transfer rates via rpm	Uptake rates
Other nutrients	Initial medium Feed systems Volume control Sterilization methods	Uptake rates
Product concentrations	Harvest methods	Productivity rates
Biological		
Cell mass	Environmental control via the above	Growth rates
Strain stability	Inoculum methods	Reversion, selection

A scaleup strategy must be determined that is based upon previous knowledge of how well the process scales up, the targeted equipment, and constraints of timing and cost of the scaleup work. If little is known about the process in question, scaleup must be approached empirically by testing the process at several intermediate scales. In contrast, for a well-understood process, changes should be predictable over a scale change of 1000 or more.

The discussion in this section begins by following the outline of Table 3. Oxygen transfer and mixing effects are emphasized in the discussion of the physical environment. After reviewing the chemical environment and inoculum buildup procedures, the biological control mechanisms and microbial kinetics are discussed. Since most actinomycetes are used for secondary metabolite production, an appreciation of specific product formation rates and product yields are important. The scaleup procedure is then discussed using literature examples and previously unpublished results. Finally, novel bioreactor designs are briefly presented. The theory behind most of the discussion is limited. The reader is encouraged to consult other references for biochemical engineering principles and scaleup (Bailey and Ollis, 1986; Aiba et al., 1973; Trilli, 1986).

B. Physical Environment

The physical aspects of the microbial environment are characterized by temperature, shear forces, and chemical concentrations, which in turn are controlled by heat, momentum, and mass transfer. In practice, temperature control is not a problem for actinomycete fermentations except in extreme cases of high metabolic activity coupled with a limited cooling capacity. Heat transfer is relatively straightforward to predict by calculating the cooling and/or heating supply versus the heating load from metabolism and mixing.

Shear forces and mass transfers to and away from the cell surface are controlled by the manner and degree of liquid mixing and by the rheological properties of the fermentor broth, which can be greatly affected by the microorganism. Sufficient oxygen transfer is the most critical. Turbulence is needed to generate gas-liquid surface area to maximize oxygen transfer. Yet the cost of high turbulence may be mechanical destruction of the organism. A compromise must often be made between oxygen transfer and shear.

With the current state of understanding, it is not possible to predict large-scale oxygen transfer and shear effects in a specific actinomycete fermentation without substantiating data at smaller scales or performance of a similar fermentation in similar equipment. A complete model does not exist that can describe the local shear and oxygen transfer at every point in a fermentor, but one can gain insight into scaleup by considering (1) the unique viscosity behavior of actinomycete fermentation broths, (2) empirical relations that describe oxygen transfer, (3) bulk mixing throughout the fermentor, and (4) shear effects and oxygen transfer at the mycelia scale. The following discussion examines this sequence.

1. Viscosity

The rheological properties of the fermentor broth will be a major factor in determining the amount of mixing and shear inside a fermentor. Shear is caused by two surfaces moving past each other. In the case of an impeller, it is the blade moving past the liquid. Viscosity can be defined as the resistance of a fluid to shear when a force is applied. For simple fluids, such as water, the viscosity is independent of the force applied and the fluid is referred to as Newtonian.

For more complex fluids, the viscosity itself becomes a function of the force applied to it. The mathematical relationships for non-Newtonian fluids can become extremely complicated. Actinomycete fermentor broths are highly non-Newtonian. In practice, an apparent viscosity is measured as if the fermentor broth is Newtonian and used as a crude measure of growth. The power-law model has been found to be adequate for describing fermentation broths. The apparent viscosity for a power law fluid is defined as

$$\eta = m \left(\frac{dv}{dy} \right)^{n-1} \tag{1}$$

The constants m and n are referred to as the consistency index and behavior index, respectively; dv/dy represents the velocity gradient in the fluid. It is a maximum at the impeller tip.

The importance of Eq. (1) is twofold. First, it predicts that the apparent viscosity in the fermentor is a function of the distance from the high shear environments of the impeller, baffles, and sparger. Second, to accurately study the viscosity in the fermentor, it is necessary to duplicate in the laboratory the range of shear rates in the fermentor, which is impractical and also typically much greater than the capabilities of a laboratory viscometer.

Table 4 gives published results for the consistency index and the behavior index. The behavior index can be expected to vary from 1.0 (Newtonian behavior) to 0.4. Broth rheology has been reviewed by Blanch and Bhavaraju (1976). Roels et al. (1974) suggested a more complicated model based upon the Casson equation:

$$\tau^{1/2} = \tau_0^{1/2} + m_1 \left(\frac{dv}{dy} \right)^{1/2} \tag{2}$$

In this model a minimum force, the yield stress τ_0, must be applied to the liquid before any shear is developed. τ_0 and m_1 were derived in terms of morphological considerations. At high shear rates such as those encountered in a fermentor, Eq. (2) reduces to the power-law model with a behavior index of 1.0, whereas at low shear rates an estimated behavior index is consistent with the values reported in Table 4.

2. Oxygen Transfer

A generalized transfer equation is

$$\text{Uptake or production rate} = \text{transfer coefficient} \times \text{driving force} \tag{3}$$

The driving force for transfer is the gradient of fluid velocity, temperature, and concentration for momentum, heat, and mass transfer, respectively. Often a simple difference is used to represent the force. For oxygen transfer, the oxygen uptake rate at a point in

Table 4 Values Reported in the Literature of Oxygen Transfer Coefficients Versus Broth Rheology for Actinomycetes Fermentations

Organism	Product	K_La	Consistency index	Flow behavior index	References
S. noursei	Cycloserine	40-180	<0.1-3.5	0.25-0.5	Járai et al., 1969
S. lavendulae	Nystatin	50-180	0.1-1.5	0.4-0.9	Járai et al., 1969
Streptomyces sp.	Fumagillin	120-420	0.5-1.5	0.5-0.8	Járai, 1972
Unknown[a]	Unknown	80-320	10-30	0.35-0.75	Tuffile and Pinho, 1970
Unknown[b]	Unknown	NA[c]	13-92	0.25-1.1	Tuffile and Pinho, 1970

[a]Typical viscosity.
[b]Very high viscosity.
[c]Not applicable.

the fermenter is written as

$$OUR = k_L a (c^* - c) \quad (4)$$

where k_L represents the liquid film diffusion rate, a represents the bubble surface area, c* represents the saturation concentration of oxygen, and c the actual oxygen concentration. The transfer coefficient k_La varies throughout the fermentor because the bubble surface area varies owing to coalescence. In addition, the apparent viscosity affects both k_L and a.

Because k_L and a cannot be studied independently except in model systems in which great care is taken to control a, a global transfer coefficient K_La is usually used. The fermentation reaches a steady state at which the oxygen transfer rate supplied by diffusion from the air/liquid interface equals the oxygen uptake rate of the organism. Most studies of transfer coefficients result in empirical correlations that generally reflect bulk mixing properties. Correlations must be applied between fermentations with great care because of the possibility of differences on the mixing microscale, which Oldshue (1981) defined as 200 μm or less. In thick streptomycete fermentations, transfer rates of 20-60 mmol/liter per hr with a K_La of 20-80 hr^{-1} are typical from a power input of 1-3 kW per 1000 liters for a mechanically agitated tank. Oxygen uptake, K_La, and dissolved oxygen are shown in Fig. 2 from a streptomycete fermentation. As the cell mass increases, the transfer coefficient drops by a factor of 8 due to increases in the apparent viscosity.

A constant mass transfer coefficient has often been suggested as a scaleup criterion. Correlations for K_La in mechanically agitated fermentors are usually reported in the form

$$K_L a = K_1 (P_g/V)^{\alpha} v_s^{\beta} \quad (5)$$

which takes into account both the gassed mechanical power P_g, and energy from air expansion in terms of the superficial air velocity v_s. Bartholomew (1960) found the trend toward smaller values of α as the fermentor size increased whereas the exponent for air velocity remained fairly constant. K_1, α, and β have been found to be functions of impeller type and geometry and other variables. Murphy et al. (1984) found values of 4.28, 0.59, and 0.4 for K_1, α, and β, respectively, for a 100,000 liter fermentation in a model that includes fermentor height. Taguchi et al. (1968) found values of 0.33 and 0.56 for α and β, respectively, for a glucamylase fermentation by an *Endomyces* on a scale of 30-50,000 liters.

Ryu and Humphrey (1972) determined a correction for the apparent viscosity effect on K_La

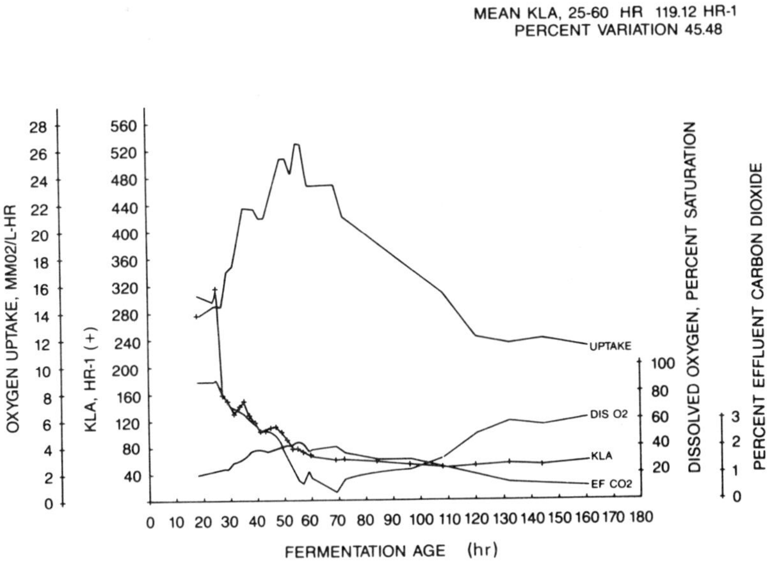

Figure 2 Time course of k_La for a 300 liter streptomycetes fermentation. (From Tuffile and Pinho, 1970; reprinted from *Biotechnology and Bioengineering*.)

$$K_La \alpha \eta^{-.86} \tag{6}$$

Chemical constituents can affect the mass transfer coefficient. Robinson and Wilke (1972) correlated the effect of ionic strength on K_La using a model system, Their results indicated that there is no effect above an ionic strength of 0.4. At lower ionic strengths there is an effect on K_1, α, and β in Eq. (5), the most pronounced effect being on K_1. From zero to 0.4 ionic strength, K_1 (and the corresponding mass transfer coefficient) decreases by a factor of 10. Defoamers are notorious for lowering oxygen transfer rates. Páca et al. (1978) demonstrated the effect of sperm whale oil on K_La for water and a simulated medium for erythromycin synthesis by *S. erythreus*. The decrease in K_La was a function of power but ranged from 5 to 80% at 1.5% (v/v) sperm whale oil.

Oxygen transfer in shake flasks can also be characterized by Eq. (4), although the effect of viscosity and power on transfer rates is drastically different. Van Suijdam et al. (1978) found oxygen transfer to be a function of the shaker rpm to the 1.4 power. In shake flasks, oxygen transfer takes place by diffusion through the liquid surface, which is relatively independent of viscosity.

Although oxygen transfer in shake flasks is less than in fermentors, as viscosity increases in shake flasks, oxygen transfer is much less affected. Bartholomew et al. (1950) reported oxygen transfer rates of 6-12 mmol/liter per hr for *S. griseus* in a 220 rpm shaker with a 1.5 inch throw.

3. Mixing Time

Uniformity of mixing is another concern during scaleup. Imperfect mixing in a fermentor may be caused by "dead spots" located behind baffles or between impeller mixing zones in a multiimpeller fermentor. Mixing is typically studied by pulsing a solute into a fermentor and following the solute concentration approach to homogeneity. Wang et al. (1979) used dimensional analysis to demonstrate that for geometrically similar tanks with constant power per unit volume, mixing times on two different sales are related by

$$\frac{t_{ml}}{t_{ms}} = \left(\frac{D_{il}}{D_{is}}\right)^{11/18} \qquad (7)$$

where t_m refers to the mixing time and D_i the impeller diameter, and the subscripts l and s refer to the large and small scale, respectively. Instead of using a single value for mixing time, Mukataka et al. (1980) suggested the use of a circulation time distribution to characterize the fermentor.

As the scale increases, so does the mixing time. Lower yields in large tanks than in smaller tanks are often attributed to longer mixing times, which result in local gradients in temperature or nutrient levels in the fermentor, although this is difficult to prove. Manfredini et al. (1983) measured the temperature profile in a 112,000 liter tetracycline fermentor and found variations of less than 0.08°C.

4. Microscale

The preceding discussion of oxygen transfer and mixing does not explicitly address microscale diffusion or shear effects. Actinomycetes can be viewed as growing in pellets or mats of interlaced mycelium. For mycelial pellets, van Suijdam et al. (1982) defined a critical pellet radius for oxygen uptake of

$$R_c = (6D_e c_s / r_{o_2})^{1/2} \qquad (8)$$

Pellets of greater radius have a dissolved oxygen concentration of zero near the center. Under typical conditions the critical pellet radius is less than 250 μm. Other geometries have a similar relationship.

High shear turbulence can affect pellet size by causing mycelial fragmentation. Bewick et al. (1976) demonstrated the differing fragmentation between shake flasks and fermentors on *S. venezuelae* in a variety of media. Power, shear, and other mixing parameters cannot be maintained in the same ratio upon scaleup because of their nonlinear and differing dependence upon impeller diameter and rpm (Oldshue, 1981). As mixing intensity is increased, a series of states has been observed. Very low power inputs result in insufficient oxygen transfer and limiting dissolved oxygen. A point is reached at which the impeller blade size design is important. Maxon (1959) found that a large impeller resulted in reduced novobiocin yields even though the power input seemed sufficient. As agitation is further increased, a point is reached at which mycelial destruction occurs.

Shear, through its effect on pellet size, combined with intraparticle diffusional effects, can result in a pseudocritical dissolved oxygen level. Dissolved oxygen measurement is a bulk fluid property. As the diffusional limitation increases, the dissolved oxygen level at the surface of the mycelial particle must increase to maintain adequate intraparticle dissolved oxygen concentrations. Steel and Maxon (1962) in a 20 liter novobiocin fermentation found higher oxygen uptake rates at constant power input with smaller diameter impellers. This result can be explained by the higher shear resulting in smaller mycelial aggregates. If the size of the aggregates is in the range of the critical size for dissolved oxygen limitation, smaller aggregates have a higher uptake rate.

Mass transfer, mixing time, and shear effects were combined by Wang and Fewkes (1977) for the novobiocin fermentation at the 6 liter scale. They show the dependence of the pseudocritical dissolved oxygen concentration versus a dimensional number of

$$\text{SHEAR TO FLOW RATIO} = N/D_i \tag{9}$$

Also, they concluded that the product of specific oxygen uptake rate and mixing time may be an indication of the importance of mixing times in a specific system.

Most of the literature is concerned with gas-liquid mass transfer due to the importance of oxygen. However, other transport behavior must be considered. In many media liquid-liquid or solid-liquid mass transfer may be important. Moo-Young et al. (1972) determined the effect of power on a yeast-dodecane fermentation and expressed the specific growth rate as a function of power. In this case, increasing power increased the oil droplet surface area for improved diffusion. Solid-liquid mass transfer may be important in complex media in which the fermentation is limited by the solubilization of an insoluble nutrient.

C. Chemical Environment

Perhaps surprisingly, nutrient levels can be affected indirectly and in some cases directly by scale due to differences in their preparation, maintenance, or selection. Complex nutrients, such as vegetable proteins and starches, are often chosen for large-scale operations based upon cost and availability. Unfortunately, their composition is not well defined and can be quite variable, leading to variable fermentor results. Also, to further increase raw material cost savings, by-products of other manufacturing processes are used. These materials may not be subject to the tight quality control of the primary material, resulting in increased fermentor variability. A suspected advantage of complex nutrients for secondary metabolite fermentations is the possibility that the nutrients may undergo slow rates of hydrolysis, which will maintain a fermentation, although probably not at optimum productivity rates. Typically, a complex material is chosen for its major constituent and then its minor components are reviewed for their effects. Thus, cornmeal may be chosen as a carbohydrate source but one must question the effects of its protein and minerals.

Different sterilization methods can affect media components. Large-scale sterilization often involves high temperatures and heat transfer rates. Also, the techniques for cooling large liquid volumes may result in the media remaining at an elevated temperature for a substantial period of time. The net result is a high temperature condition that can accelerate reaction rates. Reactions can be of the form of solubilization or insolubilization and alteration to either an inert or possibly a toxic species. The potential complexity of the kinetics was illustrated by Barton et al. (1985) for the precipitation of calcium phosphate on heated surfaces. The browning or Maillard reaction of glucose is usually one to be avoided by sterilizing glucose separately from the other raw materials. When glucose is sterilized with proteins or phosphates, a number of oxidation products are formed that can be inhibitory.

Dissolved gases are an important component of the chemical environment of the fermentation. Dissolved gases must be continuously replenished or, in the case of inhibitory components, continuously removed. Their transport is governed by Eq. (4). It is imperative that oxygen does not become a limiting nutrient. Although the effect on cell mass may be proportional to the degree of oxygen limitation, the effect on primary or secondary metabolites may be disastrous. Goodman (1985) described the effect of oxygen limitation on tetracycline fermentation. Brief periods without aeration resulted in a substantial yield loss. Sensi and Thiemann (1967) obtained an increase in rifamycin yields by *S. fradiae* from 440 to 1200 mg/liter by increasing the oxygen adsorption rate (sulfite method) from 42 to 72 mmol/liter per hr.

Dissolved gas concentrations are a function of pressure and vary with fermentor height, as illustrated by Manfredini et al. (1983). The model of Murphy et al. (1984) showed that the dissolved oxygen concentration at the top of the fermentor may be below a critical value, whereas it is substantially greater near the tank bottom. Certain organisms may show a critical dissolved oxygen concentration for optimum productivity, as demonstrated by Buckland et al. (1976) for *Nocardia*. In a 3 liter fermentor, cholesterol oxidase production by *Nocardia rhodocrous* fell 20% if the dissolved oxygen level fell below 30%.

Dissolved carbon dioxide has been shown to have inhibitory effects on a number of secondary metabolites. In shake flasks, Bylinkina and Birukov (1972) found inhibition in streptomycin synthesis by *S. griseus* at levels greater than 8%. At 14% carbon dioxide, the antibiotic synthesis was inhibited 50%. Carbon dioxide has also been shown by Nash (1974) to inhibit erythromycin synthesis by *S. erythreus*. He sparged 11% carbon dioxide beginning at 15 hr and found no effect on cell growth but a 40% decrease in antibiotic synthesis rate, possibly due to feedback inhibition of methylmalonic acid formation. Chemical species that act directly on metabolic regulation can be indirectly affected by scale. These effects are often not predictable. An example is given by Young (1979), who found that a rate-limiting component was removed by higher ventilation efficiencies on a small scale as opposed to the production scale. Ōmura et al. (1980) demonstrated that the addition of magnesium phosphate to a leucomycin medium caused the precipitation of ammonia with the result of higher antibiotic productivity due to less repression by nitrogen. Changes in the amounts and rates of solubility can be a function of mixing conditions, so one might imagine that upon scaleup, the degree of ammonia removal would change. Other media components may have more obscure behavior. Burg et al. (1979) found that the defoamer P-2000 increased avermectin synthesis by *Streptomyces avermitilis* by a factor of 3. McCann-McCormick et al. (1981) could not duplicate this stimulating effect by other defoamers. The role of P-2000 in this fermentation is unknown.

D. Inoculum Preparation

Little work has been reported on optimizing inoculum or seed stages. In a large fermentor, the organism must go through more generations than in a laboratory fermentor. Perhaps more importantly, it may involve several more transfer steps. For instance, a 100,000 liter fermentor has 13 more doublings than a 10 liter fermentor and, assuming a 5% inoculum level at each step, requires three more transfers. For an organism that has genetic instability, this may result in substantial reversions. Strains that have been highly mutated for secondary metabolite production can be genetically unstable. Reusser (1963) reported culture degeneration for *S. fradiae* (neomycin production) and *S. niveus* (novobiocin production). Antibiotic production was lowered to less than 15% of the initial yield after eight transfers of *S. fradiae* and 20 transfers of *S. niveus*.

A highly mutated strain may not grow as fast as the wild type. Selection can take place if the initial inoculum contains a small percentage of the wild type or other genotype. Yoshikawa et al. (1982) studied herbicidin A production by *Streptomyces saganonensis* and found a nonproducer to have a growth rate of 1.4 times the producing strain. If an initial inoculum (a spore suspension or frozen mycelium) contained 0.01% of the nonproducer with this growth rate, the percentage would increase to 0.2% at the 10 liter level and 7% at the 100,000 liter level. A general equation derived by Imanaka and Aiba (1981) is

$$F = \frac{\gamma + p - 1}{(\gamma - 1) + p2^{g(\gamma + p - 1)}} \tag{10}$$

where F is the biomass fraction of the parent remaining after g generations, γ is the ratio of the specific growth rate of the variant to the specific growth rate of parent, and p is the rate of appearance of the variant. Ollis and Chang (1982) extended this model to product formation by a genetically unstable organism.

The higher growth rate of a wild-type is not necessarily caused by an increase in μ_{max}. A similar effect is predicted if the nonproducer has a higher substrate affinity at low growth rates.

Table 5 illustrates the range of inoculum development sequences. Stock cultures are usually maintained as spores or frozen mycelium in liquid nitrogen. The germination medium, seed medium, and final fermentor medium serve different purposes and may be substantially different. One must be concerned about the carryover effects of nutrients between the stages. The number of steps is a function of the growth behavior of the organism, and ideally, transfer between stages is based upon the optimum cell concentration at each stage. Typically, inoculum age is used even though other parameters may be more appropriate. Orlova and Prokofieva-Belgovskaya (1961) found the amount of inoculum to be more important than age for oxytetracycline production. Buckland et al. (1985) used carbon dioxide production as a means of determining the end of the seed

Table 5 Inoculum Development Sequence

Step	Function	Inoculum percentage	Volume
Spore or mycelial stock	Long-term storage	NA[a]	NA
Flask	Germination	NA	50 ml in 250 ml flask
Seed(s)	Rapid growth, preparation for fermentor medium	0.0001-10	1 liter flask to 10,000 liter tank
Fermenter	Production	5	10-500,000 liters

[a]Not applicable.

growth phase for the thienamycin fermentation. Omstead et al. (1985) found low inoculum levels resulted in pelleted growth and depressed yields for cephamycin fermentation.

E. Microbial Kinetics

Microbial kinetics are a global approximation of the underlying physiology and are used to attempt to describe cell growth, substrate utilization, and product formation. The success or failure of a scaleup effort often depends directly upon adequate understanding of the kinetics. The kinetic model may initially be qualitative, but as a process becomes optimized, it must become increasingly quantitative. Models can be divided as to whether they assume that all cells are equivalent (distributed) or assume a population distribution (segregated) and as to whether they are based upon intracellular events (structured) or totally on extracellular conditions (unstructured). For practical purposes only distributed unstructured models are used. More research is needed before structured segregated models are used on a common basis. For a general review of models, see Roels and Kossen (1978).

1. Growth

For unicellular organisms the Monod equation is commonly referenced to describe cell growth and the Ludeking and Pirt model for product formation. The equations for cell mass, growth rate, substrate uptake, and product formation are then

$$\frac{dX}{dt} = \mu X \tag{11a}$$

$$\mu = \frac{\mu_{max} S}{K_m + S} \tag{11b}$$

$$\frac{dS}{dt} = -\frac{1}{Y_{x/s}} \frac{dX}{dt} - m_c X - \frac{q_p X}{Y_{p/s}} \tag{11c}$$

$$\frac{dP}{dt} = b_o \mu X + b_1 X \tag{11d}$$

In contrast to unicellular growth, the growth of actinomycetes takes place by tip elongation, branching, and fragmentation. The last contributes to the growth rate by creating more growing tips and is caused by fermentor shear or the normal growth cycle of the organism. Exponential growth in filamentous organisms only occurs because of branching and fragmentation. The complexity of a segregated model is illustrated by Megee et al. (1970), who divide mold growth into five states: an active growing tip, growth-associated mycelium, and three non-growth-associated states including sporulation.

A distributed model that has been successfully applied to mycelial growth begins by assuming that the radius of a mycelial pellet increases at a constant rate. The cell mass X as a function of time is given by

$$X^{1/3} = k_o t + X_o^{1/3} \tag{12}$$

For a small number of doublings, this equation is not easily distinguished from exponential growth especially if it is modified to include any degree of fragmentation or branching. Bosnjak et al. (1978) extended Eq. (12) to model the growth of *S. erythreus* to

$$\frac{dX}{dt} = k_1 X^{1/3} - k_2 X - k_3 X t \tag{13}$$

The first term in Eq. (13) corresponds to Eq. (12), whereas the last two terms are assumed to account for product inhibition and cell death. Cheruy and Durand (1979) tested various unstructured models with the best fit given by

$$\frac{dX}{dt} = k_4 X \left[1 - \left(\frac{X_o}{X_F} \right)^{k_5} \right] \tag{14}$$

More research is needed into the effect of physical and chemical parameters on the constants used in such models as Eqs. (12) through (14). In the references cited, nutrient levels are not explicitly included in the models, so the ability to use them as a function of substrate concentrations is questionable. Another difficulty is in the proper measurement of cell mass. In complex media with insoluble components, dry weight or optical density measurements are not feasible for determining cell mass. DNA is often used, but Martín and McDaniel (1975) showed that DNA and dry weight do not follow the same trends for *S. griseus* producing candicidin. They attributed the difference to accumulation of intracellular lipids or other secondary metabolites.

2. Product Formation

Qualitatively, the fermentation can be divided into the growth phase or tropophase, the stationary phase or idiophase, and the death phase or lytic phase. This distinction is especially important in product formation because most systems do not produce material until the idiophase. Martín and McDaniel (1975) attempted to quantitate this by using the cell maturation time. The maturation time was defined as the time between cell formation and the onset of product formation. The growth phase distinction is one of the more common methods of initial optimization. By using "shift" experiments, the effect of environmental parameters, such as temperature and pH, on product formation are studied as a function of age. Age is often used as an estimate of the transition from the tropophase to the idiophase. Thus, Lee et al. (1983) reported the optimization of rifamycin B synthesis by a pH shift from 6.5 in the tropophase to 7.0 in the idiophase.

Equation (11c) described the volumetric substrate uptake rate, and Eq. (11d) described the volumetric productivity rate. Of more value for scaleup are the specific rates q, defined as

$$q = \text{VOLUMETRIC RATE / CELL CONCENTRATION} \tag{15}$$

A q value represents uptake or production rates on a per cell basis. Specific rates can be approximated from batch fermentation data, but the accuracy is questionable owing to the rates of change of concentrations of interest and the accuracy of the analytical method. Also, the specific rate is a function of fermentor age and other environmental variables. Continuous culture has proved an invaluable technique for studying specific rates, as illustrated for tylosin production by Gray and Bhuwapathanapun (1980). One can see in Fig. 3 the deviation in product formation rates from the equation. They demonstrate that the maximum specific rate occurs at a growth rate of 30% of the maximum. Table 6 lists other products for actinomycetes studied by continuous culture. By allowing the system to reach a steady state, specific rates and yield coefficients can be

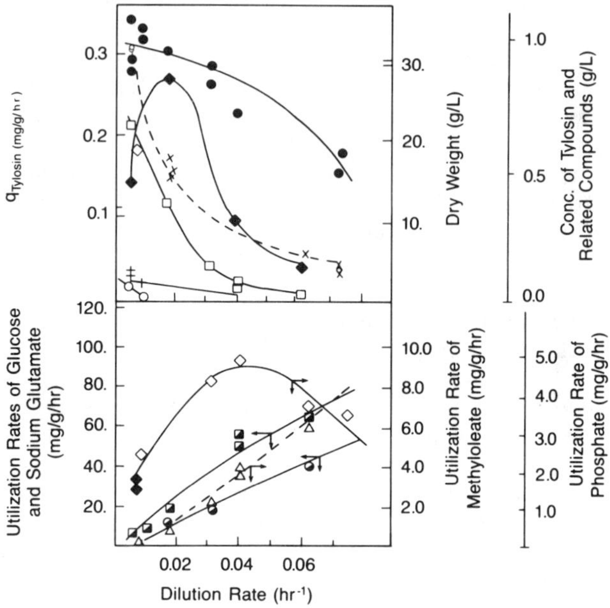

Figure 3 Effect of dilution rate on the synthesis of tylosin and related compounds and on the utilization of nutrients in chemostat culture: (●) cell concentration, (x) concentration of total tylosin-like compounds, (□) tylosin concentration (TLC), (o) macrocin concentration (TLC), (+) relomycin concentration (TLC), (◆) $q_{tylosin}$ (milligrams tylosin synthesized per gram dry weight per hour), (◇) specific uptake rate of methyloleate, (◪) specific uptake rate of glucose, (◕) specific uptake rate of sodium glutamate, (△) specific uptake rate of phosphate. (From Gray and Bhuwapathanapun, 1980; reprinted from *Biotechnology and Bioengineering*.)

measured with a high degree of accuracy. The potential drawbacks of this technique include deviations from perfect mixing, such as wall growth and genetic instability of the organism. Also, the age distribution of the mycelia is drastically different in a continuous culture as opposed to a batch culture. Finally, to reach a true steady state requires three fermentor dilutions. At a dilution rate of 0.02, 150 hr is required to reach a steady state, which can present experimental difficulties. The media used for continuous culture studies should be synthetic in order to be highly reproducible and to simplify the experimental technique.

Table 6 Selected Continuous Culture Studies of Actinomycetes

Organism/product	Dilution rate	Parameter studied	Reference
S. griseus/ streptomycin	0.047-0.098	PO_4 limitation Glucose limitation	Inoue et al., 1982
S. fradiae/ tylosin	0.005-0.08	Carbon source Nitrogen source Phosphate effects	Gray and Bhuwapathanapun, 1980
N. lactamdurans/ cephamycin C	0.025-0.045	Plasmid stability	Kirpekar et al., 1985
S. cattleya/ thienamycin	0.013-0.044	Glucose, ammonia, phosphate limitation	Lilly et al., 1981

F. Scaleup and Optimization

Early in the scaleup process, one must determine the path to large-scale implementation. There is no standard procedure for scaleup, nor should there be, given the range of potential processes. Costs, final product quality, downstream processing, and other pragmatic concerns may dictate alternatives. Even the cost of developing relevant data must be balanced against the risk of scaling up without the data. Yet there are standard steps that should be at least considered, beginning with an estimate of production costs based upon a design target. If the preliminary cost estimate looks promising, the process proceeds from shake flasks to pilot plant fermentors and, if it continues to look viable, to the production scale. Optimization of the production scale through studies at smaller scales require much more care in experimental design because of the need for greater precision in "fine-tuning" the fermentation.

Manufacturing costs are mainly a function of fermentor operation and, with the exception of carbon source or precursor feeding, usually are not directly related to culture productivity. Swartz (1979) estimated a cost of $1311 per thousand gallons of fermentor broth for penicillin production. Ryu and Oldshue (1977) calculated an 8-25% cost savings by increasing oxygen transfer through the use of a higher horsepower motor. The higher oxygen transfer would support a higher cell mass. The cost analysis highlights the most profitable areas for experimentation and suggests the basic feasibility of the process. The costs must be continually refined as the experimental work proves or disproves the initial assumptions.

From the initial cost analysis, experimental work invariably begins with shake flasks. Basic information with regard to allowable temperature ranges, inoculum levels, and acceptable media ingredients can be obtained from shake flasks. For the purposes of scaleup, it is important to keep in perspective that the shake flasks are only a step toward the larger scale. It is a common mistake to optimize shake flask yields without regard to how the process would be implemented in large equipment. For instance, culture degeneration caused by the additional steps at the large scale should at least be addressed in shake flasks. Also, the medium ingredients that are available on the production scale should be used, as opposed to reagent-grade chemicals or exotic protein hydrolysates, even though their use may add variability to shake flask results.

Medium optimization has been reviewed by Greasham and Inamine (1986). Examples of media screening techniques in shake flasks have been reported by Sensi and Thiemann (1967) for rifamycin B synthesis by *Streptomyces meditteranei*, by Stark et al. (1976) for nebramycin synthesis by *Streptomyces tenebrarius*, and by McDaniel et al. (1976) for polyene antibiotics. Each report used a different approach to media studies. Sensi and Thiemann (1967) studied the effect of addition of over 35 different nitrogen source supplements to a basic cornsteep-soybean meal medium. The nitrogen sources included inorganic nitrogen, individual amino acids, and complex nitrogen. Stark et al. (1976) discussed stepwise media improvement combined with temperature effects. McDaniel et al. (1976) demonstrated the response surface technique of optimization. This technique has the advantage of showing interactions between the media ingredients that may be significant for complex nutrients.

Statistical methods are generally most helpful early in a medium development program. The Plackett-Burman screen (1946) can be used to identify the key components. Greasham and Inamine (1986) recommended its use when more than five independent variables are to be examined. Once the key factors have been identified, optimization can be continued using a response-surface statistical method. The optimization of polyene macrolide production, performed by McDaniel et al. (1976), is an often cited example of the application of the response-surface technique. More recently, a Plackett-Burman screen and limited response-surface experiments were used for the development of a defined medium for tylosin-producing *S. fradiae* (Metzger et al., 1984).

Medium optimization should be considered a perpetual process in fermentation development, especially as new strains are introduced. Each new strain can potentially benefit from a reformulation of the medium. Between new strain optimizations, a fine-tuning of the medium can be accomplished by a theoretical fractionation of specific medium components. For example, a common component in actinomycete fermentations is soybean meal. Soybean meal is composed of approximately 30% carbohydrate and 40-50% protein (Zabriskie et al., 1980). Moreover, the mineral composition is also known. By reconstructing the medium with the theoretical components and/or more refined soymeal products, a further optimization may be obtainable. In any event, basic insights into the physiology of the organism are gained from this exercise.

The disadvantage of shake flask studies certainly relates to mixing. Shear rates in flasks are not comparable to those in fermentors, so shear effects are not representative. Oxygen transfer rates are also less, with the result that a less concentrated medium is used with resulting lower productivity. Any effect that is concentration dependent may not scale well.

From shake flasks the process is first scaled to pilot plant fermentors, typically 10-100 liter volume. At this scale, environmental variables of temperature, pH, and chemical concentrations can be controlled and studied. A series of experiments should be done to characterize the fermentation. Typical results from Proper (unpublished results) for a *Nocardia* species producing a tetracycline like antibiotic are presented in Table 7. One can see the similarity between Tables 3 and 7. The effects of the physical environment, temperature and shear, and the chemical environment must be documented. This type of data represents the first step in establishing the large-scale process. Also, optimization is a natural result of early studies. For instance, in the course of studying shear and temperature, yield increases were obtained. One might imagine further optimization by re-examining the conclusions in Table 7 or by using continuous culture to determine the specific rates, although this was not one of the scaleup objectives of the work referred to in the table.

Scaledown experiments also can be used for further optimization studies. Bošnjak et al. (1985) transferred cultures of *Streptomyces rimosus* producing oxytetracycline from production fermentors to pilot plant fermentors to study temperature, agitation, and fed-batch operation. This technique removes inoculum and media as variables. A similar technique was used by Murphy (1983) for scaling down a streptomycete fermentation from the pilot plant to shake flasks, as illustrated in Fig. 4. The difference in antibiotic potency was attributed to shear conditions. Additional work in the pilot plant may be needed if directed fermentations are involved. For instance, Sensi and Thiemann (1967) reported that sodium diethylbarbiturate causes *S. mediterranei* to synthesize only rifamycin B as opposed to rifamycin A, B, C, D, and E. However, on a large scale, it may be impractical to handle large amounts of barbiturates so a substitute must be found. Also, inducer and precursor addition times and feed rates can be more tightly controlled. As an example, Buckland et al. (1976) added cholesterol to induce cholesterol oxidase synthesis by *N. rhodocrous* based upon cell mass measurements.

Once a reliable process is defined in the pilot plant, it is ready for large-scale implementation. For proprietary reasons, this type

Table 7 Initial Pilot Plant Studies of an Antibiotic Producing *Nocardia*

Parameter studied	Pilot plant result
Temperature	Temperature profile of 30°C from 0 to 42 hr and 26°C thereafter gave a 10% yield increase over 28°C profile used in shake flasks
Shear	A 26% yield increase occurred in lowering agitation rpm by 15%
Sterilization method	Continuous sterilization did not show any advantage over batch sterilization; glucose sterilized with media as opposed to separately resulted in a 33% yield depression
Trace minerals	Low levels of minerals resulted in a 20% yield loss and abnormal fermentation broth
Phosphate	Less than 100 ppm phosphate caused an almost 100% yield depression
Ammonia	Near the end of the fermentation ammonia is depleted; addition stimulates yields
Glucose	High glucose levels depress synthesis

Source: From Proper (unpublished).

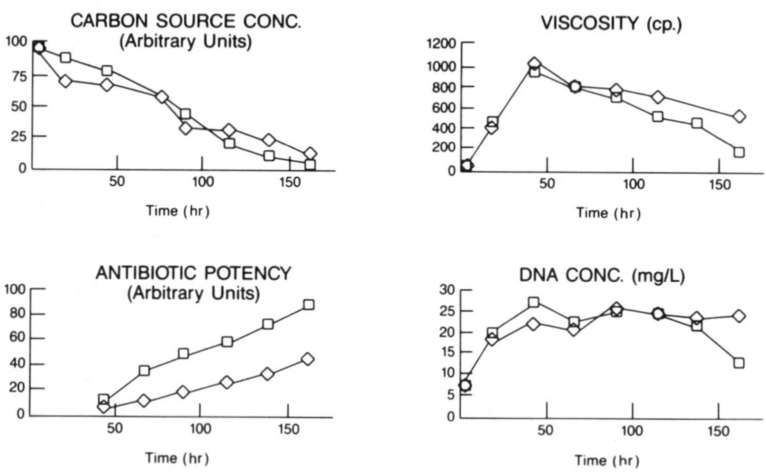

Figure 4 Shake flask (□) versus tank (◇) fermentation. (From Murphy, 1983.)

of data is not available. Certainly, the scaleup process does not stop with the first production batch. Results from fermentor performance can serve to indicate new directions for optimization. Also, the information base obtained during scaleup is needed for troubleshooting problems. Finally, an approach similar to the surface-response technique of McDaniel et al. (1976) has been applied to chemical production processes for some time. Evolutionary operations (EVOP) described by Box and Hunter (1959) can be applied to production fermentations. In an EVOP scheme, conditions are continually varied to find optimum performance.

G. Bioreactors

Although it is beyond the scope of this review to discuss fermentor design, it is nevertheless worthwhile to present some of the newer fermentor and bioreactor configurations reported for actinomycetes. Immobilized cells, airjet fermentors, and a wide range of biotransformation reactors have been described in the literature. Their application has been limited mainly to the laboratory and pilot plant because of technical problems that must be solved and a reluctance to invest significant capital in unproven technology. Yet in a competitive marketplace, the technical problems will be solved and cost advantages will be realized.

The major goal of using newer fermentor designs is to reduce the cost of oxygen transfer. In a classic stirred tank fermentor this amounts to the capital for air compressors and agitators and the resulting maintenance requirements. "Mixerless" fermentors in which air sparging supplies all the energy for oxygen transfer and mixing have been used for some time on the large scale but are limited to a few companies that have developed this technology. In rare instances the kinetics may dictate an unusual fermentor design. For instance, to avoid carbon dioxide inhibition, a "horizontal" fermentor is suggested with a low height-diameter ratio to avoid the increase in hydrostatic pressure and the corresponding increase in carbon dioxide concentration.

Airjet fermentors have been described for actinomycetes fermentations. Figure 5 illustrates the system used by Meyer and Charles (1982) for a cellulase-producing *Thermomonospora*. In this design, which is commonly called a Vogelbush fermentor, the fermentation broth is recirculated by an external pump. Fermentor mixing is caused by the velocity of the liquid jet entering the fermentor. Aeration is also due to the velocity of the liquid jet aspirating air into the system.

Increasing the velocity of the liquid jet further resulted in the high-turbulence plunging jet (HTPJ) studied by Eliezer and Jones (1984) using a mycelial fermentation. In this system, the velocity

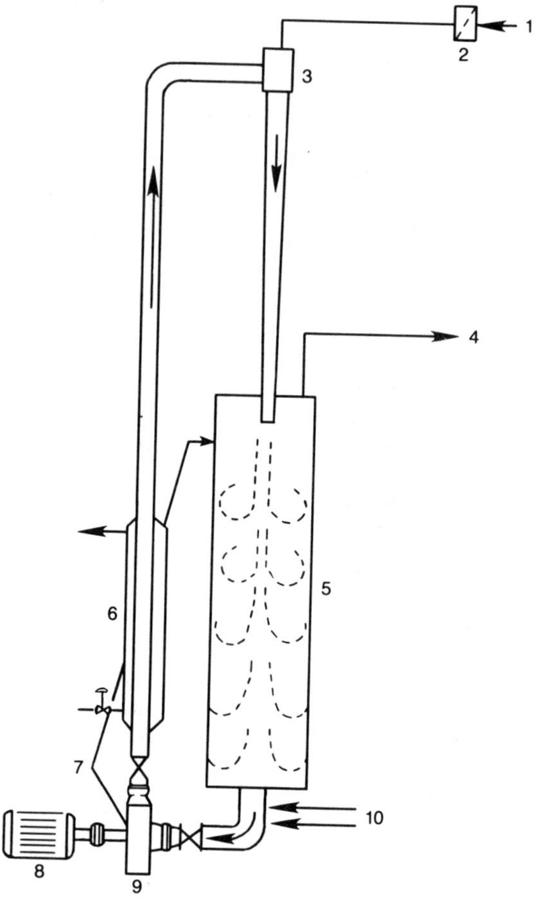

Figure 5 Schematic diagram of a deep jet fermentor: (1) air inlet; (2) air filter; (3) aeration box; (4) exhaust gas; (5) fermentation tank; (6) heat exchanger; (7) exhaust gas recirculation; (8) motor; (9) pump; (10) feed, acid, base, and so on. (From Meyer and Charles, 1982; reprinted from *Biotechnology and Bioengineering*.)

of the jet entrains the air from the headspace of the fermentor. They found a 24% increase in productivity and a 26% reduction in power cost for the HTPJ when compared with a conventional stirred tank reactor. However, their data showed the stirred tank reactor was clearly dissolved oxygen limited, so it may not be appropriate to directly compare the two fermentations.

One of the concerns of the recirculating fermentors is the design of the recirculating pumps. Meyer and Charles (1982) observed fragmentation of the organism as the pump rpm was increased. Also, to maintain dissolved oxygen levels above zero, the volume of the fermentor must be recirculated several times a minute. Pumps are not available for large-scale systems that can operate aseptically and have low internal shear at flow rates exceeding 100,000 liters/min.

Another alternative bioreactor design that has been used on an industrial scale is immobilized cells. Chibata et al. (1986) reported the use of glucose isomerase heat fixed in a *Streptomyces* species that was immobilized. Kokubu et al. (1981) used polyacrylamide-entrapped *S. fradiae* to produce protease. The rate of production of this primary metabolite was four times higher than the rate in a submerged culture and had a half-life of 30 days.

Immobilized cells have also been used for antibiotic production. They suffer from the disadvantage that the volumetric rate of production is lower although they seem to offer longer stability than the fermentation counterpart. Also, immobilization in beads or other geometrics may present increased diffusional resistances. Veelken and Pape (1982) immobilized *Streptomyces* in calcium alginate for tylosin and nikkomycin production. They measured the same productivity rate per unit cell but lower volumetric rates. Arcuri et al. (1983) described an application of immobilization for the production of thienamycin by adsorbing the organism on celite beads. Surface adsorption largely eliminates the concern of increased diffusion resistance. In this particular case the antibiotic is unstable, so that having an immobilized cell reactor allows rapid product removal.

Steroid biotransformation systems seem to be limited only by the ingenuity of the investigator. Often the hydrophobic nature of the substrate and product are utilized. Fukui and Tanaka (1981) entrapped *N. rhodochrous* in hydrophobic gels and found the rates of dehydrogenation of steroids to be a strong function of the gel hydrophobicity. Saunders et al. (1985) used resin particles as a hydrophobic product reservoir for 1-dehydrodiosgenone. Yoshida et al. (1981) used a mixed culture of *Arthrobacter simplex* and *Streptomyces roseochromogenus* for the conversion of 9α-fluorohydrocortisone to Δ'-dehydro-16α-hydroxy-9α-fluorohydrocortisone.

V. FUTURE TRENDS

Industrial actinomycete fermentations will be profoundly affected by the advent of genetic engineering and the renewed interest in biotechnology. The products, the process, and process development will all be affected. Hybrid antibiotics produced by genetically engineered organisms will place new demands on fermentation development. The types of compounds obtained from actinomycetes have also expanded to include antitumor agents, therapeutic and diagnostic enzymes, and agricultural chemicals. However, the one application in which actinomycetes may not be the organism of choice is the production of a genetically engineered protein because of the difficulty in handling mycelial organisms compared with unicellular organisms.

Perhaps the biggest impact of molecular biology may be in the development process itself. With the increasing ability to genetically modify the organisms, specific traits may be easily added instead of using labor-intensive selection and screening methods. The ability to use a standard defined medium for antibiotic production or even to use a standard organism would greatly simplify the production process. Productivity may be increased by amplification of biosynthetic genes or by insertion of unstable plasmid genes into the chromosome.

Scaleup practices and large-scale equipment design will also change but probably not at the rate of change of more research-oriented activities. Costs will remain the controlling factor, but new bioreactor and fermentor designs will be introduced as the processes dictate. Modeling of scaleup parameters and knowledge of microbial kinetics could improve to the point at which pilot-scale testing becomes a formality. A situation that will not change in the foreseeable future is the amount of information that remains proprietary. Published descriptions of fermentations will continue to lag behind the actual process in sophistication.

ACKNOWLEDGMENTS

The authors wish to thank Douglas A. Jaeger (Manager; Fermentation Process Improvement, Abbott Laboratories) for critical reading of the manuscript and helpful discussion and gratefully acknowledge Janlyn Lahti for typing and manuscript preparation.

APPENDIX: SYMBOLS

Symbol	Definition	Units
a	Gas/liquid interfacial area	cm^2/cm^3
b_0, b_1	Product formation kinetic constants	various
c	Oxygen concentration	mol/liter
c_s	Oxygen concentration at surface	mol/liter
c*	Oxygen concentration at saturation	mol/liter
dv/dy	Shear rate (i.e., velocity gradient)	sec^{-1}

Symbol	Definition	Units
D_e	Effective diffusivity	cm^2/sec
D_i	Impeller diameter	cm
D_{il}	Impeller diameter at large scale	cm
D_{is}	Impeller diameter at small scale	cm
F	Fraction of biomass that is parent strain	Dimensionless
g	Number of generations	None
k_L	Liquid film mass transfer coefficient	cm/sec
k_La	Local oxygen transfer coefficient	sec^{-1}
k_0-k_5	Growth kinetic constants [e.g., Eqs. (12) through (14)	Various
K_La	Global or volumetric oxygen transfer coefficient	hr^{-1}
K_m	Monod constant	mol/liter
K_1	Constant for K_La correlation	hr^{-1}-(W/liter)$^{-\alpha}$ (cm/sec)$^{-\beta}$
m	Consistency index for power-law fluid	g/cm-(sec)$^{2-n}$
m_1	Constant for Casson fluid model	$g^{1/2}/cm^{1/2}$-$sec^{1/2}$
m_c	Maintenance coefficient	g substrate/(g cell)-hr
n	Behavior index for power-law fluid	None
N	Impeller rpm	min^{-1}
OUR	Oxygen uptake rate	mol/liter-sec
p	Probability per generation of appearance of variants	None
P	Product concentration	mol/liter
Pg	Gassed power	W
q	Specific rate of substrate uptake or product formation	g/g cell per hr
q_p	Specific product formation rate	g product/(g cell)-hr
R_c	Critical pellet radius	cm
r_{O_2}	Rate of oxygen uptake by pellet	mol/liter-sec
S	Substrate concentration	mol/liter

Symbol	Definition	Unit
t	Time	sec
t_{ml}	Large-scale mixing time	sec
t_{ms}	Small-scale mixing time	sec
v_s	Superficial gas velocity	cm/sec
V	Volume	liters
X	Cell mass	g/liter
X_F	Final cell mass	g/liter
X_0	Initial cell mass	g
$Y_{p/s}$	Product yield coefficient	g product/ g substrate
$Y_{x/s}$	Cell yield coefficient	g cell/g substrate
	Greek Letters	
α	Exponent for P_g/V in K_La correlation	None
β	Exponent for v_s in K_La correlation	None
γ	Ratio of specific growth rates	None
η	Apparent viscosity	Pseudopoise
μ	Specific growth rate	hr^{-1}
μ_{max}	Maximum specific growth rate	hr^{-1}
τ	Shear stress	$g/cm\text{-}sec^2$
τ_0	Yield stress	$g/cm\text{-}sec^2$

REFERENCES

Aguilar, A., and Hopwood, D. (1982). Determination of methylenomycin A synthesis by the pSV1 plasmid from *Streptomyces violaceus-ruber* SANK 95570. *J. Gen. Microbiol. 128*: 1893-1901.

Aharonowitz, Y. (1979). Regulatory interrelationships of nitrogen metabolism and cephalosporin biosynthesis. In *Genetics of Industrial Microorganisms*. Sebek, O. K., and Laskin, A. I. (Eds.). American Society for Microbiology, Washington, D.C., pp. 210-217.

Aharonowitz, Y. (1980). Nitrogen metabolite regulation of antibiotic biosynthesis. *Annu. Rev. Microbiol. 34*: 209-233.

Aharonowitz, Y., and Demain, A. L. (1979). Nitrogen nutrition and regulation of cephalosporin production in *Streptomyces clavuligerus*. *Can. J. Microbiol. 25*: 61-67.

Aiba, S., Humphrey, A. E., and Millis, N. F. (1973). *Biochemical Engineering*, 2nd ed. Academic Press, New York.
Alikhanian, S. I. (1962). Induced mutagenesis in the selection of microorganisms. *Izv. Akad. Nauk SSSR [Biol.]*: 544-575.
Altenbuchner, J., and Cullum, J. (1984). DNA amplification and an unstable arginine gene in *Streptomyces lividans* 66. *Mol. Gen. Genet. 195*: 134-138.
Altenbuchner, J., and Cullum, J. (1985). Structure of an amplifiable DNA sequence in *Streptomyces lividans* 66. *Mol. Gen. Genet. 201*: 192-197.
Arcuri, E. J., Nichols, J. R., Brix, T. S., Santamaniam, V. G., Buckland, B. C., and Drew, S. W. (1983). Thienamycin production by immobilized cells of *Streptomyces cattleya* in a bubble column. *Biotechnol. Bioeng. 25*: 2399-2411.
Bailey, J. E., and Ollis, D. F. (1986). *Biochemical Engineering Fundamentals*, 2nd ed. McGraw-Hill, New York.
Baltz, R. H. (1978). Genetic recombination in *Streptomyces fradiae* by protoplast fusion and cell regeneration. *J. Gen. Microbiol. 107*: 93-102.
Baltz, R. H. (1986a). Mutagenesis in *Streptomyces* spp. In *Manual of Industrial Microbiology and Biotechnology*, A. L. Demain and N. A. Solomon (Eds.). American Society for Microbiology, Washington, D.C.
Baltz, R. H. (1986b). Mutation in *Streptomyces*. In *The Bacteria*, vol. 9. *Antibiotic-Producing Streptomyces*. Academic Press, New York.
Baltz, R. H., and Matsushima, P. (1981). Protoplast fusion in *Streptomyces*: conditions for efficient genetic recombination and cell regeneration. *J. Gen. Microbiol. 127*: 137-146.
Baltz, R. H., and Stonesifer, J. (1985). Mutagenic and error-free DNA repair in *Streptomyces*. *Mol. Gen. Genet. 200*: 351-355.
Bartholomew, W. H. (1960). Scale-up of submerged fermentations. *Adv. Appl. Microbiol. 2*: 289-300.
Bartholomew, W. H., Karow, E. O., Sfat, M. R., and Wilhelm, R. H. (1950). Oxygen transfer and agitation in submerged fermentations. Mass transfer of oxygen in submerged fermentations by *Streptomyces griseus*. *Ind. Eng. Chem. 42*: 1801-1809.
Barton, K. P., Chapman, T. W., and Lund, D. (1985). Rate of precipitation of calcium phosphate on heated surfaces. *Biotechnol. Prog. 1*: 39-45.
Berdy, J. (1974). Recent developments of antibiotic research and classification of antibiotics according to chemical structures. *Adv. Appl. Microbiol. 18*: 309-406.
Berdy, J. (1980). Recent advances in and prospects of antibiotic research. *Proc. Biochem. Oct./Nov.*: 28-35.

Bewick, M. W. M., Williams, S. T., and Veltkamp, C. (1976). Growth and ultrastructure of *Streptomyces venezuelae* during chloramphenicol production. *Microbios. 16*: 191-199.

Bibb, M. J., Ward, J. M., and Hopwood, D. A. (1978). Transformation of plasmid DNA into *Streptomyces* at high frequency. *Nature 284*: 398-400.

Bibb, M. J., Shottel, J. L., and Cohen, S. N. (1980). A DNA cloning system for interspecies gene transfer in antibiotic-producing *Streptomyces*. *Nature 284*: 526-531.

Blanch, H. W., and Bhavaraju, S. M. (1976). Non-newtonian fermentation broths: Rheology and mass transfer. *Biotechnol. Bioeng. 18*: 745-790.

Bošnjak, M., Topolovec, V., and Vrana, M. (1978). Growth kinetics of *Streptomyces erythreus* during erythromycin biosynthesis. *J. Appl. Chem. Biotechnol. 28*: 791-798.

Bošnjak, M., Stroj, A., Ćurčić, M., Adamović, V., Glunčič, Z., and Bravar, D. (1985). Application of scale-down experiments in the study of kinetics of oxytetracycline biosynthesis. *Biotechnol. Bioeng. 17*: 398-408.

Box, G. E. P., and Hunter, J. S. (1959). Condensed calculations for evolutionary operation programs. *Technometrics 1*: 77-95.

Braña, A. F., Wolfe, S., and Demain, A. L. (1985). Ammonium repression of cephalosporin production by *Streptomyces clavuligerus*. *Can. J. Microbiol. 31*: 736-743.

Buckland, B. C., Lilly, M. D., and Dunnill, P. (1976). The kinetics of cholesterol oxidase synthesized by *Nocardia rhodocrous*. *Biotechnol. Bioeng. 18*: 601-621.

Buckland, B. C., Omstead, D. R., and Santamarina, V. (1985). Novel β-lactam antibiotics. In *Comprehensive Biotechnology*, H. W. Blanch, S. Drew, and D. I. C. Wang (Eds.). Pergamon Press, New York, pp. 49-68.

Burg, R. W., Miller, B. M., Baker, E. E., Birnbaum, J., Currie, S. A., Hartman, R., Kong, Y., Monaghan, R., Olson, G., Putter, I., Tunac, J., Wallick, H., Stapley, E., Oiwa, R., and Ōmura, S. (1979). Avermectins, new family of potent anthelmintic agents: Producing organism and fermentation. *Antimicrob. Agents Chemother. 15*: 361-367.

Bylinkina, E. S., and Birukov, V. V. (1972). The problem of scale-up in antibiotic biosynthesis. In *Proc. IV IFS: Ferment. Technol. Today*, G. Terui (Ed.), pp. 105-115.

Byrne, K. M., Hilton, B. D., White, R. J., Misra, R., and Pandey, R. C. (1985). Biosynthesis of fredericamycin A, a new antitumor antibiotic. *Biochemistry 24*: 478-486.

Cain, R. B. (1981). Regulation of aromatic and hydroaromatic catabolic pathways in *Nocardioform actinomycetes*. In *Actinomycetes*, K. P. Schaal and G. Pulverer (Eds.). Gustav Fischer Verlag, New York, pp. 335-354.

Calam, C. (1970). Improvement of microorganisms by mutation, hybridization and selection. *Methods Microbiol. 3A*: 435-439.

Calam, C. T. (1986). Shake-flask fermentations. In *Manual of Industrial Microbiology*, A. L. Demain and N. A. Solomon (Eds.). American Society for Microbiology, Washington, D.C., pp. 59-65.

Campbell, I. M. (1984). Secondary metabolism and microbial physiology. *Adv. Microb. Physiol. 25*: 1-60.

Cane, D. E., Hasler, H., and Liang, T. C. (1981). Macrolide biosynthesis. Origin of the oxygen atoms in the erythromycins. *J. Am. Chem. Soc. 103*: 5960-5962.

Castro, J. M., Liras, P., Cortes, J., and Martín, J. F. (1985). Regulation of α-aminoadiphy-cysteinyl-valine, isopenicillin N synthetase, isopenicillin N isomerase, and diacetoxycephalosporin C synthetase by nitrogen sources in *Streptomyces lactamdurans*. *Appl. Microbiol. Biotechnol. 22*: 32-40.

Chang, L. T., and Elander, R. P. (1979). Rational selection for improved cephalosporin C productivity in strains of *Acremonium chrysogenum* Gams. *Dev. Ind. Microbiol. 20*: 367-379.

Chater, K. F., and Bruton, C. J. (1985). Resistance, regulatory and production genes for the antibiotic methylenomycin are clustered. *EMBO J. 4*: 1893-1897.

Chater, K. F., Hopwood, D. A., Kieser, T., and Thompson, C. J. (1982). Gene cloning in *Streptomyces*. *Curr. Top. Microbiol. Immunol. 96*: 69-95.

Chatterjee, S., and Vining, L. C. (1982). Catabolite repression in *Streptomyces venezuelae*. Induction of β-galactosidase, chloramphenicol production, and intracellular cyclic adenosine 3',5'-monophosphate concentrations. *Can. J. Microbiol. 28*: 311-317.

Chen, W. P. (1980). Glucose isomerase (a review). *Proc. Biochem.* 30-41.

Cheruy, A., and Durand, A. (1979). Optimization of erythromycin biosynthesis by controlling pH and temperature: Theoretical aspects and practical applications. *Biotechnol. Bioeng. Symp. 9*: 303-320.

Chibata, I., Tosa, T., and Sato, T. (1986). Methods of cell immobilization. In *Manual of Industrial Microbiology and Technology*, A. L. Demain and N. A. Solomon (Eds.). American Society of Microbiology, Washington, D.C., pp. 217-229.

Chung, S. T., and Molvar, N. B. (1982). Characterization and development of *Streptomyces* gene-cloning vectors. I. *Streptomyces fradiae*. In *Trends in Antibiotic Research*, H. Umezawa, A. Demain, T. Tota, and C. R. Huchinson (Eds.). Japan Antibiotics Research Association, Tokyo, pp. 42-51.

Clark, C. H., and Hopwood, D. A. (1976). Ultraviolet mutagenesis in *Streptomyces coelicolor*: Induction of reversions in a polyauxotrophic strain. *Mutat. Res. 41*: 201-207.

Cox, K., Fishman, S., Hershberger, C., and Seno, E. (1987). Improvements in or relating to antibiotic-producing microorganisms. European Patent Application 87302318.8.

Crameri, R., and Davies, J. E. (1986). Increased production of aminoglycosides associated with amplified antibiotic resistance genes. *J. Antibiot.* *39*: 128-135.

Dahl, S. G., Waaler, T., Thomassen, S., Oystese, B., and Hoagland, T. (1972). Testing error in the screening of bacitracin producing strains for high yielding mutants. *Pharm. Acta Helv.* *47*: 424-432.

Delavier-Klutchko, C., and Flavin, M. (1965). Enzymatic synthesis and cleavage of cystathionine in fungi and bacteria. *J. Biol. Chem.* *240*: 2537-2549.

Delic, V., Hopwood, D. A., and Friend, E. J. (1970). Mutagenesis by *N*-methyl-*N'*-nitro-*N*-nitrosoguanidine (NTG) in *Streptomyces coelicolor*. *Mutat. Res.* *9*: 167-182.

Demain, A. L., Aharonowitz, Y., and Martín, J. F. (1983). Metabolic control of secondary biosynthetic pathways. In *Biochemistry and Genetic Regulation of Commercially Important Antibiotics*, L. C. Vining (Ed.). Addison-Wesley, Reading, Massachusetts, pp. 49-72.

DeWitt, J. P. (1985). Evidence for a sex factor in *Streptomyces erythreus*. *J. Bacteriol.* *164*: 969-971.

Dotzlaf, J. E., Metzger, L. S., and Foglesong, M. A. (1984). Incorporation of amino acid-derived carbon into tylactone by *Streptomyces fradiae* GS14. *Antimicrob. Agents Chemother.* *25*: 216-220.

Drew, S. W., and Demain, A. L. (1977). Effect of primary metabolites on secondary metabolism. *Annu. Rev. Microbiol.* *31*: 343-356.

Dulaney, E. L., and Dulaney, D. D. (1967). Mutation populations of *Streptomyces viridifaciens*. *Trans. N.Y. Acad. Sci.* *29*: 782-799.

Elander, R. P., and Chang, L. T. (1979). Microbial culture selection. *Microb. Technol.* *2*: 243-302.

Eliezer, E., and Jones, R. T. (1984). The application of a novel bioreactor design to mass transfer limited fermentations. *World Biotech. Rep.* *2*: 303-314.

Eritt, I., Gräfe, U., and Fleck, W. F. (1984). Inducers of both cytodifferentiation and anthracycline biosynthesis of *Streptomyces griseus* and their occurrence in actinomycetes and other microorganism. *Z. Allg. Mikrobiol.* *24*: 3-12.

Fairbairn, D. A., Priest, F. G., and Stark, J. R. (1986). Extracellular amylase synthesis by *Streptomyces limosus*. *Enzyme Microb. Technol.* *8*: 89-92.

Feller, K., and Träger, L. (1985). Steroid-induced regulation of 3α,20β- and 3β,17β-hydroxysteroid dehydrogenase activity in wild type and mutants of *Streptomyces hydrogenans*. *Arch. Microbiol. 143*: 256-259.

Fishman, S., and Hershberger, C. (1983). Amplified DNA in *Streptomyces fradiae*. *J. Bacteriol. 155*: 459-466.

Florent, J. (1985). Streptomycin and commercially important aminoglycoside antibiotics. *Compr. Biotechnol. 3*: 137-162.

Foster, J. W., and Katz, E. (1981). Control of actinomycin D biosynthesis in *Streptomyces parvulus*: Regulation of tryptophan oxygenase activity. *J. Bacteriol. 148*: 670-677.

Freese, E. (1981). Initiation of bacterial sporulation. In *Sproulation and Germination*, H. S. Levinson, A. L. Sohenshein, and D. J. Tipper (Eds.). American Society for Microbiology, Washington, D.C., pp. 1-12.

Friedrich, W., Bormann, E. J., and Gräfe, U. (1984). Isolation and biological properties of arsenate-resistant strains of *Streptomyces noursei*. *Z. Allg. Mikrobiol. 24*: 13-19.

Friend, E. J., Warren, M., and Hopwood, D. A. (1978). Genetic evidence for a plasmid controlling fertility in an industrial strain of *Streptomyces rimosus*. *J. Gen. Microbiol. 98*: 453-465.

Fukui, S., and Tanaka, A. (1981). Synthetic prepolymer methods of biocatalysts. Application of gel-entrapped microbial cells to bioconversions of lipophilic compounds in hydrophobic conditions. *Adv. Biotechnol. 3*: 343-348.

Gallo, M., and Katz, E. (1972). Regulation of secondary metabolite synthesis: Catabolite repression of phenoxazinone synthase and actinomycin formation by glucose. *J. Bacteriol. 109*: 659-667.

Gersch, D., Skurk, A., and Romer, W. (1979). Phosphate inhibition of secondary metabolism in *Streptomyces hygroscopicus* and its reversal by cyclic AMP. *Arch. Microbiol. 121*: 91-96.

Gibb, G. D., and Strohl, W. R. (1987). Physiological regulation of protease activity in *Streptomyces peucetius*. *Can. J. Microbiol. 34*: 187-190.

Ginther, C. L. (1979). Sporulation and the production of serine protease and cephamycin C by *Streptomyces lactamdurans*. *Antimicrob. Agents Chemother. 15*: 522-526.

Godfrey, O. (1973). Isolation of regulatory mutants of the aspartic and pyruvic acid families and their effect on antibiotic production in *Streptomyces lipmanii*. *Antimicrob. Agents Chemother. 4*: 73-79.

Godfrey, O., Ford, L., and Huber, M. L. B. (1978). Interspecies matings of *Streptomyces fradiae* with *Streptomyces bikiniensis* mediated by conventional and protoplast fusion techniques. *Can. J. Microbiol. 24*: 994-997.

Goodman, J. J. (1985). Fermentation and mutational development of the tetracyclines. *Handb. Exp. Pharmacol. 78*: 5-57.

Gräfe, U. (1982). Relevance of microbial nitrogen metabolism to production of secondary metabolites. In *Overproduction in Microbial Products*, V. Krumphanzl, B. Sikyta, and Z. Vanek (Eds.). Academic Press, New York, pp. 63-75.

Gray, P. P., and Bhuwapathanapun, S. (1980). Production of the macrolide antibiotic tylosin in batch and chemostat cultures. *Biotechnol. Bioeng. 22*: 1785-1804.

Greasham, R., and Inamine, E. (1986). Nutritional improvement of processes. In *Manual of Industrial Microbiology*, A. L. Demain and N. A. Solomon (Eds.). American Society for Microbiology, Washington, D.C., pp. 41-48.

Gregory, K. F., and Huang, J. (1964). Tyrosinase inheritance in *Streptomyces scabies*. I. Genetic recombination. *J. Bacteriol. 87*: 1281-1294.

Hamilton, B., White, R., McGuire, J., Montgomery, P., Stroshane, R., Kalite, C., and Pandey, R. (1980). Improvement of the daunorubicin fermentation realized at 10,000 liter fermentor scale. *Adv. Biotechnol. 1*: 63-68.

Hänel, F., Gräfe, U., Friedrich, W., and Bormann, E. J. (1984). ATP levels in phosphate-deregulated and arsenate-resistant mutants of *Streptomyces noursei*. *Z. Allg. Mikrobiol. 24*: 239-245.

Hanlon, G. W., and Hodges, N. A. (1981). Bacitracin and protease production in relation to sporulation during exponential growth of *Bacillius licheniformis* on poorly utilized carbon and nitrogen sources. *J. Bacteriol. 147*: 427-431.

Hara, O., and Beppu, T. (1982). Induction of streptomycin-inactivating enzyme by A-factor in *Streptomyces griseus*. *J. Antibiot. 35*: 1208-1215.

Harold, F. M. (1966). Inorganic polyphosphates in biology: Structure, metabolism and function. *Bacteriol. Rev. 30*: 772-794.

Hasegawa, M., Hintermann, G., Simonet, J. M., Crameri, R., Piret, J., and Hutter, R. (1985). Certain chromosomal regions in *Streptomyces glaucescens* tend to carry amplifications and deletions. *Mol. Gen. Genet. 200*: 375-384.

Hirsch, C. F., and McCann-McCormick, P. A. (1985). Biology of *Streptomyces*. In *Biology of Industrial Microorganisms*, A. L. Demain and N. A. Solomon (Eds.). Benjamin/Cummings, Menlo Part, California, pp. 291-314.

Hodgson, D. A. (1982). Glucose repression of carbon source uptake and metabolism in *Streptomyces coelicolor* A3(2) and its perturbation in mutants resistant to 2-deoxyglucose. *J. Gen. Microbiol. 128*: 2417-2430.

Holt, G., and Saunder, G. (1986). Genetic modification of industrial microorganisms. In *Comprehensive Biotechnology, vol. 1, The Principles of Biotechnology: Scientific Fundamentals*, A. T. Bull and H. Dalton (Eds.). Pergamon Press, Oxford, pp. 51-76.

Hopwood, D. A. (1976). Genetics of antibiotic production in *Streptomyces*. In *Microbiology 1976*. American Society for Microbiology, Washington, D.C.

Hopwood, D. A. (1978). Extrachromosomally determined antibiotic production. *Annu. Rev. Microbiol. 32*: 373-392.

Hopwood, D. A. (1981). Genetic studies with bacterial protoplasts. *Annu. Rev. Microbiol. 35*: 237-272.

Hopwood, D. A. (1983). Actinomycete genetics and antibiotic production. In *Biochemistry and Genetic Regulation of Commercially Important Antibiotics*, L. C. Vining (Ed.). Addison-Wesley, Reading, Massachusetts, pp. 1-23.

Hopwood, D. A., and Chater, K. F. (1984). Streptomycetes. In *Genetics and Breeding of Industrial Microbiology*, C. Ball (Ed.). CRC Press, Boca Raton, Florida, pp. 7-42, 51-76.

Hopwood, D. A., and Merrick, M. J. (1977). Genetics of antibiotic production. *Bacteriol. Rev. 49*: 595-635.

Hopwood, D. A., and Sermonti, G. (1962). The genetics of *Streptomyces coelicolor*. *Adv. Genet. 11*: 273-342.

Hopwood, D. A., and Wright, H. M. (1972). Transformation in *Thermoactinomyces vulgaris*. *J. Gen. Microbiol. 71*: 383-398.

Hopwood, D. A., Chater, K. F., Dowding, J. E., and Vivian, A. (1973). Advances in *Streptomyces coelicolor* genetics. *Bacteriol. Rev. 31*: 373-403.

Hopwood, D. A., and Wright, H. M. (1978). Bacterial protoplast fusion: Recombination in fused protoplasts of *Streptomyces coelicolor*. *Mol. Gen. Genet. 162*: 307-317.

Hopwood, D. A., Wright, H. M., Bibb, M. J., and Cohen, S. N. (1977). Genetic recombination through protoplast fusion in *Streptomyces*. *Nature 268*: 171-174.

Hopwood, D. A., Kieser, T., Wright, H. M., and Bibb, M. J. (1983a). Plasmids, recombination and chromosome mapping in *Streptomyces lividans* 66. *J. Gen. Microbiol. 129*: 2257-2269.

Hopwood, D. A., Bibb, M. J., Bruton, C. J., Chater, K. F., Feitelson, J. S., and Gil, J. A. (1983b). Cloning*Streptomyces* genes for antibiotic production. *Trends Biotechnol. 1*: 42-48.

Hopwood, D. A., Bibb, M. J., Chater, K. F., Kieser, T., Bruton, C. J., Kieser, H. M., Lydiate, J., Smith, C. P., Ward, J. M., and Schrempf, H. (1985). *Genetic Manipulation in Streptomyces: A Lab Manual*. John Innes Foundation, Norwich, U.K.

Hopwood, D. A., Bibb, M. J., Chater, K., and Kieser, T. (1987). Plasmid and phage vectors for gene cloning and analysis in *Streptomyces*. *Methods Enzymol.* *153*(D): 116-166.

Hranueli, D., Pigac, J., Smokvina, T., and Alacevic, M. (1983). Genetic interactions in *Streptomyces rimosus* mediated by conjugation and by protoplast fusion. *J. Gen. Microbiol.* *129*: 1451-1422.

Hutter, R., Kieser, T., Crameri, R., and Hintermann, G. (1981). Chromosomal instability in *Streptomyces glaucescens*. *Zastralbl. Bakteriol.* [*Suppl.*] *11*: 551-559.

Ichikawa, T., Date, M., Ishikura, I., and Ozaki, A. (1971). Improvement of kasugamycin-producing strain by the agar piece method and the prototroph method. *Folia Microbiol. (Praha)* *16*: 218-224.

Ikeda, N., Inoue, M., and Ōmura, S. (1983). Improvement of macrolide antibiotic-producing streptomycete strains by the regeneration of protoplasts. *J. Antibiot.* *36*: 283-288.

Ikeda, H., Seno, E. T., Bruton, C. J., and Chater, K. F. (1984). Genetic mapping, cloning, and physiological aspects of the glucose kinase gene of *Streptomyces coelicolor*. *Mol. Gen. Genet.* *196*: 501-507.

Imanaka, T., and Aiba, S. (1981). A perspective on the application of genetic engineering: Stability of recombinant plasmid. *Ann. N.Y. Acad. Sci.* *369*: 1-14.

Inoue, S., Nishizawa, Y., and Nagai, S. (1982). Stimulation of streptomycin formation by *Streptomyces griseus* grown in a phosphate deficient media. *J. Ferment. Technol.* *60*: 417-422.

Ivanitskaya, L. P., Bibikova, M. V., Gromova, M. N., Zhadanovich, Y. V., Vagina, I. M., and Ishatov, E. N. (1980). Selective media with lincomycin for directed screening of antibiotic producing microorganisms. *Antibiotiki* *26*: 83-86.

Iwamoto, T., Nakaya, T., Inaoka, M., and Naka, H. (1983). Induction of β-*N*-acetyl-glucosaminidase and uptake of *N*-acetylglucosamine by *Streptomyces* sp. *J. Ferment. Technol.* *61*: 499-503.

Járai, M., Győry, E., and Tombor, J. (1969). Oxygen transfer in *Streptomyces* fermentation broths. *Biotechnol. Bioeng.* *11*: 605-622.

Járai, M. (1972). Oxygen transfer in the fermentations of primary and secondary metabolites. In *Proc. IV IFS: Ferment. Technol. Today*, G. Terui (Ed.), pp. 97-103.

Kacser, H., and Burns, J. A. (1973). The control of flux. In *Rate Control of Biological Processes*, D. D. Davies (Ed.). Cambridge University Press, London, pp. 65-104.

Kalk, J. P., and Langlykke, A. L. (1986). Cost estimation for biotechnology projects. In *Manual of Industrial Microbiology*, A. L. Demain and N. A. Solomon (Eds.). American Society for Microbiology, Washington, D.C., pp. 363-385.

Kawaguchi, T., Asahi, T., Satoh, T., Uozumi, T., and Beppu, T. (1984). B-factor, an essential regulatory substance inducing the production of rifamycin in a *Nocardia* sp. *J. Antibiotics 37*: 1587-1595.

Keller, U., Poschmann, S., Krengel, U., Kleinkauf, H., and Kraepelin, G. (1983). Studies of protoplast fusion in *Streptomyces chrysomallus*. *J. Gen. Microbiol. 129*: 1725-1731.

Kendall, K., and Cullum, J. (1984). Cloning and expression of an extracellular-agarase gene from *Streptomyces coelicolor* A3(2) in *Streptomyces lividans* 66. *Gene 29*: 315-321.

Kendrick, K. E., and Ensign, J. C. (1983). Sporulation of *Streptomyces griseus* in submerged culture. *J. Bacteriol. 155*: 357-366.

Kern, B. A., and Inamine, E. (1981). Cystathionine-γ-lyase activity in the cephamycin C producer *Streptomyces lactamdurans*. *J. Antibiot. 34*: 583-589.

Khoklov, A. S. (1982). Low molecular weight microbial bioregulators of secondary metabolism. In *Overproduction of Microbial Metabolites*, V. Krumphanzl, B. Sikyta, and Z. Vanek (Eds.). Academic Press, London, pp. 97-109.

Khoklov, A. S., and Tovarova, I. I. (1979). Autoregulator from *Streptomyces griseus*. *Regulation of Secondary Product and Plant Hormone Metabolism*, M. Luckner and K. Schreiber (Eds.). Pergamon Press, New York, pp. 133-145.

Kirby, R., and Hopwood, D. A. (1977). Genetic determination of methylenomycin synthesis by the SCP1 plasmid of *Streptomyces coelicolor* A3(2). *J. Gen. Microbiol. 98*: 239-252.

Kirby, R., Wright, L., and Hopwood, D.A. (1975). Plasmid-determined antibiotic synthesis and resistance in *Streptomyces coelicolor*. *Nature 254*: 265-267.

Kirpekar, A. C., Kirwan, D. J., and Stieber, R. W. (1985). Modeling the stability of cephamycin C-producing *N. lactamdurans* during continuous culture. *Biotechnol. Prog. 1*: 231-236.

Kitagiri, I. (1954). Improvement of chlorotetracycline producing strain by several kinds of methods. *J. Antibiot. 7*: 45-52.

Kitano, K., Nozaki, Y., and Imada, A. (1985). Strain improvement of a carbapenem antibiotic producer, *Streptomyces incarnatus* NRRL8089. *J. Antibiot. 38*: 1204-1210.

Kleiner, E. M., Pliner, S. A., Soifer, V. S., Onoprienko, V. V., Balashova, T. A., Rozynov, B. V., and Khoklov, A. S. (1976). Structure of the A-factor, a bioregulator from *Streptomyces griseus*. *Bioorg. Khim. 2*: 1142-1147.

Koepsel, R., and Ensign, J. C. (1984). Microcycle sporulation of *Streptomyces viridochromogenes*. *Arch. Microbiol. 140*: 9-14.

Kokubu, T., Karube, I., and Suzuki, S. (1981). Protease production by immobilized mycelia of *Streptomyces fradiae*. *Biotechnol. Bioeng. 23*: 29-39.

Kominek, L. A. (1972). Biosynthesis of novobiocin by *Streptomyces niveus*. *Antimicrob. Agents Chermother.* *1*: 123-134.

Krugel, H., Fiedler, G., and Noack, D. (1980). Transformation of protoplasts from *Streptomyces lividans* 66 with actinophage SH10 DNA. *Mol. Gen. Genet.* *177*: 297-300.

Kuklin, V., Emelyanova, L., Zhdanov, V., and Yustrotova, L. (1983). Protoplast fusion for improvement of *Streptomyces griseus* producing grisin, a streptothricin antibiotic. *Antibiotiki* *28*: 883-889.

Lechevalier, M. P., and Lechevalier, H. (1980). The chemotaxonomy of actinomycetes. In *Actinomycete Taxonomy*, A. Dietz and D. W. Thayer (Eds.). SIM Spec. Pub. No. 6, Arlington, Virginia, pp. 227-291.

Lechevalier, M. P., and Lechevalier, H. (1985). Biology of actinomycetes not belonging to the genus *Streptomyces*. In *Biology of Industrial Microorganisms*, A. L. Demain and N. A. Solomon (Eds.). Benjamin/Cummings, Menlo Park, California, pp. 315-358.

Lee, J. G., Choi, C. Y., Seong, B. L., and Han, M. H. (1983). Optimal pH profile in rifamycin B fermentation. *J. Ferment. Technol.* *61*: 49-53.

Lilly, G., Clark, A. E., and Lawrence, G. C. (1981). Control of the production of cephamycin C and thienamycin by *Streptomyces cattleya* NRRL 8057. *J. Chem. Tech. Biotechnol.* *31*: 127-134.

L'vova, N. A., and Lapchinskaya, O. A. (1981). Selection of productive variants from cultures of olivomycin producing *Streptomyces olivoreticuli* 16749. *Antibiotiki* *26*: 890-893.

Madry, N., and Pape, H. (1982). Formation of secondary metabolism enzymes in the tylosin producer *Streptomyces* T59-235. *Arch. Microbiol.* *131*: 170-173.

Madry, N., Sprinkmeyer, R., and Pape, H. (1979). Regulation of tylosin synthesis in *Streptomyces*: Effects of glucose analogs and inorganic phosphate. *Eur. J. Appl. Microbiol. Biotechnol.* *7*: 365-370.

Makins and Holt (1981). Liposome mediated transformation of streptomycetes by chromosomal DNA. *Nature* *293*: 671-673.

Malina, H., Tempete, C., and Robert-Gero, M. (1985). Enhanced sinefungin production by medium improvement, mutagenesis, and protoplast regeneration of *Streptomyces incarnatus* NRRL8089. *J. Antibiot.* *38*:1204-1210.

Manfredini, R., Cavallera, V., Marini, L., and Donati, G. (1983). Mixing and oxygen transfer in conventional stirred fermentors. *Biotechnol. Bioeng.* *25*: 3115-3131.

Martín, J. F. (1977). Control of antibiotic synthesis by phosphate. *Adv. Biochem. Eng.* *6*: 105-127.

Martín, J. F., and Demain, A. L. (1976). Control by phosphate of candicidin production. *Biochem. Biophys. Res. Commun. 71*: 1103-1109.

Martín, J. F., and Demain, A. L. (1980). Control of antibiotic biosynthesis. *Microbiol. Rev. 44*: 230-251.

Martín, J. F., and McDaniel, L. E. (1974). The submerged culture production of the polyene antifungal antibiotics candicidin and candihexin. *Dev. Ind. Microbiol. 15*: 324-337.

Martín, J. F., and McDaniel, L. E. (1975). Kinetics of biosynthesis of polyene macrolide antibiotics in batch cultures: Cell maturation time. *Biotechnol. Bioeng. 17*: 925-938.

Martín, J. F., Naharro, G., Liras, P., and Villanueva, J. R. (1979). Isolation of mutants deregulated in phosphate control of candicidin biosynthesis. *J. Antibiot. 32*: 600-606.

Maxon, W. D. (1959). Aeration-agitation studies on the novobiocin fermentation. *J. Biochem. Microbiol. Technol. Eng. 1*: 311-324.

McCann-McCormick, P. A., Monaghan, R. L., Baker, E. E., Goegelman, R. T., and Stapley, E. O. (1981). Studies on the avermectin fermentation. *Adv. Biotechnol. 1*: 69-74.

McDaniel, L. E., Bailey, E. G., Ethiraj, S., and Andrews, H. P. (1976). Application of response surface optimization techniques to polyene macrolide fermentation studies in shake flasks. *Dev. Ind. Microbiol. 17*: 91-98.

Megee, R. D., III, Kinoshita, S., Fredrickson, A. G., and Tsuchiya, H. M. (1970). Differentiation and product formation in molds. *Biotechnol. Bioeng. 12*: 771-801.

Metzenberg, R. L. (1979). Implications of some genetic control mechanisms in *Neurospora*. *Microbiol. Rev. 43*: 361-383.

Metzger, L. S., Dotzlaf, J. E., and Foglesong, M. A. (1984). Development of a defined medium for tylosin producing strains of *Streptomyces fradiae*. Abstr. Annu. Meet. Am. Soc. Microbiol. 061, p. 199.

Meyer, H., and Charles, M. (1982). Cultivation of a filamentous bacterium in a deep jet fermentor. *Biotechnol. Bioeng. 24*: 1905-1909.

Miller, T. L. (1985). Steroid fermentations. In *Comprehensive Biotechnology*, vol. 3, *The Practice of Biotechnology: Current Commodity Products*, H. W. Blanch, S. Drew, and D. I. C. Wang (Eds.). Pergamon Press, New York, pp. 297-318.

Miller, T. L., and Churchill, B. W. (1986). Substrates for large-scale fermentation. In *Manual of Industrial Microbiology*, A. L. Demain and N. A. Solomon (Eds.). American Society for Microbiology, Washington, D.C., pp. 122-136.

Modarski, M., Kuryłowicz, W., and Jeljaszewicz, J. (Eds.). *Nocardia and Streptomyces*. Gustav Fisher Verlag, New York, 1978.

Moo-Young, M., Whitworth, D. A., and Shimizu, T. (1972). Effects of agitation power on cell productivity in hydrocarbon fermentation. In *Proc. IV IFS: Ferment. Technol. Today*, G. Terui (Ed.), pp. 91-96.

Mracek, M., Blumauerova, J., Paleckova, F., and Hostalek, Z. (1969). Regulation of biosynthesis of secondary metabolites, XI. Induction of variants in *Streptomyces aureofaciens* and the specificity of mutagens. *Mutat. Res. 7*: 19-24.

Mukataka, S., Kataoka, H., and Takahashi, J. (1980). Effects of vessel size and rheological properties of suspensions on the distributions of circulation times in stirred vessels. *J. Ferment. Technol. 58*: 155-161.

Müller, P. J., Haubold, G., Menner, M., Grosse, H. H., Ozegowski, J. H., and Bocker, H. (1984). Effect of phosphate on the biosynthesis of nourseothricin in *Streptomyces noursei*, J. A. 3890b. *Z. Allg. Mikrobiol. 24*: 555-564.

Murphy, T. F. (1983). Fermentation scale-up: A challenge in biotechnology. Presented at 83rd Annual Meeting American Society of Microbiology.

Murphy, T. F., Wald, S. A., and Jackson, J. V. (1984). Oxygen transfer in a 100,000 liter fungal fermentation, presented at 188th National ACS Meeting, August 1984.

Nagasawa, T., Kanzaki, H., and Yamada, H. (1984). Cystathionine-γ-lyase of *Streptomyces phaeochromogenes*. *J. Biol. Chem. 259*: 10393-10403.

Nakano, M. M., Ishihara, H., and Ogawara, H. (1982). Fusion of protoplasts of *Streptomyces lavendulae*. *J. Antibiot. 35*: 359-363.

Nash, C. H., III (1974). Effect of carbon dioxide on synthesis of erythromycin. *Antimicrob. Agents Chemother. 5*: 544-545.

Novotna, J., and Hostalek, Z. (1985). Phosphorylation of hexoses in *Streptomyces aureofaciens:* Evidence that the phosphoenolypyruvate:sugar phosphotransferase system is not operative. *FEMS Microbiol. Lett. 28*: 347-350.

Ochi, K. (1986). A decrease in GTP content is associated with aerial mycelium formation in *Streptomyces* MA406-A-1. *J. Gen. Microbiol. 132*: 299-305.

Ochi, K. (1987). Changes in nucleotide pools during sporulation of *Streptomyces griseus* submerged culture. *J. Gen. Microbiol. 133*: 2787-2795.

Ochi, K., Hitchcock, M., and Katz, E. (1979). High-frequency fusion of *Streptomyces parvulus* or *Streptomyces antibioticus* protoplasts induced by polyethylene glycol. *J. Bacteriol. 139*: 984-992.

Ogata, S., Yamada, S., and Hayashida, S. (1985). Genetic recombination in *Streptomyces azureus* by protoplast fusion and high production of thiostrepton by recombinants. *J. Gen. Appl. Microbiol. 31*: 187-191.

Ogawa, H., Imai, S., Shimizu, T., Satoh, A., and Kojima, M. (1983). Cosynthesis and protoplast fusion by mutants of bialophos (AMPBA) producing *Streptomyces hygroscopicus. J. Antibiot. 36*: 1040-1044.

Oh, Y. K., Speth, J. L., and Nash, C. (1980). Protoplast fusion with *Streptosporangium viridogriseum. Dev. Ind. Microbiol. 21*: 219-226.

Okanishi, M. (1979). Plasmids and antibiotic synthesis in streptomycetes. In *Genetics of Industrial Microorganisms*, O. K. Sebek and A. I. Laskin (Eds.). American Society for Microbiology, Washington, D.C., pp. 134-140.

Oldshue, J. Y. (1981). Let's understand mixing. *Chemtechnology 11*: 554-561.

Ollis, D. F., and Chang, H. (1982). Batch fermentation kinetics with (unstable) recombinant cultures. *Biotechnol. Bioeng. 24*: 2583-2586.

Omstead, D. R., Hunt, G. R., and Buckland, B. C. (1985). Commercial production of cephamycin antibiotics. In *Comprehensive Biotechnology*, vol. 3, H. W. Blanch, S. Drew, and D. I. C. Wang (Eds.). Pergamon Press, New York, pp. 187-210.

Ōmura, S., Nakagawa, A., Takeshima, H., Miyazawa, J., Kitao, C., Piriou, F., and Lukacs, G. (1975). A ^{13}C nuclear magnetic resonance study of the biosynthesis of the 16-membered macrolide antibiotic tylosin. *Tetrahedron Lett. 50*: 4503-4506.

Ōmura, S., Takeshima, H., Nakagawa, A., Kanemoto, N., and Lukacs, G. (1976). Studies on carboxylic acid metabolism in a macrolide-producing microorganism using carbon-13 magnetic resonance. *Biorg. Chem. 5*: 451-454.

Ōmura, S., Takeshima, H., Nakagawa, A., Miyazawa, J., Piriou, F., and Lukacs, G. (1977). Studies on the biosynthesis of 16-membered macrolide antibiotics using carbon-13 nuclear magnetic resonance spectroscopy. *Biochemistry 16*: 2860-2866.

Ōmura, S., Tanaka, Y., Tanaka, H., Takahishi, Y., and Iwai, Y. (1980). Stimulation of the production of macrolide antibiotics by magnesium phosphate and related insoluble materials. *J. Antibiot. 33*: 1568-1569.

Ōmura, S., Tanaka, Y., Tanaka, H., Takahashi, Y., Iwai, Y., and Kitao, C. (1981). Stimulation of leucomycin production by magnesium phosphate and its relevance to nitrogen catabolite regulation. *Adv. Biotechnol. 3*: 181-186.

Ōmura, S., Tanaka, Y., Mamada, H., and Masuma, R. (1983). Ammonium ion suppresses the biosynthesis of tylosin aglycone by interference with valine catabolism in *Streptomyces fradiae. J. Antibiot. 36*: 1792-1794.

Ōmura, S., Tanaka, Y., Mamada, H., and Masuma, R. (1984a). Effect of ammonium ion, inorganic phosphate and amino acids on the biosynthesis of protylonolide, a precursor of tylosin aglycone. *J. Antibiot.* *37*: 494-502.

Ōmura, S., Taki, A., Matsuda, K., and Tanaka, Y. (1984b). Ammonium ions suppress the amino acid metabolism involved in the biosynthesis of protylonolide in a mutant of *Streptomyces fradiae*. *J. Antibiot.* *37*: 1362-1369.

Ōmura, S., Ikeda, H., Malpartida, F., Kieser, H. M., and Hopwood, D. A. (1986). Production of new hybrid antibiotics, Mederrhodins A and B, by a genetically engineered strain. *Antimicrob. Agents Chemother.* *29*: 13-19.

Ono, H., Hintermann, G., Crameri, R., Wallis, G., and Hutter, R. (1982). Reiterated DNA sequences in a mutant strain of *Streptomyces glaucescens* and cloning of the sequence in *Escherichia coli*. *Mol. Gen. Genet.* *186*: 106-110.

Orlova, N. V., and Prokofieva-Belgovskaya, A. A. (1961). The effect of amount and age of culture material on the development of *Actinomyces rimosus* and on production of oxytetracycline. *Antibiotiki* *6*: 15-20.

Páca, J., Ettler, P., and Grégr, V. (1978). Oxygen transfer rate in media used for erythromycin biosynthesis. *J. Ferment. Technol.* *56*: 144-151.

Palleroni, N. J. (1983). Genetic recombination in *Actinoplanes brasiliensis* by protoplast fusion. *Appl. Environ. Microbiol.* *45*: 1865-1869.

Pastan, I., and Adhya, S. (1976). Cyclic adenosine 5'-monophosphate in *Escherichia coli*. *Bacteriol. Rev.* *40*: 527-551.

Payne, G. F., and Wang, H. Y. (1988). Phosphate feeding to permit growth while maintaining secondary product synthesis. *Appl. Microbiol. Biotechnol.* *27*: 572-576.

Petty, T. M., and Crawford, D. L. (1984). Enhancement of lignin degradation in *Streptomyces* spp. by protoplast fusion. *Appl. Environ. Microbiol.* *47*: 439-440.

Plackett, R. L., and Burman, J. P. (1946). The design of optimum multifactorial experiments. *Biometrika* *33*: 305-325.

Puglia, A. M., Spada-Sermonti, I., Basile, S., Misuraca, F., and Sermonti, G. (1973). Infectious transfer of a fertility factor in *Streptomyces coelicolor*. *Mutat. Res.* *21*: 107-113.

Queener, S. W., and Lively, D. H. (1986). Screening and selection for strain improvement. In *Manual of Industrial Microbiology and Biotechnology*, A. L. Demain and N. A. Solomon (Eds.). American Society for Microbiology, Washington, D.C., pp. 155-169.

Raczyńska-Bojanowska, K., Ruczaj, Z., Sawnor-Korszyńska, D., and Rafalski, A. (1973). Limiting reactions in activation of acyl units in biosynthesis of macrolide antibiotics. *Antimicrob. Agents Chemother.* *3*: 162-167.

Reusser, F. (1963). Stability and degeneration of microbial cultures on repeated transfer. *Adv. Appl. Microbiol. 5*: 189-215.
Riesenberg, D., Bergter, F., and Kari, C. (1984). Effect of serine hydroxamate and methyl-α-D-glucopyranoside treatment on nucleoside polyphosphate pools, RNA and protein accumulation in *Streptomyces hydroscopicus. J. Gen. Microbiol. 130*: 2549-2558.
Rimsay, R. L., Murphy, G. W., Martin, C. J., and Orr, J. C. (1988). The 20 -hydroxysteroid dehydrogenase of *Streptomyces hydrogenans. Eur. J. Biochem. 174*: 437-442.
Rinehart, K. L., Jr. (1983). Newer methods for characterization of antibiotics. I. Stable isotopes in the study of antibiotic synthesis. *Biotechnology August*: 495-502.
Robinson, C. W., and Wilke, C. R. (1972). Oxygen volumetric mass transfer coefficients in agitated basal salt solutions simulating fermentation media. In *Proc. IV IFS: Ferment. Technol. Today*, G. Terui (Ed.), pp. 73-82.
Robinson, M., Lewis, E., and Napier, E. (1981). Occurrence of reiterated DNA sequences in strains of *Streptomyces* produced by an interspecific protoplast fusion. *Mol. Gen. Genet. 182*: 336-340.
Roels, J. A., and Kossen, N. W. F. (1978). On the modeling of microbial metabolism. *Prog. Ind. Microbiol. 14*: 95-203.
Rolleston, F. (1972). A theoretical background to the use of measured concentrations of intermediates in the study of the control of intermediary metabolism. *Curr. Top. Cell Regul. 5*: 47-75.
Rowlands, R. T. (1984a). Industrial strain improvement: Mutagenesis and random screening procedures. *Microbiol. Technol. 6*: 3-10.
Rowlands, R. T. (1984b). Industrial strain improvement: Rational screens and genetic recombination techniques. *Enzyme Microb. Technol. 6*: 290-300.
Ryu, D. Y., and Humphrey, A. E. (1972). A reassessment of oxygen transfer rates in antibiotic fermentations. *J. Ferment. Technol. 6*: 424-431.
Ryu, D. Y., and Oldshue, J. Y. (1977). A reassessment of mixing cost in fermentation processes. *Biotechnol. Bioeng. 19*: 621-629.
Ryu, D., Kim, K. S., Cho, N. Y., and Pai, H. S. (1983). Genetic recombination in *Micromonospora rosaria* by protoplast fusion. *Appl. Environ. Microbiol. 45*: 1854-1858.
Sabater, B., and Asensio, C. (1973). Transport of hexoses in *Streptomyces violaceoruber. Eur. J. Biochem. 39*: 201-205.
Sabater, B., Sebastian, J., and Asensio, C. (1972a). Identification and properties of an inducible mannokinase from *Streptomyces violaceoruber. Biochem. Biophys. Acta 284*: 406-413.

Sabater, B., Sebastian, J., and Asensio, C. (1972b). Identification and properties of an inducible and highly specific fructokinase from *Streptomyces violaceoruber*. *Biochim. Biophys. Acta 284*: 414-420.

Saier, M. H., Jr. (1977). Bacterial phosphoenolpyruvate:sugar phosphotransferase systems: Structural, functional, and evolutionary interrelationships. *Bacteriol. Rev. 41*: 856-871.

Satoh, A., Ogawa, H., and Satomura, Y. (1976). Regulation of *N*-acetylkanamycin amidohydrolase in the idiophase in kanamycin fermentation. *Agr. Biol. Chem. 40*: 191-196.

Saunders, R. P., Hardman, R., and Cheetham, P. S. J. (1985). The use of a hydrophobic resin as a product reservoir in steroid transformations. *Biotechnol. Bioeng. 27*: 825-831.

Schaal, K. P., and Pulverer, G. (Eds.) (1981). *Actinomycetes*. Gustav Fischer Verlag, New York.

Schoemer, U., and Martin, C. K. A. (1980). Microbial transformation of sterols. *Biotechnol. Bioeng. 22*: 11-25.

Schrempf, H. (1983). Deletion and amplification of DNA sequences in melanin-negative variants of *Streptomyces reticuli*. *Mol. Gen. Genet. 189*: 501-505.

Seno, E. T., and Baltz, R. H. (1982). S-adenosyl-L-methionine: Macrocin *O*-methyltransferase activities in a series of *Streptomyces fradiae* mutants that produce different levels of the macrolide antibiotic tylosin. *Antimicrob. Agents Chemother. 21*: 758-763.

Seno, E. T., and Chater, K. F. (1983). Glycerol catabolic enzymes and their regulation in wild-type and mutant strains of *Streptomyces coelicolor* A3(2). *J. Gen. Microbiol. 129*: 1403-1413.

Seno, E. T., Bruton, C. J., and Chater, K. F. (1984). The glycerol utilization operon of *Streptomyces coelicolor:* Genetic mapping of *gyl* mutations and the analysis of cloned *gyl* DNA. *Mol. Gen. Genet. 193*: 119-128.

Sensi, P., and Thiemann, J. E. (1967). Production of rifamycins. *Prog. Ind. Microbiol. 6*: 21-60.

Stark, W. M., Knox, N. G., Wilgus, R., and DuBus, R. (1976). The nebramycin fermentation: Culture and fermentation development. *Dev. Ind. Microbiol. 17*: 61-77.

Steel, R., and Maxon, W. D. (1962). Some effects of turbine size on novobiocin fermentations. *Biotechnol. Bioeng. 4*: 231-240.

Stonesifer, J., and Baltz, R. H. (1985). Mutagenic DNA repair in *Streptomyces*. *Proc. Natl. Acad. Sci. USA 82*: 1180-1183.

Streicher, S. L., and Tyler, B. (1981). Regulation of glutamine synthetase activity by adenylylation in the gram-positive bacterium *Streptomyces cattleya*. *Proc. Natl. Acad. Sci. USA 78*: 229-233.

Studdard, C. (1979). Transduction of auxotrophic markers in a chloramphenicol-producing strain of *Streptomyces*. *J. Gen. Microbiol. 110*: 479-483.

Suarez, J. E., and Chater, K. F. (1980a). Polyethylene glycol-assisted transfection of *Streptomyces* protoplasts. *J. Bacteriol. 142*: 8-14.

Suarez, J. E., and Chater, K. F. (1980b). DNA cloning in *Streptomyces*: A bifunctional replicon comprising pBR322 inserted into a *Streptomyces* phage. *Nature 286*: 527-529.

Surowitz, K. G., and Pfister, R. M. (1985). Glucose metabolism and pyruvate excretion by *Streptomyces alboniger*. *Can. J. Microbiol. 31*: 702-706.

Suss, F., and Klaus, S. (1981). Transduction in *Streptomyces hygroscopicus* mediated by temperate bacteriophage SH110. *Mol. Gen. Genet. 181*: 552-555.

Suter, M., Hutter, R., and Leisinger, T. (1978). Mutants of *Streptomyces glaucescens* affected in the production of extracellular enzymes. In *Genetics of the Actinomycetales*, E. Freerksen, I. Tarnok, and J. Thumin (Eds.). Gustav Fischer Verlag, Stuttgart, pp. 61-64.

Svensson, M.-L., Roy, P., and Gatenbeck, S. (1983). Glycerol catabolite regulation of D-cycloserine production in *Streptomyces garyphalus*. *Arch. Microbiol. 135*: 191-193.

Swartz, R. W. (1979). The use of economic analysis of penicillin G manufacturing costs in establishing priorities for fermentation process/production. *Annu. Rep. Ferment. Proc. 3*: 75-110.

Szvoboda, G., Lang, T., Gado, I., Ambrus, G., Kari, C., Fodor, K., and Alfoldi, L. (1980). Fusion of *Micromonospora*. *Symp. Biol. Bulg. 22*: 235-240.

Taguchi, H., Imanaka, T., Teramoto, S., Takatsu, M., and Sato, M. (1968). Scale up of glucamylase fermentation by *Endomyces* sp. *J. Ferment. Technol. 46*: 823-828.

Thompson, C. J., Ward, J. M., and Hopwood, D. A. (1980). DNA cloning in *Streptomyces*: Resistance genes from antibiotic-producing species. *Nature 286*: 525-527.

Townsend, M. E., Wright, H. M., and Hopwood, D. A. (1971). Efficient mutagenesis by near ultraviolet light in the presence of 8-methoxypsoralen in *Streptomyces*. *J. Appl. Bacteriol. 34*: 799-801.

Trilli, A. (1986). Scale-up of fermentations. In *Manual of Industrial Microbiology and Technology*, A. L. Demain and N. A. Solomon (Eds.). American Society for Microbiology, Washington, D.C., pp. 277-307.

Trilli, A., Costanzi, I., Lamanna, F., and Di Dio, N. (1982). Development of the agar-disc method for the rapid screening of strains with increased productivity. *J. Chem. Tech. Biotechnol. 32*: 281-291.

Tuffile, C. M., and Pinho, F. (1970). Determination of oxygen-transfer coefficients in viscous streptomycete fermentations. *Biotechnol. Bioeng. 12*: 849-871.

Unowsky, and Hoppe, (1978). Increased production of the antibiotic Aurodox (X5108) by Aurodox-resistant mutants. *J. Antibiot. 31*: 662-666.

Valerianova, Z. T., Todonov, T., Naydenova, M., Kozarova, S., Minchov, D., and Gancheva, V. (1981). UV irradiation of *Streptomyces aureofaciens* and isolation of chlorotetracycline highly productive mutants. *Acta Microbiol. Bulg. 8*: 25-29.

van Suijdam, J. C., Kossen, N. W. F., and Joha, A. C. (1978). Model for oxygen transfer in a shake flask. *Biotechnol. Bioeng. 20*: 1695-1709.

van Suijdam, J. C., Hols, H., and Kossen, N. W. F. (1982). Unstructured model for growth of mycelial pellets in submerged cultures. *Biotechnol. Bioeng. 24*: 177-191.

Veelken, M., and Pape, H. (1982). Production of tylosin and nikkomycin by immobilized *Streptomyces* cells. *Eur. J. Appl. Microb. Biotechnol. 15*: 206-210.

Verhoff, F. H., Boguslawski, G., Lantero, O. J., Schlager, S. T., and Jao, Y. C. (1985). Glucose isomerase. In *Comprehensive Biotechnology, vol. 3, The Practice of Biotechnology: Current Commodity Products*, H. W. Blanch, S. Drew, and D. I. C. Wang (Eds.). Pergamon Press, New York, pp. 837-860.

Vezina, C., Bolduc, C., Kudelski, A., and Sehgal, S. N. (1976). Antimycin A fermentation. I. Production and selection of strains. *J. Antibiot. 29*: 248-264.

Vu-Trong, K., and Gray, P. P. (1982). Continuous culture studies on the regulation of tylosin biosynthesis. *Biotechnol. Bioeng. 24*: 1093-1103.

Vu-Trong, K., and Gray, P. P. (1987). Influence of ammonium on the biosynthesis of the macrolide tylosin. *Enzyme Microb. Technol. 9*: 590-593.

Vu-Trong, K., Bhuwapathanapun, S., and Gray, P. P. (1980). Metabolic regulation in tylosin-producing *Streptomyces fradiae*: Regulatory role of adenylated nucleotide pool and enzymes involved in biosynthesis of tylonolide precursors. *Antimicrob. Agents Chemother. 17*: 519-525.

Vu-Trong, K., Bhuwapathanapun, S., and Gray, P. P. (1981). Metabolic regulation in tylosin-producing *Streptomyces fradiae*: Phosphate control of tylosin biosynthesis. *Antimicrob. Agents Chemother. 19*: 209-212.

Waksman, S. A. (1959). *The Actinomycetes*, vol. I, *Nature Occurrence, and Activities*. Williams & Wilkins, Baltimore.

Walsh, K., and Koshland, D. E., Jr. (1984). Determination of flux through the branch point of two metabolic cycles. *J. Biol. Chem. 259*: 9646-9654.

Wang, D. I. C., and Fewkes, R. C. J. (1977). Effect of operating and geometric parameters on the behavior of non-newtonian, mycelial, antibiotic fermentations. *Dev. Ind. Microbiol. 18*: 39-56.

Wang, D. I. C., Cooney, C. L., Demain, A. L., Dunnill, P., Humphrey, A. E., and Lilly, M. D. (1979). *Fermentation and Enzyme Technology.* John Wiley and Sons, New York.

Wanner, B. L. (1983). Overlapping and separate controls on the phosphate regulon in *Escherichia coli* K12. *J. Mol. Biol. 166*: 283-308.

Wax, R., Snyder, L., and Kaplan, L. (1982). Inactivation of glutamine synthetase by ammonia shock in the gram-positive bacterium *Streptomyces cattleya. Appl. Environ. Microbiol. 44*: 1004-1006.

Weinberg, E. D. (1970). Biosynthesis of secondary metabolites: Role of trace metals, *Adv. Microb. Physiol. 4*: 1-44.

Wesseling, A. C., and Lago, B. D. (1981). Strain improvement by genetic recombination of cephamycin producers, *Nocardia lactamdurans* and *Streptomyces griseus. Dev. Ind. Microbiol. 22*: 641-651.

Westerhoff, H. V., Groen, A. K., and Wanders, R. J. A. (1984). Modern theories of metabolic control and their applications. *Biosci. Rep. 4*: 1-22.

Williamson, J. M., Meyer, R., and Inamine, E. (1985). Reverse transsulfuration and its relationship to thienamycin biosynthesis in *Streptomyces cattleya. Antimicrob. Agents Chemother. 28*: 478-484.

Yamashita, F., Hotta, K., Kurasawa, S., Okami, Y., and Umezawa, H. (1985). New antibiotic-producing streptomycetes, selected by antibiotic resistance as a marker. I. New antibiotic production generated by protoplast fusion treatment between *Streptomyces griseus* and *S. tenjimariensis. J. Antibiot. 38*: 58-63.

Yoshida, T., Taguchi, H., Kulprecha, S., and Nilubol, N. (1981). Kinetics and optimization of steroid transformation in a mixed culture. *Adv. Biotechnol. 3*: 501-506.

Yoskikawa, H., Takiguchi, Y., and Terao, M. (1982). Degeneration of Herbicidin A-Producer, *Streptomyces saganonensis*, and strain improvement. *J. Ferment. Technol. 60*: 385-391.

Young, M. D., Kempe, L. L., and Bader, F. G. (1985). Effects of phosphate, glucose and ammonium on cell growth and linocomycin production by *Streptomyces lincolnensis* in chemically defined medium. *Biotechnol. Bioeng. 27*: 327-333.

Young, T. B., III (1979). Fermentation scale-up: Industrial experience with a total environmental approach. *Ann. N.Y. Acad. Sci. 326*: 165-180.

Zabriskie, D. W., Armiger, W. B., Phillips, D. H., and Albano, P. A. (1980). *Traders Guide to Fermentation Media Formulation.* Traders Protein, Memphis, Tennessee.

2

Fermentation of Bacilli

FERGUS G. PRIEST

Heriot-Watt University, Edinburgh, Scotland

RICHARD J. SHARP

Public Health Laboratory Service, Centre for Applied Microbiology and Research, Porton Down, Salisbury, England

I. INTRODUCTION

The genus *Bacillus* comprises rod-shaped bacteria that differentiate into heat-resistant endospores under aerobic conditions. Although they all possess cell walls of a typical gram-positive structure, strains of several species, in particular *Bacillus macerans* and *Bacillus polymyxa*, react gram negatively in the traditional gram staining procedures. Similarly, although they are often described as the aerobic endospore formers, it should be remembered that many strains are facultative anaerobes and some, such as *Bacillus coagulans*, are best described as microaerophilic.

The genus is markedly heterogeneous and contains a diversity of bacteria; this is best illustrated by the variation in base composition (mol% guanine plus cytosine) of the chromosomal DNA from representative strains, which may be as low as 33% for some strains of *Bacillus cereus* and *Bacillus thuringiensis* or more than 60% for some extreme thermophiles. Most strains, however, lie in the region of 36-54%, which is still considerably broader than the 10% variation in GC content that is usually considered to represent the maximum genetic diversity that should be accommodated in a single genus (Priest, 1981). This genetic diversity has been confirmed by ribosomal (r) RNA analyses, which revealed that the genus *Bacillus* has equivalent evolutionary divergence to the combined familes of Enterobacteriaceae and Vibrionaceae in which there are over 20 genera (Stackebrandt and Woese, 1981). It is therefore not surprising that a tremendous diversity of ecological and physiological types is found in *Bacillus*: alkaliphiles, acidophiles, psychrophiles and thermophiles are represented. Physiologically they vary in metabolic pathways and end products of metabolism, giving rise to a diversity of potential products. Most strains are robust, grow rapidly on inexpensive media, and are nonpathogenic; indeed, *Bacillus subtilis* has GRAS (generally regarded as safe) clearance in the United States. Genetically, *B. subtilis* is well characterized, and increasingly *Bacillus megaterium, Bacillus thuringiensis, Bacillus licheniformis*, and *Bacillus amyloliquefaciens* are being studied in more detail. These are then in most respects ideal industrial organisms.

II. PRODUCTS FROM BACILLI

A. Enzymes

Bacilli secrete a variety of enzymes, including polysaccharases, proteases and nucleic acid-hydrolyzing enzymes. Amylase from *B. subtilis* was considered an industrial enzyme as early as 1917, but commercial production of *Bacillus* enzymes began in the late 1940s with the introduction of submerged culture borrowed from the antibiotic industry. Today, strains of perhaps 8 or 10 species are used to prepare seven enzymes on a large scale (Table 1).

α-Amylase hydrolyzes the α(1-4)-glycosidic linkages in starch and glycogen. Its major use is for the liquefaction of starch (reduction of high molecular mass to about 10 or 11 glucose units) in the starch-processing and related industries and for the desizing of textiles. Originally prepared from *B. amyloliquefaciens*, the enzyme from *B. lichenformis* is gaining market share because of its greater thermostability (Madsen et al., 1973; Kindle, 1983). The exo-acting β-amylase has a limited market for the preparation of high-maltose syrups from liquefied starch and perhaps in the brewing industry as a replacement for barley β-amylase, but at present it is not produced to any extent.

Pullulanase is a debranching enzyme that hydrolyzes the α(1,6)-bonds in amylopectin, glycogen, and pullulan. It is produced commercially from *Klebsiella pneumoniae* and *"Bacillus acidopullulyticus"* and is used to assist the total saccharification of starch. β-Glucanase preparations from *B. subtilis* and *Bacillus circulans* are used in the brewing and related industries for the removal of plant cell β-glucans, which, owing to their viscosity, cause filtration problems.

Of the proteases, the serine or alkaline protease dominates the market. This enzyme, produced from *B. licheniformis* or alkaliphilic bacilli, has a high pH optimum, is relatively thermostable, has no

Table 1 Some Industrial Enzymes Produced from *Bacillus*

Enzyme	Organism	Comments
α-Amylase	*B. amyloliquefaciens*	Traditional enzyme
	B. licheniformis	Thermostable enzyme
Maltose-producing exoamylase	*B. stearothermophilus*	Produced from a recombinant organism
Glucose isomerase	*B. coagulans*	Intracellular enzyme; used in immobilized form
β-Glucanase	*B. amyloliquefaciens*	
	B. circulans	
	B. subtilis	
Alkaline protease	*B. licheniformis*	"Detergent enzyme"
	Alkaliphilic bacilli	Improved stability at high pH
Neutral protease	*B. amyloliquefaciens*	
	"B. thermoproteolyticus"	Temperature stable
Pullulanase	*"B. acidopullulyticus"*	

metal ion requirements, and is ideally suited to inclusion in household washing detergents, its major application, and for use in the leather industry (Ward, 1983).

The industrial enzyme industry is large, currently estimated at around $500 million per annum, much of which is derived from *Bacillus* and consequently accounts for much of the industrial application of these bacteria.

B. Insecticides

Although many bacteria are pathogenic toward insects, few have been exploited as biological control agents. Insect pathogenic bacilli include strains of *B. thuringiensis*, *Bacillus lentimorbus*, *Bacillus popilliae*, *Bacillus sphaericus*, and *Bacillus larvae*, a pathogen of honeybees that is discouraged rather than promoted. *B. thuringiensis*, first isolated in 1902, is a close relative of *B. cereus* from which it can only be reliably distinguished by serology or by virtue of its entomopathogenicity. This was first discovered in the 1940s, and subsequently the organism considered as a biological control agent. Today it is the most effective bacterial insect control agent (Bulla et al., 1985).

B. thuringiensis is pathogenic by virtue of an intracellular crystalline protein (or δ-endotoxin) that is synthesized during the early stages of sporulation and consequently became known as the parasporal body. When ingested by susceptible larvae, the protoxin (M_r 130,000-160,000) is cleaved by proteolysis to toxic polypeptides of M_r 30,000-80,000 that cause gut paralysis and the gut cells to swell, vacuolate, and burst. The spore penetrates the weakened midgut and multiplies in the hemolymph. Death results from either intoxication or septicemia (Deacon, 1983). Different strains of *B. thuringiensis* are classified by serotyping and given varietal names (reviewed by Dean, 1985). Different serotypes produce toxins of slightly different specificity, but most are pathogenic to lepidopteran larvae, such as the tobacco hornworm, and it is these strains that have been successful commercially. In particular, the subspecies *thuringiensis* (*berliner;* serotype 1), *galleriae* (serotype 5a, 5b), and *kurstaki* (3a, 3b) are used (Bulla et al., 1985).

In 1977, Goldberg and Margalit isolated a strain of *B. thuringiensis* (subsequently designated serotype 14 or subsp. *israelensis*) that was pathogenic for mosquito and blackfly larvae. Thereafter other serotypes (10, 11a, and 11c) were also found to be toxic to mosquito larvae, and the exciting possibility that an insect vector of human and animal disease might be controlled biologically was realized. Mosquito-pathogenic strains also produce crystal toxins that are active on the larval midgut, but these proteins differ in size and composition from the lepidopteran toxins (see Aronson et al., 1986, for review). More recently, strains toxic for Coleopteran (beetle) larvae have been isolated (Herrnstedt et al., 1986).

Some strains of *B. sphaericus* are also toxic to mosquito larvae, in particular those of DNA homology group IIA (Krych et al., 1980). These bacteria also contain crystalline parasporal bodies, but it is not yet certain that this is solely responsible for the insecticidal activity since vegetative cells are also toxic. Although no commercial preparations based on *B. sphaericus* are currently available, this organism has considerable potential as a mosquito and blackfly control agent (Bulla et al., 1985).

Finally, *B. popilliae* and *B. lentimorbus*, the milky disease agents of scarabaeid beetle larvae, have been used for the control of the Japanese beetle for many years. Although strains of *B. popilliae* form a crystalline inclusion, these bacteria are infectious rather than toxic and cause death through septicemia. Thus as the larvae die, spores are liberated and reinfect other larvae, giving rise to a long-lasting, chronic infestation of the treated grassland.

C. Antibiotics

Bacilli are prolific producers of antibacterial compounds, but few are of clinical value. This is largely because of limited antibacterial activity and in many cases toxicity to the patient. Nevertheless, some are produced on a commercial scale, for example, polymyxin and tyrocidine, and some, such as bacitracin, are used in cattle feeds (Katz and Demain, 1977). Antibiotics of bacilli can be grouped in four classes: (1) cyclic oligopeptides that inhibit cell wall synthesis include bacitracin from *B. licheniformis*, a linear chain of 12 amino acids linked to a heptapeptide ring; (2) linear or cyclic peptides that act on the membrane: tyrocidines and linear gramicidins from *Bacillus brevis* are included here as well as gramicidin S and polymyxin; (3) basic peptides, such as the edeines, also from *B. brevis*, that inhibit protein synthesis; and (4) aminoglycoside antibiotics from some strains of *B. circulans*. The antibiotics of bacilli have been reviewed recently (Kleinkauf and von Dohren, 1983).

D. Other Products

Bacilli are used and have potential for the manufacture of various other products. Purine nucleotides, of which 5'-guanosine monophosphate and 5'-inosine monophosphate are the major examples, are used as flavor enhancers in the food industry (Demain, 1978). *Corynebacterium glutamicum* can be used for the production of the nucleotides, but a commonly used route is to prepare guanosine from *B. subtilis* followed by chemical phosphorylation (Enei et al., 1985).

Suitably modified *Bacillus* strains can be used for the manufacture of some amino acids, such as phenylalanine (Enei and Hirose, 1984), tryptophan, and arginine (Kinoshita and Nakayama, 1978), but this industry is dominated by the coryneforms. Surface-active agents are also liberated by some strains of *B. subtilis* (Cooper and Zajic, 1980). The metabolic versatility and potential biotechnological uses of the bacilli have been reviewed by Debabov (1982) and Priest (1989).

E. Heterologous Gene Products

B. subtilis has many potential advantages for use as a host for the manufacture of cloned gene products. It is nonpathogenic and grows readily on inexpensive media, there is a great deal of established knowledge concerning the large-scale growth of this organism, and it possesses an efficient protein secretion system that might be amenable to adaption for the secretion of foreign proteins. The considerable interest in *B. subtilis* is illustrated by the range of foreign gene products that have been successfully synthesized by this bacterium, a selection of which is shown in Table 2 and includes both bacterial and eukaryotic proteins.

Table 2 Some Foreign Genes That Have Been Cloned and Expressed in *Bacillus subtilis*

Protein	Origin	Secretion system	Reference
Protein A	*Staphylococcus aureus*	α-Amylase *B. amyloliquefaciens*	Fahnestock and Fisher, 1986
Protein A	*S. aureus*	S. protease[a] *B. amyloliquefaciens*	Vasantha and Thompson, 1986
Protein A	*S. aureus*	N. protease[b] *B. amyloliquefaciens*	
Proteins E1 and E2	Semliki Forest virus	α-Amylase *B. amyloliquefaciens*	Lundström et al., 1985 Ulmanen et al., 1985
G protein	Vesicular stomatitis virus	α-Amylase *B. amyloliquefaciens*	Lundström, 1984
Insulin	Human	ß-Lactamase *B. licheniformis*	Chang et al., 1982
Interferon	Human	α-Amylase *B. amyloliquefaciens*	Palva et al., 1983
Interferon	Mouse	α-Amylase *B. subtilis*	Shiroza et al., 1985
β-Lactamase	*E. coli*	α-Amylase *B. amyloliquefaciens*	Palva et al., 1982
β-Lactamase	*E. coli*	α-Amylase *B. subtilis*	Ohmura et al., 1984

[a]S, serine (alkaline) protease.
[b]N, neutral (metal) protease.

III. CONTROL OF SYNTHESIS

A. Carbon Catabolite Repression

Catabolite repression is the repression of the synthesis of enzymes concerned with the metabolism of alternative carbon sources that occurs when a microorganism is provided with a rapidly catabolized carbon source. The rapidly utilized carbon source is often glucose; hence it is commonly described as glucose repression, but this obscures the fact that any "good" carbon source has the same effect, for example, succinate in pseudomonads or glycerol in *B. subtilis* but not in *Escherichia coli*. Catabolite repression offers the organism a means to coordinate its carbon and energy metabolism. In their natural environments, bacteria are surrounded by a variety of carbon sources suitable for growth, and it would be extremely wasteful for the organisms to synthesize induced catabolic enzymes for all the available sugars. Instead they tend to use them in the order of the rate at which they will support growth, the "best" ones being used first.

E. coli and related bacteria achieve this by modulating the transcription of catabolite-sensitive operons in a positive fashion. The principal effector molecule is adenosine 3',5'-cyclic monophosphate (cAMP), which, in conjunction with its receptor protein (CRP), is required for the efficient transcription of these operons. In the absence of cAMP, transcription is effectively prevented since the CRP alone cannot promote transcription.

Regulation is achieved by altering the intracellular level of cAMP. Adenylate cyclase, the enzyme responsible for cAMP synthesis, is regulated via the phosphoenolpyruvate-sugar phosphotransferase uptake system, probably through the steady-state level of phosphorylation of enzyme IIIGlc (for review, see Postma and Nelson, 1985). Thus the bacterium can sense the catabolic status of the cell and affect the cAMP concentration accordingly. In particular, when growing slowly on a poor carbon source, the cAMP level is maintained at a high level, and in the presence of glucose, adeylate cyclase activity is repressed and the cAMP concentration is reduced.

However, the cAMP system does not explain all aspects of catabolite repression in *E. coli*; indeed, catabolite repression can occur in cells under conditions in which cAMP cannot be synthesized or in cells that lack a functional CRP (Ullman and Danchin, 1983). Thus some other factors seem to be involved (see Postma, 1986, for review).

In many gram-positive bacteria, particularly *Bacillus*, there is no detectable cAMP (Setlow, 1973; Botsford, 1981). Several alternative molecular mechanisms for catabolite repression have been suggested, including the involvement of cGMP (Bernlohr et al., 1974; Majumdar and Bose, 1985), but there is little experimental support for these various schemes. However, recent observations indicate a more complex role for RNA polymerase in catabolite repression (Doi et al., 1985). RNA polymerase in bacilli can combine with various σ factors, each of which directs the start of transcription to particular sites or promoters on the chromosome. These σ factors are classified according to their molecular masses; thus the major vegetative cell polypeptide is referred to as σ^{43} (previously σ^{55}) denoting a σ factor of 43 kD. It is thought that the process of sporulation is at least partly brought about by the sequential replacement of σ factors and hence the sequential transcription of sporulation genes (see Sec. III.C.1).

A connection between catabolite repression and sporulation was revealed by the characterization of certain mutants able to sporulate in the presence of a high concentration of glucose (Takahashi, 1979). These catabolite-resistant sporulation (CRS) mutations have been mapped to several loci (Sun and Takahashi, 1984, 1986), and some will suppress certain sporulation mutants blocked at stage zero (*spoOF* mutants), thus allowing them to sporulate normally. Moreover, *crsA* mutations have been shown to lie in the *rpoD* locus, which codes for the σ^{43} polypeptide, thus connecting catabolite repression with the major σ factor (Doi et al., 1985).

It should be remembered that both nitrogen and phosphorus metabolism are also subject to regulation according to the availability of substrates. In *B. subtilis* the extracellular proteases and glutamate and glutamine synthetases are regulated by nitrogen source availability (Fisher et al., 1984). Glutamine synthetase seems to be autogenously controlled since the enzyme governs its own synthesis and probably others in a trans fashion (Sonenshein, 1985). Little is known about the control of metabolism by phosphorus (see Sonenshein, 1985, for review).

Despite catabolite repression remaining very much a mystery in these bacteria, it is a most important consideration for the fermentation technologist since many of the major products of bacilli are regulated by this process.

1. Extracellular Enzymes

Virtually all extracellular enzymes are controlled by carbon catabolite repression. This is particularly evident in the amylases but is also encountered in other carbohydrases and enzymes, such as proteases. Catabolite repression has been observed for all the common commercial enzymes, including the amylases of *B. licheniformis* (Thirunavukkarasu and Priest, 1980) and *B. amyloliquefaciens* (Ingle and Boyer, 1976) and the proteases of *B. licheniformis*, *B. amyloliquefaciens*, and *B. subtilis* (see Ingle and Boyer, 1976, for review).

There are two solutions to catabolite repression in industrial fermentations: the preparation of resistant mutants (Eveleigh and Montenencourt, 1979; Nicholson and Chambliss, 1986; see also Sec. IV.B.1) or the use of slowly metabolized carbon sources, such as starch, starch hydrolysates, or lactose in the medium (see Sec. V. A.3). In practice both approaches are usually used since industrial producer strains are generally extensively mutated and catabolite repression has been intentionally or incidentally destroyed in these bacteria and starchy materials are among the cheapest available carbon sources.

2. Antibiotics

The synthesis of many peptide antibiotics is inhibited when glucose is used as the carbon source, thus implicating catabolite repression in the regulation of synthesis of these molecules (see Demain et al., 1979, for review). It should be remembered that these peptides are generally not ribosomal but synthesized on multifunctional enzymes using the protein thiotemplate mechanism (Kleinkauf and von Dohren, 1981, 1983) and therefore the synthesis of the peptide can be regulated by the synthesis and activity of the synthetase. In the case of gramicidin S synthesis in *B. brevis*, the antibiotic becomes detectable several hours after the appearance of GS-synthetase activity and production continues for a short time after the disappearance of the enzyme (Vandamme, 1982). When grown in defined medium, GS-synthetase formation occurs during a short lag phase following the initial metabolism of glutamine and arginine and before the utilization of fructose. This led to the suggestion that catabolite repression, effected by amino acid metabolism, might be responsible for the regulation of GS-synthetase formation.

Catabolite repression has also been implicated in bacitracin synthesis (Haavik, 1974), but in this system the inhibitory effect of glucose was later related to the increased excretion of acetic and pyruvic acids that occurred in the presence of glucose. As with enzymes, catabolite repression of antibiotic synthesis has been overcome by the use of resistant strains, slowly metabolized carbon sources, or the slow feeding of rapidly metabolized sugars.

3. Miscellaneous

Since catabolite repression is a means by which the organism coordinates its metabolism, it is not surprising that it is implicated in the synthesis of most products from bacilli. However, in many instances this is not a primary effect. For example, synthesis of crystal protein in *B. thuringiensis* accompanies sporulation and is repressed by those conditions that repress sporulation, such as the presence of glucose. However, it seems unlikely that crystal protein synthesis itself is subject to catabolite repression. This distinction is important and will be relevant when considering medium development and strain improvement so that the appropriate strategy is adopted.

B. Induction

Induction typically refers to the increase in synthesis of specific catabolic enzymes that occurs in the presence of the relevant substrate. The *lac* operon of *E. coli* is the archetypal system and shows a thousandfold increase in β-galactosidase activity when the organism is grown in the presence of the substrate lactose. With regard to the products of bacilli, most are constitutive; that is, synthesis of antibiotics or toxins is not inducible. A possible exception is the extracellular enzymes, some of which may be inducible.

There is much controversy about the inducibility of the extracellular enzymes of bacilli (reviewed by Ingle and Erickson, 1978; Priest, 1983). It has been suggested that, because the synthesis of amylases, for example, is slightly greater in the presence of maltose or starch than in the presence of some unrelated sugar, synthesis of these enzymes is being induced. This is well substantiated in some cases; indeed the induction of β-glucanase synthesis in *B. circulans* WL-12 has been studied extensively (Tanaka et al., 1981). This bacterium secretes numerous enzymes, including β(1-3)-, β(1-6)-, and α(1-3)-glucanases and β-xylanase, in response to various macromolecules in the environment (Esteban et al., 1984).

However, the majority of the industrially important enzymes, including the amylases and proteases of *B. licheniformis, B. amyloliquefaciens* and *B. subtilis*, are synthesized constitutively (Priest, 1977; Priest and Thirunavukkarasu, 1985), and any increased levels of activity observed after growth on α-glucans is probably due to release from catabolite repression. Nevertheless, it is important to remember during medium formulation that a particular product may be inducible.

C. Temporal Regulation

When the supply of readily metabolizable carbon or nitrogen is exhausted, bacilli differentiate into heat-resistant endospores. The synthesis of many products by bacilli, including extracellular enzymes, antibiotics, and toxins, occurs during the late exponential phase as sporulation is initiated or during sporulation itself. Before considering how the regulation of synthesis of these products may operate, it is necessary to outline briefly current ideas on the regulation of sporulation. This has been considered in detail in many excellent reviews, such as those by Losick and Youngman (1984) and Piggot (1985).

1. Transcriptional Regulation During Sporulation

RNA polymerase, the enzyme responsible for transcription, in *B. subtilis* can associate with several σ factors. The two key recognition sequences for RNA polymerase of some 6-10 bases are situated 35 and 10 bases upstream from the starting point of the mRNA. Different σ factors recognize different -35 and -10 sequences and are responsible for directing RNA polymerase to different promoters. Therefore, by the progressive replacement of σ factors during sporulation, sequential transcripts are produced that are at least partly responsible for the process of differentiation (Losick and Pero, 1981; Losick and Youngman, 1984).

Table 3 Recognition Sequences for Sigma Factors from *Bacillus subtilis*[a]

σ Factor	Stage of growth	Consensus sequences -35	gap	-10
43	Vegetative	TTGACA	17-19	TATAAT
37	Minor components in vegetative cells	AGGATTTNA	11-15	GGAATTNTTT
32		AAATC	14-15	TANTGNTTNTA
29	Sporulation	TTNAAA	14-17	CATATT
28	Minor component in vegetative cells	CTAAA	16	CCGATAT
E. coli σ^{70}		TTGACA	16-18	TATAAT

[a] *Source:* Data from Doi and Wang (1986).

The major σ factor is σ^{43} (previously σ^{55}). This is responsible for the transcription of vegetative genes and recognizes -35 and -10 sequences that are very similar to those recognized by the RNA polymerase from *E. coli* (Table 3). Largely for this reason, vegetative *Bacillus* genes are usually transcribed in *E. coli*. Although σ^{43} accounts for 90-95% of RNA polymerase holoenzyme in vegetative cells, three other sigma factors (σ^{37}, ^{32}, and ^{28}) are present in small amounts and increase in activity toward the end of exponential growth (reviewed by Doi and Wang, 1986). An involvement of at least two of these in sporulation is implicated by the effect of certain *spo0* mutations that block sporulation at a very early stage (stage zero) and reduce transcription from σ^{28} and σ^{37} promoters.

A fourth σ factor, σ^{29}, is the product of the *spoIIG* gene and was the first protein associated with RNA polymerase that was shown to be essential for sporulation.

2. Sporulation and the Synthesis of Enzymes, Antibiotics, and Toxins

That sporulation was accompanied by the synthesis of extracellular enzymes, antibiotics, and toxins led many people to suspect that the synthesis of these products was actually involved in sporulation and that they were essential to the process of differentiation (reviewed by Schaeffer, 1969). Moreover, many asporogenous mutants were deficient in the synthesis of these molecules, and vice versa;

for example, protease mutants were deranged in sporulation. However, in most cases, analysis of further mutants has clarified the role, if any, of these products in differentiation.

Mutants of *B. subtilis* that are deficient in the synthesis of α-amylase, neutral and serine protease, and DNase and yet sporulate normally have been isolated, thus excluding extracellular enzymes from any essential role in sporulation. It would seem that these are catabolic enzymes that are secreted at the end of exponential growth in response to the changing environment and their synthesis accompanies the early stages of sporulation but the enzymes are not involved in sporulation. In most cases the promoters of these genes have typical σ^{43} recognition sequences (Table 4). but the serine protease of *B. subtilis* is an exception and the gene is transcribed in vivo by the σ^{37} form of RNA polymerase. This seems to be an important aspect of its synthesis since replacement of the promoter by a vegetative promoter initiates synthesis at a much earlier stage in the growth cycle (Wong and Doi, 1986).

With regard to antibiotics, a close relationship with sporulation has again been demonstrated by biochemical and genetic analyses (Katz and Demain, 1977; Marahiel et al., 1979; Nam and Ryu, 1985). Although the regular isolation of antibiotic-deficient mutants that sporulate normally excludes an essential role for antibiotics in sporulation, spores from tyrocidine- and gramicidin-negative mutants of *B. brevis* can show reduced dipicolinate content and heat sensitivity and the organisms synthesize reduced levels of serine protease. This phenotype could be reversed by the addition of antibiotic (reviewed by Kleinkauf and von Dohren, 1981; Malik, 1982). An involvement of tyrocidine and gramicidin in transcription has also been suggested since tyrocidine forms a complex with DNA, preventing the initiation of transcription, and this effect can be at least partially relieved by gramicidin (Ristow et al., 1979). More recent in vitro studies suggest that gramicidin S does not function during growth and sporulation but has a role during germination and outgrowth (Frangou-Lazaridis and Seddon, 1985). Gramicidin S synthetase (Krause et al., 1985) has now been cloned, and it seems likely that it is transcribed from a sporulation-specific promoter.

The δ-endotoxin of *B. thuringiensis* is also a sporulation-associated product. Vegetative cells grown on rich medium make neither spores nor parasporal bodies. Exhaustion of carbon and nitrogen, on the other hand, leads to sporulation, and some 2 hr after the end of exponential growth crystal protein antigen can be detected. Synthesis then continues for several hours, after which crystal protein accounts for 20–30% of the total protein of the sporangium (see Dean, 1985, for review). This parasporal body is dispensable for sporulation since cells that do not synthesize it sporulate normally.

Table 4 Promoter Sequences for Some Extracellular Enzyme Genes and δ-Endotoxin

Organism	Protein[a]	Promoter sequence[b] -35	gap	-10	σ-Factor	Reference
B. amyloliquefaciens	Amylase	TTGTTA	17	TATAAT	43	Lehtovaara et al., 1984
	S. protease	GGTCTA	16	TACTAT	37	Wells et al., 1983
B. licheniformis	Amylase	TTGTTA	17	TACAAC	43	Gray et al., 1986
B. stearothermophilus	Amylase	TTGAAA	18	TATAAT	43	Gray et al., 1986
B. subtilis	Amylase	TTGATA	17	TAAAAT	43	Yamazaki et al., 1983
	N. protease[c]	TTGAGT	17	TAATAT	43	Henner et al., 1985
	S. protease	AGTCTT	16	TGAATT	37	Wong et al., 1984
	β-Glucanase	TTGACC	18	AATCAT	43	Murphy et al., 1984
	Levansucrase	TTGCAA	17	TAGAAT	43	Steinmetz et al., 1985
B. thuringiensis subsp. *berliner*	δ-Endotoxin	CCTACG	15	TACAAT	?	Klier et al., 1983
Consensus		TTGACA	17/18	TATAAT	43	Moran et al., 1982

[a]S. protease, serine protease; N. protease, neutral protease.
[b]Many of these sequences are presumptive recognition sites.
[c]May also contain a σ-37 sequence.

There is now overwhelming evidence that in most cases the δ-endotoxin gene is naturally plasmid borne. Curing and gene transfer experiments both support this view (see Clark et al., 1984), and more recently toxin genes from several strains of *B. thuringiensis*, including subspp. *kurstaki*, *berliner*, *sotto*, *subtoxicus*, and *aizawai*, and *B. sphaericus* have been cloned (reviewed by Aronson et al., 1986). The chromosomal gene from *B. thuringiensis* subsp. *berliner* seems to be unique, but the plasmid genes from subspp. *kurstaki* and *sotto* are virtually identical. Promoter sites have been established in the gene from subsp. *kurstaki* by Tn*5* insertions and S1 nuclease mapping (Wong et al., 1983). Three different promoter sites have been defined, one used exclusively by *E. coli* and two by *B, thuringiensis*. In *B. thuringiensis*, expression from the downstream promoter begins at stage II and continues until midsporulation after which transcription from the upstream site begins. The downstream promoter is similar to that for *spoVG* and *spoVC*, both of which are activated early in sporulation and are recognized by σ-37. In *B. subtilis*, on the other hand, expression begins much earlier in the growth cycle and it seems likely that the *E. coli* promoter is being recognized by the vegetative σ-43 enzyme (Shivakumar et al., 1986). These studies do not reveal the reasons for overproduction of this protein, however, since the sequence data do not tell us how strong the promoters are, merely the stage of growth at which they are activated. Nevertheless, this is an important consideration in strain, fermentation, and medium development.

IV. STRAIN IMPROVEMENT

The development of high-yielding strains is generally crucial to the commercialization of a microbially derived product. Moreover, improvements in the output and efficiency of production must be regularly achieved to maintain a competitive edge on rival manufacturers. Strain development encompasses several aspects of the fermentation process: (1) higher product yield, (2) more rapid and higher biomass yield, (3) growth on less expensive substrates, and (4) efficient secretion of product, where appropriate, to enable simpler processing and purification. With regard to bacilli, since sporulation terminates cell metabolism it is often desirable to select asporogenous mutants; this also eliminates spores that may also survive the recovery process, resulting in microbial contamination during downstream processing. The occurrence of polypeptide antibiotics has been widely exploited, but their production as secondary products in a process in which the primary product is intended for use in the food or pharmaceutical industry is undesirable. Selection of suitable nonantibiotic-producing mutants is therefore necessary in these cases.

Traditionally, most of these objectives were achieved by mutation and selection programs. Since such approaches were successful and are still used today they are described briefly here, but recombinant DNA techniques are now superseding the classic mutation programs in most areas of biotechnology.

A. Mutation and Screening Programs

Mutation of a producer strain and screening or preferably selection for a hyperproducing phenotype has proved successful for the improvement of yields of enzymes and primary and secondary metabolites (for reviews see Eveleigh and Montenencourt, 1979; Malik, 1979; Yamane and Maruo, 1980). In the case of extracellular enzymes the approach was exemplified by the work of Maruo's group with the α-amylase of *B. subtilis*. Several independent mutations that increased amylase synthesis in the parental strain were obtained either by screening for increased zones of starch hydrolysis following mutagenesis or by selecting for antibiotic-resistant mutants, such as tunicamycin- or cycloserine-resistant strains that might be pleiotropically affected in enzyme secretion. These individual mutations were then combined in a single strain using DNA-mediated transformation. In this way a 1500-fold increase in amylase productivity was obtained (Yamane and Maruo, 1980).

A second area in which traditional mutation programs have been successful is in the production of purine nucleotides. For example, removal of feedback inhibition and/or end-product repression has been obtained by isolating strains resistant to toxic purine analogs, and combinations of mutations in single strains has again given rise to increased yields (Demain, 1978). But both these examples have their limitations today. The main shortcomings are those of time and effort. The increase in amylase productivity, impressive though it is, represents a formidable investment in money and labor, and a similar improvement in yield could probably be obtained today in a fraction of the time using genetic engineering. Similarly, removal of a rate-limiting step in the nucleotide biosynthetic pathway by cloning the appropriate gene onto a multicopy plasmid could improve purine biosynthesis dramatically as has been achieved for amino acid biosynthesis (Enei and Hirose, 1984). So, unless there is some extenuating circumstance, strain improvement will almost invariably center on recombinant DNA and genetic engineering.

B. Recombinant DNA Routes to Strain Improvement

Developments in molecular biology have not only indicated the points at which gene expression can be increased but also provided some of the techniques and information that allow this to be achieved. It should be remembered that the molecular biology of the bacilli is not as well developed as for *E. coli*; nevertheless, the gap continually closes and much can be done with *B. subtilis* and its relatives.

1. Transcription

As described in Sec. III.C, there are several types of promoter in *Bacillus*, each recognized by its own form of RNA polymerase. With regard to strain improvement, the promoter recognition sequence should be optimized to provide the strongest form of promoter consistent with general cell health. Too few promoters from *Bacillus* have been sequenced and related to in vivo expression to indicate the structure of a strong promoter in *B. subtilis*, but similarity to the consensus sequences at -35 and -10 for the vegetative σ factor is presumably necessary. Moreover, the gap between these two sequences should be optimized. By replacing promoters, differences in temporal expression can be achieved. For example, when the σ-37 promoter of serine protease from *B. subtilis* was replaced with a σ-43 promoter, expression was activated much earlier in the batch culture growth cycle (Wong and Doi, 1986). Such a scheme could perhaps be adopted for antibiotic synthetase genes or those for insect toxins.

A further improvement could be obtained by providing two promoters. The rRNA genes of *B. subtilis* have tandem promoters for σ-43 RNA polymerase that constitute very powerful promoters (Stewart and Bott, 1983). Alternatively, two promoters with different recognition sequences could be provided either adjacent or in an overlapping configuration. Some natural promoters, such as that for serine protease (σ-32 and σ-37), have this structure. The P43 gene, a cryptic gene, has σ-37 and σ-43 sequences, thus allowing expression during the exponential and early stationary phases of growth (Wang and Doi, 1984).

A second feature of strong promoters seems to be an AT-rich region upstream from the -35 sequence (Moran et al., 1982; Stewart and Bott, 1983). This may be important for efficient transcription. These major requirements for effective transcription in *B. subtilis* have been reviewed by Doi (1985) in detail.

For expression of heterologous genes the amylase promoters from *B. subtilis* and *B. amyloliquefaciens* seem to be the most popular. These employ σ-43 sequences and are constitutive. In many circumstances it is preferable to use a regulated promoter since this gives the opportunity to grow large quantities of biomass in the absence of massive synthesis of a foreign and perhaps toxic protein and subsequently to switch synthesis on, either by the addition of inducer or by a temperature change. Of secreted proteins the β-lactamase from *B. licheniformis* is inducible and has been used to secrete foreign proteins from *B. subtilis* (Chang et al., 1982). Levansucrase is also an inducible extracellular enzyme (Shimotsu and Henner, 1986) and might be useful in this connection. Yansura and Henner (1984) adopted the *E. coli lac* operon system for regulating transcription in *B. subtilis*. They placed the *lac* operator between the promoter and structural gene and regulated expression using a cloned *lacI* gene The system was effective and expression could be controlled using the gratuitous inducer IPTG. With the analysis of catabolic and anabolic pathways progressing at an increasing rate in *B. subtilis*, it seems likely that other regulated promoters, such as the *trp* (Kuroda et al., 1986) or *suc* (Fouet et al., 1984) promoters, might provide useful material for the construction of regulated expression vectors.

In addition to regulated expression, freedom from catabolite repression is a valuable advantage since it allows the inclusion of glucose- or glycerol-based feedstocks in fermentation media and generally leads to improved yields. Mutants of *B. subtilis* that synthesize amylase in the presence of glucose have been isolated and contain a cis-acting alteration to the promoter region that confers resistance to catabolite repression (Nicholson and Chambliss, 1986). This might form the basis of an effective expression vector.

2. Translation

Efficient translation requires a strong ribosome binding site (RBS) and a mRNA that is stable. The RBS has three critical features: (1) the Shine-Dalgarno (SD) sequence that interacts with the 16S rRNA to effect ribosome binding, (2) the initiation codon, usually AUG or GUG, and (3) a space between the SD sequence and the initiation codon.

The SD sequences of bacilli, like those of other gram-positive bacteria, are highly complementary to the respective rRNA sequences. Initial studies with phage ϕ29 of *B. subtilis* revealed that for two genes the SD sequences GGAGG were correct in all five positions with the CCUCC of the 16S rRNA (Murray and Rabinowitz, 1982). In *E. coli*, on the other hand, complementarity in several bases is seen, but seldom in all five. This suggested that bacilli required more stringent homology than *E. coli* and explained why *E. coli* would translate gram-positive mRNA but not vice versa. Numerous mRNAs have now been characterized. The calculated free energies of interaction of the SD region of rRNA and mRNA are -11.7 kcal/mol for *E. coli* and -16.7 kcal/mol for gram-positive bacteria, thus supporting the stringency hypothesis for translational specificity (Hager and Rabinowitz, 1985). However, it seems that a strong SD sequence complementarity is a necessary but not sufficient criterion for translation by *B. subtilis* ribosomes. The late message from coliphage T7 has a very powerful RBS with considerable homology, but it is not translated by *B. subtilis* ribosomes, suggesting that there are additional requirements for translation in *B. subtilis* or that there is a secondary structure block (Hager and Rabinowitz, 1985). Indeed there may be a region up to 25 bases each side of the SD sequence that influences translation.

Optimization of the SD region should therefore enhance translation. Band and Henner (1984) explored this assumption using a cloned leukocyte interferon A gene in an *E. coli*/*B. subtilis* shuttle vector as a model. A range of plasmids was prepared differing in free energy of binding to the *B. subtilis* 16S rRNA between -9.4 and -29.4 kcal/mol. Expression ranged 400-fold, and a direct correlation between complementarity and expression was found. No such correlation was observed in *E. coli*.

The space between the SD sequence and the initiation codon and its composition are important for efficient translation in *E. coli*. A space of 9-12 bases seems to be optimal, with a preference for A or U residues. This has received little attention in *B. subtilis*. Band and Henner (1984) concluded that the size of the space region was less important than its composition for maximal translation with a requirement for A and U residues, but Chang et al. (1983) noted that the space distance was important for efficient translation of the interferon gene from a β-lactamase promoter.

Finally, the stability of the message will have important implications for translation. Although mRNA in *B. subtilis* is generally short-lived, with a half-life of 1-2 min at 37°C, crystal protein synthesis in *B. thuringiensis* subsp. *berliner* appears to be coded by a message with an exceptionally long half-life (Klier and Lecadet, 1976). This may be responsible in part for the massive expression of this gene observed in stationary phase and suggests that mRNAs may be engineered for increased stability. Indeed this possibility has recently become a reality. A positive-acting retroregulation that enhances mRNA stability has been detected (Wong and Chang, 1986). Derived from the crystal protein gene of *B. thuringiensis* subsp. *kurstaki*, this regulator resides in a 381 bp fragment located at the distal end of the gene. When fused to the distal ends of *penP* or interleukin-2 cDNA, it increased the mRNA half-life from 2 to 6 min in *B. licheniformis* and *E. coli*. The increased stability is probably due to the incorporation of a stem and loop structure at the 3' end of the mRNA and gave rise to increased product yield. In conclusion, maximal translation can be achieved by paying attention to the SD sequence, the composition of the bases each side of the SD sequence, providing a spacer region of the optimal size and composition, and increasing the stability of the message.

3. Secretion

One of the great potential advantages of using bacilli as a host for recombinant DNA is the opportunity of secreting the product, thus providing the possibility of obtaining higher yields of product together with ease of recovery and purification and the avoidance of denaturation caused by the aggregation of proteins sometimes seen in *E. coli*.

The most popular secretion vectors are based on the amylase genes of *B. subtilis* or *B. amyloliquefaciens* (Table 2). By fusing foreign genes to the signal peptides of these secreted genes, export of the hybrid proteins and correct processing have been achieved. However, it seems that proteins that are not fully exported in their natural host cannot be efficiently secreted by *B. subtilis* despite the presence of a *Bacillus* signal peptide. Thus, β-lactamase, interferon, and insulin can be secreted, but viral envelope glycoproteins and the *ompA* and *ompF* products of the *E. coli* outer membrane are not secreted (Table 2). A second problem is that the yield of secreted heterologous proteins is generally very low. For example, yields of secreted interferon-α-2 using the amylase signal peptide from *B. amyloliquefaciens* were never greater than 1-2 mg/liter, well below the 250-400 mg/liter of α-amylase produced by this bacterium (Palva, 1982; Palva et al., 1983). This could be due to proteolytic degradation of the protein in the culture fluid or inefficient secretion or poor expression. Schein et al. (1986) addressed this problem, again using the interferon system as a model. They created an accurate fusion in which the interferon gene was adjacent to the signal peptide and compared expression and secretion with the cloned amylase. Secretion of interferon was poor (1-3% of the amylase yield), and pulse-labeling experiments showed that this was largely the result of inefficient secretion. Messenger RNA levels were similar for both interferon and amylase, but significant amounts of interferon precursor were associated with the cells. Proteolytic degradation, although significant in stationary phase cells, was negligible in exponential phase cells and thus the lower yield of interferon was attributed to inefficient transport or processing of the precursor. Thus there is a need for strains of *Bacillus* that secrete heterologous proteins more efficiently for commercial exploitation of protein secretion.

4. Stability

Recombinant plasmids are susceptible to two forms of instability. Defective partitioning at cell division gives rise to segregational instability in which daughter cells lacking the plasmid arise. Since these cells generally have a growth advantage compared with the producer cells, they quickly predominate, particularly in continuous fermentation (Nugent et al., 1983). Such a loss of chimeric plasmids from *B. subtilis* has been observed (Uhlen et al., 1981). Low copy number plasmids must be accurately partitioned at cell division, and genetic loci responsible for partitioning (*par*, *stb*, or *sop*) have been described for several plasmids from gram-negative systems (Sherratt, 1985).

Although plasmids from gram-positive sources are not so well understood, those used in *B. subtilis* are almost invariably derived from *Staphylococcus aureus*, and it may be that indigenous plasmids would be more stable. Indeed, Chang et al. (1983) stabilized a *B. subtilis* cloning vector by inserting a fragment of DNA from a naturally occurring *B. subtilis* plasmid and explained this as the insertion of a *B. subtilis par* function. Alternatively, an indigenous *B. subtilis* plasmid, such as pMI3 (Monod et al., 1986), which replicates with a very high copy number approaching 200, should be considered as a potential cloning vector.

Multicopy plasmids, such as pMI3, do not appear to have active partitioning systems and rely on probability for segregation at cell division (Sherratt, 1985). This is obviously reliable under natural conditions, but during prolonged growth in a chemostat the copy number may decline with the associated problems of segregational instability. In these circumstances it is advisable to clone a *par* locus into the plasmid.

An alternative approach to stabilizing foreign genes in *B. subtilis* is to engineer integration of the gene into the chromosome (Young, 1980). Saunders et al. (1984) overcame the problem of instability of a cloned protein A gene from *S. aureus* in *B. subtilis* by integrating the gene into the chromosome using an integration vector. Recombinational instability is also a problem. Having established a foreign gene in *B. subtilis*, rearrangements and deletions often occur, particularly after prolonged cultivation (Grandi et al., 1981; Lopez et al., 1984). Such recombinational events probably take place between short regions of homology that are flanked by regions of dyad symmetry and are independent of the host recombination system (Hahn and Dubnau, 1985).

Since little is known in this area it is difficult to overcome these problems. Nevertheless, Ostroff and Pene (1984) described a mutant of *B. subtilis* MI 112 (r^-m^- *recE4*) that transformed as efficiently as the parent and provided improved stability at least in the short term.

5. Proteolytic Degradation of Product

A principal reason for using *B. subtilis* as a host for the production of cloned gene products is the possibility of secreting the protein into the medium. Unfortunately, *B. subtilis* also secretes proteases into the medium and these rapidly degrade the product.

B. subtilis secretes at least three proteases (Millet, 1970): a serine or alkaline protease, a metal or neutral protease, and an esterase. These are synthesized predominantly during the late exponential and early stationary phases of growth and rapidly hydrolyze any foreign proteins. Acceptable hosts must therefore be deficient in these enzymes. Both the serine and metal enzymes have been cloned, and this has enabled the preparation of in vitro-derived deletions in the genes prior to their insertion into the genome. In this way, bacteria virtually devoid of extracellular protease activity have been developed (Stahl and Ferrari, 1984; Yang et al., 1984), and these organisms are a great improvement. Nevertheless, they display residual proteolytic activity presumably due to minor extracellular enzymes, such as the esterase, or leakage of intracellular enzymes. Thus there is still a requirement for improved host strains in this area.

C. Implications of Strain Development for the Fermentation Technologist

The remarkable achievements of the molecular biologists present the fermentation development engineer with a somewhat different set of problems from those faced 15 years ago. Traditionally, empirical and, later, statistical approaches were used to optimize media and growth conditions in order to enhance product yield. Either consciously or in ignorance, media constituents were chosen that effected induction where appropriate, relieved catabolite repression, repressed protease synthesis, and extended the late exponential-early stationary phase of growth before the onset of sporulation if this was necessary. Thus amylase is typically produced in a medium comprising high concentrations of starch and amino acids, which does not catabolite repress but reduces protease synthesis and encourages an extended stationary phase before sporulation. Moreover, for the traditional fermentation engineer it is also unlikely that the choice of continuous or batch fermentation could be entertained because most products, such as antibiotics, enzymes, or insect toxins, are synthesized in stationary phase and therefore would not be produced in a chemostat.

With the recombinant organism the fermentation engineer has a different set of problems. It is likely that the product, whether derived from a host gene or foreign gene, will be synthesized constitutively from a catabolite-resistant promoter. The host will be largely protease deficient, and product synthesis can be switched on at any stage of growth either by the addition of inducer or by a change in the physical conditions, such as a temperature shift. A major decision is therefore choice of fermentor, basically continuous culture (which with improved engineering and microprocessor control has many attractions) or traditional batch fermentor. Medium ingredients will be chosen largely on economic and availabillty grounds, and the fermentation will be optimized for strain stability. Although poorly understood at present, it is in the general area of maintenance of plasmid copy number and segregational accuracy that major developments will be made.

Despite these developments few products are manufactured on an industrial scale from recombinant organisms at present, not always because of technical limitations but also because of consumer resistance. It is therefore important to consider the more traditional approaches to fermentation optimization, and these are covered in detail in the next section.

V. DEVELOPMENT OF PRODUCTION MEDIA AND FERMENTATION PROCESSES

The greatest improvements in the productivity of fermentations are generally considered due to improvements in the microorganism following mutation and strain selection studies. However, the nutritional and physical environment used to cultivate the organism and the methods used for product recovery and purification also have a significant effect on the economics and final product yield of the process.

The medium used to isolate and screen industrial microorganisms is not usually the same as that used for development studies and production. The former often include components that are relatively expensive, support suboptimal growth, or may be regarded as undesirable for use at the production scale for other reasons. Development of an economic process requires the selection of sources of carbon, nitrogen, phosphorus, sulfur, potassium, and trace elements. An energy source is required that will maximize microbial growth and product yield, with the maximum yield of product or biomass per unit of substrate. The rate of product formation must be maximized and undesirable side products kept to a minimum. The ingredients must be cheap, of consistent quality, and readily available all year, and the costs of transport and any pretreatment should be minimal. Industrial media are generally based on the use of natural materials, although a careful balance of carbohydrate, nitrogen, and mineral salts is essential. Substrates for industrial fermentations are described by Miller and Churchill (1986), who provide tables of analysis for medium ingredients. The use of natural materials, which are generally agricultural commodity products, results in some geographic variation in availabillty, composition, and cost, which should be considered when carrying out media formulation studies.

Although industrial fermentation media are based on complex nutrients, development studies require the use of empirical procedures to assess the suitability of medium ingredients for growth and product induction and repression. Medium development is generally initiated at the level of the shake flask, but pilot-scale fermentation studies generally indicate further modifications. Statistical methods for medium optimization are discussed by Greasham and Inamine (1986).

Considerable information is available on laboratory-scale studies, but as processes become more commercially significant the availability of published literature begins to decline. The information available, therefore, on large-scale industrial practices becomes generalized, with competitors unwilling to disclose the more significant and novel details of their processes. A second problem is that a multiplicity of strains and assays has been used over the years, particularly with regard to extracellular enzymes, and this is complicated by the inability to distinguish *B. subtilis* from *B. amyloliquefaciens*. In many papers it is not clear which organism is being studied (this particular confusion has since been clarified; O'Donnel et al., 1980; Priest et al., 1987). We therefore thought it appropriate to present the available information on medium development in a fairly uncritical fashion, although we do make some attempt to highlight those principles of the processes that can be extracted from the mass of data published over the past 25 years.

A. Production of α-Amylase

The production of extracellular enzymes is a relatively simple process: microorganisms are cultured in a suitable medium and the enzyme subsequently recovered from the medium, after removal of the cells. The problems for production center on achieving a process that is economical and reliable and complies with the strict codes for safety and production procedures laid down by the regulatory authorities, such as the U.S. Food and Drug Administration or the Medicines Inspectorate or Health and Safety Executive in the United Kingdom. Process plant and equipment must be maintained in good order, and maintenance of production records and quality control data is essential.

Rather than provide a comprehensive and necessarily shallow account of all the extracellular enzymes of *Bacillus* and since the principles tend to be similar for all enzymes, we treat one fermentation in depth. α-Amylase was chosen as the example since a wealth of information is available for this fermentation.

1. Medium Development

Amylase production is based on naturally occurring materials, such as corn, wheat, malt, soy extracts, and cornsteep liquor. Fukumoto et al. (1957) described a medium for the production of α-amylase from *B. amyloliquefaciens* based on soybean cake extracts (6%), potato starch (3.5%, dextrinized with bacterial amylase), $(NH_4)_2HPO_4$ (1.2%), KCl, and $MgSO_4 \cdot 7H_2O$ (0.02%). Studies of amylase production from *B. subtilis* indicated that when carbohydrate was used as the carbon source, phosphate had a significant effect on enzyme production. The optimum phosphate concentration was in the region of 0.1 M. When casein and soybean extracts were used as nitrogen source, the choice of carbon source was not critical; starch, most sugars, and organic acids were suitable. Peptones and amino acids provided a more effective source of nitrogen than ammonia. A 300-fold increase in α-amylase production from *B. amyloliquefaciens* (ATCC 23350) was achieved when glycine was added to a chemically defined medium (Zhang et al., 1983), and yet studies by Fukumoto et al. (1957) using washed cells of *B. subtilis* (probably *B. amyloliquefaciens*) in a phosphate-buffered medium with 1% soluble starch showed that glycine and cysteine inhibited enzyme production. The concentration of Fe^{3+} ions was critical to enzyme production (Coleman and Elliott, 1962).

Ammonium hydrogen phosphate was the preferred source of nitrogen and phosphate for α-amylase production from *Bacillus apiarius* CBML-152 (Ghosh and Chandra, 1984). Enzyme production was enhanced by Mn^{2+}, Ca^{2+}, and Zn^{2+}, although there was little effect on cell growth; Mg^{2+}, Fe^{2+}, Fe^{3+}, and Na^{+} enhanced both growth and enzyme production. A total of nine amino acids (alanine, arginine, asparagine, aspartic acid, cystine, histidine, phenylalanine, lysine, and valine) stimulated growth and enzyme production when examined individually in a starch-based mineral salts medium. A combination of arginine, cystine, lysine, and valine was found to be the most effective in stimulating enzyme production, and their omission resulted in a decrease in activity. Biotin, pantothenic acid, pyridoxine, and thiamine all enhanced growth and enzyme production. Studies with naturally occurring substrates indicated enhanced production using oil seed cakes and husks of pulses. Dried and powdered mushrooms also increased production in the presence or absence of peptone and yeast extract. α-Amylase production from *B. licheniformis* CUMC-305 was supported using alanine, arginine, aspartic acid, cystine, isoleucine, phenyalanine, and serine. The omission of aspartic acid and cystine resulted in a 50% reduction in enzyme production. Biotin, pantothenic acid, and thiamine enhanced production; $(NH_4)_2HPO_4$ and peptone were the optimum sources of phosphate and nitrogen. Galactose, dextrin, glycogen, and inulin were the optimum sources of carbon; low concentrations (0.05-0.1%) of glucose, lactose, maltose, fructose, rhamnose, and sorbitol supported moderate production, but higher concentrations were inhibitory. Na^{2+}, Cu^{2+}, Mn^{2+}, Mg^{2+}, and Fe^{3+} all stimulated enzyme production (Chandra et al., 1980). The use of oil seed cakes in a basal medium containing peptone and beef extract was found to provide an ideal fermentation base for the production of amylase from this strain of *B. licheniformis* (Krishnan and Chandra, 1982). The use of sesame, linseed, and cottonseed cakes required the addition of cornsteep liquor to provide an additional nitrogen source.

A medium for the cultivation of *B. licheniformis* (ATCC 39326) was based on 10% lactose as carbon source, with K_2HPO_4 (1.4%),

KH_2PO_4 (0.6%), $(NH_4)_2SO_4$ (0.6%), sodium citrate (0.3%), whole-grain soy flour (3%), and $CaCl_2 \cdot 2H_2O$ (0.05%). Alternative carbon sources included glucose, starch, or complex carbohydrate containing such materials as cornmeal. The optimum carbohydrate concentration was considered 15% (w/v). The nitrogen source was based on soluble nitrates, ammonium salts, or organic sources, such as yeast extract, cornsteep liquor, or peanut meal (Horwath, 1984).

The production of a thermostable α-amylase from *Bacillus stearothermophilus* (strain 1503-4) was examined in a defined medium (Welker and Campbell, 1963a). The enzyme appeared constitutive since enzyme activity was observed on glycerol, glucose, and sucrose; the addition of casein hydrolysate (0.1%) permitted growth on fructose. Growth did not occur on sodium acetate and sodium succinate. Maximum production was observed on starch and maltose, which induced enzyme activity. Glycerol, glucose, and sucrose exhibited an inverse relationship between the rate of growth and total amylase production, indicative of catabolite repression. Analysis of technical-grade maltose indicated the presence of 3-5% glucose, 0.5% maltotriose, and 2.5% higher molecular weight maltodextrins. Studies with maltodextrins as individual constituents showed maximum enzyme induction with maltotetraose, followed by maltopentaose, maltohexaose, maltotriose, and maltose (Welker and Campbell, 1963b). A soil isolate of *B. stearothermophilus* with a lower optimum growth temperature than the strain studied by Welker and Campbell (1963a) produced an α-amylase with an optimum temperature for activity of 82°C (Srivastava and Baruah, 1986). Maximal production was observed on a complex medium based on 0.4% cornsteep liquor; moderate production was found with meat and beef extracts. Growth was comparable on yeast extract, but enzyme levels decreased. Starch was required to induce amylase synthesis; maximum enzyme production was observed in the late exponential and stationary phases of growth. Isoleucine, phenylalanine, cysteine, and aspartic acid were essential for enzyme production, and glutamic acid, methionine and glycine were stimulatory. Diammonium hydrogen phosphate and ammonium dihydrogen phosphate were preferable to ammonium sulfate and ammonium nitrate as a nitrogen source. Ca^{2+}, Sr^{2+}, Mg^{2+}, and Ba^{2+} slightly stimulated enzyme production, whereas Ag^{2+}, Cu^{2+}, Fe^{3+}, Sb^{3+}, As^{3+}, and Au^{3+} inhibited enzyme production and Zn^{2+} and Mn^{2+} had no effect. Growth on glucose, galactose, fructose, and sucrose suppressed enzyme production. Enzyme activity was not observed on maltose; however, amylase was induced with maltotriose, maltotetraose, maltopentaose, and maltohexaose. Maximum production was observed on starch, dextrin, and glycogen and the disaccharide cellobiose. Thermostable amylase was extracted from the culture supernatant of 10 strains of *B. stearothermophilus* isolated from heat-treated aerobic sewage sludge (Grueninger et al., 1984). In a basal salts medium of pH 7.0 supplemented with malt extract, yeast extract, and simple carbon sources, the growth of all 10 strains was observed. Amylase activity was only produced on media containing either lactose (6 strains), acetate (3 strains), or maltose (2 strains), but no activity was observed on glucose, fructose, galactose, sucrose, glycerol, and citrate. Growth on soy grist, rape grist, and molasses increased amylase production by 10- to 50-fold compared with growth on simple carbon sources. Enzyme yields were lower on peptone, casein, wheat flour, or cornsteep liquors, and no activity was found in supernatants of cells grown on soluble or insoluble starch. Combinations of complex substrates gave increased enzyme yields, and the best combination examined was malt extract-soy grist-peptone. Studies appeared to indicate that low growth rate, which could be achieved by decreasing agitation or aeration, enhanced amylase production.

Buonocore et al. (1976) studied the effect of different carbon sources on amylase production from *Bacillus acidocaldarius* using a basal salts medium supplemented with 0.5% yeast extract and 0.5% carbon source. Enzyme production occurred during the stationary phase of growth in the presence of oligo- or polysaccharides having α(1-4)-glycosidic linkages. Maximum enzyme yields were achieved after 18 hr using glycogen, although starch and maltose also produced high enzyme activity. No enzyme activity was detected following growth on glucose, glycerol, lactose, sucrose, trehalose, or cellobiose. Amylase production increased linearly with inducer concentration up to a maximum of 0.3-0.5%; above this concentration enzyme activity slowly declined. There was no significant variation in the cell growth when the inducer was varied.

In conclusion, it will be appreciated that the composition of the medium is a critical factor in obtaining maximum enzyme yields, and the optimum medium composition varies significantly with different *Bacillus* species (Table 5). Generally, starch is the preferred carbon source, although maltose and maltodextrins provide a suitable alternative. α-Amylase production is generally repressed in the presence of glucose. The nitrogen source in industrial fermentations is generally supplied by complex substrates, such as cornsteep liquor or soy extract; $(NH_4)_2HPO_4$ is generally the preferred inorganic supply of nitrogen.

2. Fermentation Studies

Laboratory- and pilot-scale fermentation studies of amylase production by a number of different species of *Bacillus* have been reported. These have in the main involved small batch and continuous fermentation studies to evaluate the control of enzyme production, optimization of media, and physical growth parameters.

Continuous culture studies of amylase and protease synthesis by *B. amyloliquefaciens* and some strains of *B. subtilis* have often, after initially exhibiting high enzyme levels in batch culture, been followed by fluctuating enzyme yields that subsequently decline during continuous culture (Heineken and O'Connor, 1972; Fencl et al., 1972; Sikyta and Fencl, 1976). This pattern of enzyme yield can be explained by the occurrence of low or nonproductive mutants (Sikyta and Fencl, 1976; Fencl et al., 1979). The regulation of synthesis of extracellular enzymes in these two organisms also provides an explanation. In *B. subtilis* Marburg and *B. amyloliquefaciens*, extracellular enzyme synthesis is repressed during exponential growth and derepressed during early stationary phase prior to sporulation (see Sec. III.C). Since the chemostat provides continuous exponential growth, extracellular enzyme synthesis would be repressed. Fencl and Pazlarova (1982) studied amylase production from *B. subtilis* using a two-stage chemostat; the enzyme concentration in the second stage reached more than 75% of the maximum concentration achieved in batch fermentation.

B. stearothermophilus (Davis et al., 1980), *B. licheniformis* (Emanuilova and Toda, 1984; Priest and Thirunavukkarasu, 1985), and *B. polymyxa* (Griffin and Fogarty, 1973), in which amylase production occurs during exponential growth, all exhibit constant levels of amylase production during continuous culture studies.

Chemostat studies of *B. licheniformis* under nitrogen limitation using glucose, glucose with glutamate, and glucose with alanine as carbon source indicated minimal production of α-amylase. Under carbon limitation with glucose, alanine, or glutamate as carbon source, amylase production was optimal at a dilution rate (D) of 0.05 hr^{-1} and decreased by more than 50% as D was increased to 0.2 hr^{-1}. Although low growth rate gave an increased rate of enzyme production per gram of substrate, the rate of enzyme production per unit fermentation volume was independent of dilution rate (Meers, 1972). In batch fermentation studies using soluble starch, a range of polymers from glucose to maltohexaose were detected; within 10 hr of inoculation the sugar concentration decreased to a level at which it was no longer detectable and α-amylase production was initiated. Increasing the concentration of glucose significantly diminished the rate of amylase production until the glucose was dissimilated. Glucose produced in the fermentor as a result of α-amylase activity was metabolized and did not accumulate to detectable levels. Repression of α-amylase activity was also observed with fructose and maltose. Similar results have been obtained in batch and continuous cultures of *B. licheniformis* NC1B 6346 (Priest and Thirunavukkarasu, 1985), and catabolite repression has been observed in *Bacillus caldolyticus* (Emanuilova and Toda, 1984), *B. stearothermophilus* (Srivastava and Mathur, 1984), and *B. acidocaldarius* (Buonocore et al., 1976).

Table 5 Constituents Used for Medium Formulation for the Production of α-Amylase

Organism	Production phase	Carbon source	Carbon repressors	Nitrogen source
B. amyloliquefaciens	Stationary	Starch Lactose	—	Soybean cakes $(NH_4)_2HPO_4$
B. acidocaldarius	Stationary	Maltose Starch Glycogen Maltotriose Maltotetraose	Glycerol Glucose Lactose Sucrose Cellobiose	Yeast extract
B. apiarius	Biphasic	Starch	—	$(NH_4)_2HPO_4$ Oil seed cakes Husks of pulses
B. licheniformis	Logarithmic	Starch Lactose Galactose Inulin Oil seed cakes	Glucose Maltose Fructose	Yeast extract Peptone Soy flour Cornsteep liquor
B. polymyxa	Logarithmic	Starch	Excess starch	Peptone
B. stearothermophilus	Logarithmic	Maltose Starch Maltotriose Maltotetraose Glycogen Cellobiose Dextrin Maltohexaose	Glucose Maltose Galactose Fructose Sucrose	Cornsteep liquor $(NH_4)_2HPO_4$ $NH_4H_2PO_4$ Tryptone
B. caldolyticus	Logarithmic	Starch	Glucose	Casitone
B. subtilis	Stationary	Starch Lactose Maltose Glucose	—	$(NH_4)_2SO_4$ Peptone

Organism	Amino acids: Stimulatory	Amino acids: Inhibitory	Metal ions: Stimulatory	Metal ions: Inhibitory	Reference
B. amyloliquefaciens	Glycine	—	—	—	Fukumoto et al., 1957 Zhang et al., 1983
B. acidocaldarius	—	—	—	—	Buonocore et al., 1976
B. apiarius	Alanine, arginine, asparagine, cystine, histidine, lysine, phenylalanine, valine	Glutamine leucine	Mn^{2+}, Ca^{2+}, Zn^{2+}, Mg^{2+}, Fe^{2+}, Fe^{3+}, Na^{+}, K^{+}	—	Ghosh and Chandra, 1984
B. licheniformis	Alanine, arginine, cysteine, isoleucine, phenylalanine, serine, aspartic acid	—	Na^{+}, Cu^{2+}, Mn^{2+}, Mg^{2+}, Fe^{3+}	Cd^{2+}, Ag^{+} Hg^{2+}, Ca^{2+}, Pb^{2+}, Se^{4+}, W^{6+}, $2m^{2+}$	Meers, 1972 Krishnan and Chandra, 1982 Horwath, 1984 Chandra et al., 1980
B. polymyxa	—	—	Ca^{2+}, Mn^{+}	—	Griffin and Fogarty, 1973
B. stearothermo philus	Isoleucine, cysteine, phenylalanine, aspartic acid, glutamic acid, methionine, glycine	—	Ca^{2+}, Sr^{2+}, Mg^{2+}, Ba^{2+}	Ag^{2+}, Cu^{2} Fe^{3+}, Sb^{3+} As^{3+}, Au^{3+}	Welker and Campbell, 1963a,b Davis et al., 1980 Sidler and Zuber, 1977 Grueninger et al., 1984 Srivastava and Mathur, 1984 Srivastava and Barush, 1986
B. caldolyticus	—	—	—	—	Emanuilova and Toda, 1984
B. subtilis	—	Glycine, cysteine	Fe^{3+}	—	Fukumoto et al., 1957 Coleman and Elliott, 1962

Davis et al. (1980) studied α-amylase production from a strain of *B. stearothermophilus* in a 1 liter fermentation vessel with the airflow at 1.5 vvm and impeller speed at 300 rpm. Studies were carried out on a tryptone-maltose-based medium at 55°C. In batch cultures the pH increased to 8.2 after 3.5 hr, yielding a maximum of 790 U of α-amylase per milliliter of culture. When the pH was controlled at pH 7.2, amylase concentration reached 470 U/ml after 3.5 hr, although the biomass yield was the same. In a chemostat under tryptone limitation at 55°C, the α-amylase concentration increased from 513 U/ml at D = 0.31 hr^{-1} to 5788 U/ml at D = 0.75 hr^{-1}. The amylase concentration remained the same up to a D = 0.95 hr^{-1}; above this and below 0.31 hr^{-1} the concentration was greatly reduced. Cell dry weight and optical density followed a similar trend. Maximum rates of both α-amylase and biomass production were reported at a D of 0.95 hr^{-1}. Studies of protease and α-amylase production from *B. stearothermophilus* strains NCIB 8924 and NRRL-B-3880 in a medium based on beef extract, tryptone, and maltose were examined in 10 and 300 liter fermentors (Sidler and Zuber, 1977). In a 300 liter batch fermentor with the pH controlled at 6.8 with 4 N NaOH, impeller speed at 200 rpm, airflow at 0.5 vvm with 0.5 atm over pressure, and temperature of 53°C, the cultivation time was 5-7 hr. The vessel was inoculated with a 5% inoculum from a preculture grown under similar conditions.

Emanuilova and Toda (1984) studied α-amylase production from *B. caldolyticus* during batch and continuous fermentation using a 2 liter fermentor with a working volume of 1.5 liters. Growth temperature was maintained at 60°C, airflow at 0.5 vvm, and impeller speed at 500 rpm. The pH was not controlled, dissolved oxygen was maintained above 1.0 ppm, and foaming controlled using KM70 antifoam. A basal salts medium with casitone (2 g/liter) was supplemented with either glucose or starch (1 or 0.1 g/liter). In batch culture with starch (1 g/liter), the pH decreased from 7.0 to 5.5 and then increased at the end of exponential phase, reaching pH 8.8 in late stationary phase. Amylase production occurred during exponential growth when the cell doubling time was approximately 15 min. Without glucose or starch as carbon source, α-amylase production was minimal on casitone; the addition of glucose (1 g/liter) resulted in amylase production after utilization of the glucose (approximately 3 hr). Addition also of starch (0.1 g/liter) led to the immediate production of α-amylase, with final yields similar to yields on glucose alone. Enzyme yields were doubled when casitone and starch (2 g/liter) were used in the absence of glucose (Table 6). In continuous culture amylase activity was low (3 U/ml) at dilution rates above 1.2 hr^{-1} when cell dry weight, glucose concentration (0.1 g/liter) and pH (5.5) remained relatively constant. The maximum specific growth rate determined by washout experiments was

Table 6 α-Amylase Production in Batch and Continuous Cultures of *B. caldolyticus*

Carbon source			Batch culture		Continuous culture		
Casitone (g/liter)	Glucose (g/liter)	Starch (g/liter)	Amylase (U/ml)	Amylase (U/g dry weight)	Dilution rate (hr^{-1})	Amylase (U/ml)	Amylase (U/g dry weight)
2.0	0	0	2.5	3.0	0.1-0.9	1.5	3.2
2.0	1.0	0	9.0	9.7	0.15	5.0	7.1
2.0	1.0	0.1	8.5	7.2	0.15	80.0	120.0
2.0	0	1.0	13.5	18.0	0.25	200.0	230.0

Source: Data from Emanuilova and Toda (1984).

2.7 hr^{-1}. Peak amylase activity was observed at a dilution rate of 0.1-0.25 hr^{-1}, one-tenth of the maximum specific growth rate when the concentration of glucose and starch were so low that accurate determination using the normal assay methods was not possible. The addition of a small amount of starch had a more significant effect on amylase production than in batch culture. The addition of starch (0.1 g/liter) to the casitone and glucose (1.0 g/liter) medium resulted in a 17-fold increase in productivity compared with batch studies. Using casitone and starch (1.0 g/liter) without glucose increased enzyme yields some 15-fold over the batch fermentation, giving a yield of 200 U/ml of culture.

Continuous culture studies of those *Bacillus* strains in which α-amylase production occurs during exponential growth indicate that as the growth rate is increased above a maximum value the level of enzyme activity begins to decrease quite rapidly. Enzyme production in batch cultures undergoes catabolite repression when glucose is present in the medium.

3. Industrial Production of α-Amylase

The industrial production of α-amylase from *B. amyloliquefaciens* was developed by Boidin and Effront (1917). The organism, which was initially named *Bacillus mesentericus*, was reclassified as *B. subtilis* and later as *B. amyloliquefaciens*. *B. amyloliquefaciens* is used for much of the commercial production of α-amylase. The enzyme has a molecular weight of about 50,000 daltons and is stable over a pH range of 5.5-8.0, with optimum activity at pH 5.9. The enzyme is stabilized by calcium ions, which are generally added to reaction mixtures, increasing stability and activity. The enzyme can be used to hydrolyze starch at temperatures approaching 90°C. Starch granules are resistant to α-amylase attack and must first be ruptured by heat before hydrolysis can occur. "Cooking" temperatures above 100°C are normally required to ensure complete breakage of corn granules and gelatinization of the starch. Enzymes able to operate at higher temperatures would enable faster, more efficient, and cheaper starch liquefaction since the material would not have to be cooled after gelatinization for enzymic hydrolysis. α-Amylase from *B. licheniformis*, which was introduced in 1973 (Madsen et al., 1973), hydrolyzes starch at 105-110°C. The enzyme has a wider pH range than the *B. amyloliquefaciens* enzyme and has a molecular weight of 60,000 daltons. Despite being more thermostable it is less dependent upon Ca^{2+}. It has the disadvantage that the reaction is more difficult to terminate by heat treatments.

Depending upon the strain being cultivated, fermentations are carried out at 30-40°C and pH 7.0. The pH may be maintained by the addition of $CaCO_3$ to the medium, which also increases enzyme stability. Enzyme activity is lost if the pH falls below 6.0. Amyl-

Table 7 Production Media for the Production of α-Amylase from *Bacillus*

Components	Amount % (w/v)
Ground soybean meal	1.85
Amber BYF (autolyzed brewer's yeast fractions, Amber Laboratories)	1.5
Distiller's dried solubles	0.76
N-Z amine (enzymic casein hydrolyzate, Sheffield Chemical Company)	0.65
Lactose	4.75
$MgSO_4$	0.04
Hodag KG-1 antifoam (Hodag Chemical Corp.)	0.05
(Underkofler, 1976)	
Cornstarch	4.0
Ground corn (hominy)	10.0
Cornsteep liquor	6.5
(Smythe et al., 1950)	
Potato starch	10.0
Ground barley	5.0
Soybean meal	2.0
Sodium caseinate	1.0
$Na_2HPO_4 \cdot 12H_2O$	0.9
(Outtrup et al., 1972)	

ase production based on complex natural ingredients (Table 7) begins as the cell count approaches 10^9-10^{10} cells ml^{-1} after approximately 10-20 hr. Production continues until the utilizable carbon source is exhausted, usually 100-150 hr (Aunstrup, 1979).

Both *B. amyloliquefaciens* and *B. licheniformis* produce a serine protease, and *B. amyloliquefaciens* also produces a neutral protease, β-glucanase, and hemicellulase. The proteases need to be removed or inactivated since their presence may lead to the formation of melaninlike products during starch hydrolysis. The proteases may be destroyed by heating, taking care to minimize the inactivation of the α-amylase. Adsorption of the protease to clay minerals (bentonite) or fractional precipitation may also be used. The use of protease-free mutants provides an alternative, but this may then require the use of prehydrolyzed protein substrate in the fermentation medium (Aunstrup, 1979).

Seed inocula for the production fermentation are generally grown in a seed fermentor and aseptically transferred to the production vessel while the culture is still in exponential growth. Depending on the size of the production fermentor, intermediate vessels may be required; a 5% v/v inoculum is generally considered desirable. The use of complex natural ingredients may require milling or pretreatment before use and separate sterilization before transfer to the production fermentor. Amylase production in the culture supernatant may be monitored using the methods of Welker and Campbell (1963a), Pfeuller and Elliot (1969), Emanuilova and Toda (1984), or Horwath (1984).

The cells and any insoluble medium ingredients remaining need to be removed from the fermentation broth by centrifugation or filtration. Flocculating agents may be necessary to settle the cells before clarification. The clarified broth is then concentrated by low-temperature high-vacuum evaporation. After dilution of the liquid preparation to give a standard activity, preservatives and calcium salts may be added to aid stabilization of the preparation.

The activities of commercial preparations vary by a factor of 100 and are adjusted to suit the required application. The production of more concentrated products leads to reduced packaging and transport costs. The most active liquid preparations may contain 2% active amylase and the most active solid preparations up to 5% active amylase. Solid preparations are obtained by further evaporation of liquid, and spray drying is used to achieve a concentrated powder that may be mixed with an inorganic salt. Further purification steps to achieve highly active preparations include precipitation of the enzyme with sodium or ammonium sulfate or solvent systems based on ethanol or isopropanol. To assist recovery, the enzyme may be isolated in the presence of an inert carrier, such as starch, lactose, calcium sulfate, or casein, depending upon the application for which the product is designated (Burbidge and Collier, 1968).

4. α-Amylase Production from Recombinant *Bacillus* Strains

The cloning of the α-amylase gene and studies of its expression in various species of *Bacillus* have been widely reported (Aiba et al., 1983; Ortlepp et al., 1983; Palva, 1982; Joyet et al., 1984; Piggott et al., 1984; Mielenz and Mickel, 1985; Aiba et al., 1986; Vehmaanperä and Korhola, 1986). Few reports, however, discuss plasmid stability and gene expression during fermentation and scaleup of laboratory studies.

Nybergh (1984) studied the expression of the α-amylase gene cloned from *B. amyloliquefaciens* to *B. subtilis* using a plasmid pUB110. The recombinant strain produced a 5-fold increase in α-amylase compared with the *B. amyloliquefaciens* strain and 2500-fold more than the wild-type *B. subtilis* strain. Comparative fermentation studies between the original *B. amyloliquefaciens* strain and the recombinant *B. subtilis* strain were carried out in 8 dm^3, 1 m^3, and 10 m^3 fermentors. The strains were maintained on L agar supplemented with 1% starch. Kanamycin (5 μg/ml) was added to plates used for the subculture of the recombinant strain. The recombinant organism produced higher levels of α-amylase than the donor strains on L broth supplemented with starch hydrolysate or L broth in which casein hydrolysate was substituted for tryptone. Growth on complex raw materials available from a grain alcohol production plant and a starch sugar plant resulted in little difference in the level of amylase production between *B. amyloliquefaciens* and the recombinant *B. subtilis*. The availability of free amino acids in these complex medium ingredients was relatively low, and the greater proteolytic activity of *B. amyloliquefaciens* was a significant advantage for growth on complex substrates. Laboratory-scale studies indicated the plasmid construction to be stable for some 50 generations; however, at the semi-industrial scale, loss of expression was observed resulting from deletion of part of the α-amylase gene.

Pinches et al. (1985) examined the expression of the α-amylase gene from an industrially important *Bacillus* strain (Bsl) cloned into *B. subtilis* IA297 on plasmid pVC102 (Corfield et al., 1984). Stocks of the recombinant organism were lyophilized and maintained at 4°C. Working stocks were subcultured on L broth agar supplemented with 2% starch and neomycin sulfate and chloramphemical (10 μg/ml). Fermentation seed cultures were prepared by transferring a few colonies from a fresh 24 hr plate culture to 20 ml of L broth supplemented with antibiotics and incubating at 37°C for 8-10 hr on a reciprocal shaker. Plasmid loss was determined by plating on L broth agar plates supplemented with starch (LS medium) and antibiotics.

Fermentations (5 liter) were carried out at 37°C and pH 7.0 controlled with H_2SO_4 and KOH. Foaming was controlled by the addition of Dow Corning DB31 silicon emulsion. The stirrer speed was maintained at 600 rpm, and sterile air was supplied at 0.4 vvm. Plasmid loss was negligible during exponential growth in L broth supplemented with either starch or glucose, only becoming significant during stationary phase. In L broth supplemented with starch, α-amylase production was growth associated, and plasmid loss, which became apparent after 50 hr, had little effect on enzyme production. Enzyme production on L broth supplemented with glucose was associated with late exponential growth and stationary phase. Loss of plasmid toward the end of the exponential phase significantly affected enzyme yields, although this was improved following the addition of antibiotics to the medium. Plasmid instability was associated with both structural instability, resulting in deletions, and total elimination of the plasmid. The cloning of a thermostable α-amylase from *B. stearothermophilus* into *B. subtilis* on plasmid pUB110 was reported by Zeaman and McCrea (1985). Vehmaanperä and Korhola (1986) studied the stability of plasmid pKTH10, which carried the α-amylase gene from *B. licheniformis* cloned into pUB110. In two strains of *B. subtilis* examined, the plasmid remained stable after 50 generations. In a third strain, *B. subtilis sacU9*, the plasmid was almost completely lost after 50 generations, indicating the importance of selecting a suitable host strain for cloned gene expression.

Studies of the α-amylase from the recombinant strain and wild-type *B. stearothermophilus* indicated them to be identical, and animal feeding studies were carried out to confirm the safety of the recombinant product for use in the food industry. Following the completion of toxicity trials, a petition was filed with the FDA seeking affirmation of GRAS status for the cloned enzyme.

B. Production of Insecticidal Endotoxin from *B. thuringiensis*

The principal source of new strains of *B. thuringiensis* is diseased or dead insects; nonspore formers may be eliminated by heating to 65°C for 15 min. Other spore formers can be reduced by the inclusion of selective antibiotics, such as polymyxin B (5 μg/ml), into the nutrient medium (Akiba and Katoh, 1986). Strains may be cultivated on nutrient agar or tryptone-soy agar. Stock cultures are generally maintained as spore suspensions or dried spores, which remain viable for several years. Alternatively, strains have been maintained on nutrient agar slopes at 4°C with monthly subculture (Dharmsthiti et al., 1985). The development of new strains requires the use of bioassays for screening activity against a range of different insects.

1. Medium Development

A defined medium for the growth of *B. thuringiensis* (Nickerson and Bulla, 1974) contained glucose (1 g/liter), $(NH_4)_2SO_4$ (2 g/liter), K_2HPO_4 (0.5 g/liter), aspartic acid (1 g/liter), glutamic acid 1 g/liter), and trace elements $MgSO_4 \cdot 7H_2O$ (0.3 g/liter), $MnSO_4 \cdot H_2O$ (0.05 g/liter), $CaCl_2 \cdot 2H_2O$ (0.08 g/liter), $ZnSO_4 \cdot 7H_2O$ (0.005 g/liter), $CuSO_4 \cdot 5H_2O$ (0.005 g/liter), and $FeSO_4 \cdot 7H_2O$ (0.0005 g/liter). The medium was adjusted to pH 7.4 with NaOH.

Studies of 21 strains of *B. thuringiensis* using $NaNO_3$ as the nitrogen source indicated growth of 19 strains using glutamate as the sole carbon source, 17 strains utilized aspartate, 16 DL-arginine, and 16 L-ornithine. A number of strains were able to utilize acetate or citrate, although spores were unable to germinate on these carbon sources (Akiba and Katoh, 1986). Seed cultures for primary-stage fermentors were prepared from spore suspensions or deep-frozen spore suspensions stored at -20°C in 10-20% (v/v) glycerol. Before use, stock cultures should be examined for evidence of crystal formation and must be phage free. Erlenmeyer flasks containing nutrient broth were inoculated with spores and incubated at 30°C on an orbital incubator shaker at 150-200 rpm for 6-8 hr. Krieg and Miltenburger (1984) recommended a seed medium containing starch, 0.5%; glucose, 0.5%; yeast extract, 0.2%; tryptone, 1.0%, K_2HPO_4, 0.001%; and KH_2PO_4, 0.001%. Smith (1982) reported studies of six different culture media for the production of mosquito toxin production using shake flask cultures of *B. thuringiensis* subsp. *israelensis*. Buffered medium gave no improvement on unbuffered medium ranging in pH from 5.7 to 8.1 at the end of the culture. A 2% (v/v) seed inoculum is used to inoculate the primary seed fermentor.

The production of bacteria for use as biological control agents is far less sophisticated than the production of chemical insecticides. Their production can therefore be undertaken in many underdeveloped countries where mosquito-borne diseases are endemic. It follows that the use of cheap, locally available substrates is an important consideration. Obeta and Okafor (1984) assessed five formulations of media for the production of spores from *B. thuringiensis* subsp. *israelensis*. The media were based on groundnuts (peanuts), soybeans, Bambara beans, and black and white cowpeas blended with dried cow's blood and mineral salts. All five media gave high levels of sporulation and total viable cell counts of $4.2\text{-}5.4 \times 10^9$ ml^{-1}. Biological assay of the spore-δ-endotoxin complexes on three types of mosquito larvae indicated high activity.

Dharmsthiti et al. (1985) developed two new media for the production of *B. thuringiensis* subsp. *israelensis* and *B. sphaericus* based on hydrolyzed liquor, a by-product of the production of monosodium glutamate. The yield of *B. thuringiensis* spores after 48 hr was 2.5×10^8 ml^{-1}. Comparative studies using a nutrient broth, glucose, and salts medium (Delafield et al., 1968) indicated that similar yields of spores and crystals were obtained on both media, although the cost of the medium based on hydrolyzed liquor was considerably cheaper.

Most production media now utilize complex natural products as sources of carbon, nitrogen, and minerals. Nitrogen sources include fishmeals, cottonseed flour, soy flour, cornsteep liquor, autolyzed yeast, and casein. Carbohydrate sources include corn products, starch dextrose, and beet molasses.

Dulmage (1970) studied the production of δ-toxin from 12 strains of *B. thuringiensis* in medium based on tryptone-cornstarch-glucose-salts and peptone-cottonseed flour-glucose-salts. Considerable strain variation was indicated between the media; high sporulation levels did not always correlate with high levels of δ-toxin production based on biological assay, and vice versa. Further studies (Dulmage, 1971) indicated problems of obtaining reproducible results when using the same medium and fermentation conditions.

To achieve maximum sporulation, Megna (1963) considered it important that the fermentable carbohydrate source and nitrogen source be depleted at approximately the same time after the onset of sporulation. If not, sporulation is slow, with the result that before an appreciable fraction of the cell population sporulates, those cells that sporulate first have fractured, freeing crystals and spores, some of which begin to germinate.

2. Fermentation Development

Holmberg et al. (1980) described a process scaled up to 1000 liters for the cultivation of *B. thuringiensis* subsp. *thuringiensis* (ATCC 10792) on soy flour and molasses (Table 8). Toxin production was not significantly affected when the pH was varied from 6.5 to 8.0 or aeration varied from 0.1 to 1 vvm. Pendleton (1969) described the production of *B. thuringiensis* in a medium based on fishmeal and starch in a 54,000 liter fermentor. The culture was grown at 30°C with high aeration to achieve maximum sporulation. Krieg and Miltenburger (1984) described a process that was scaled through 15, 300, and 3000 liter fermentors to a 30,000 liter production vessel. Transfer from the seed vessels was carried out at the end of the logarithmic phase of growth (6-9 hr). The yield was dependent upon the specific strain and medium used. Megna (1963) recommended a medium containing beet molasses and cornsteep solids for the seed stages and a more concentrated medium including cottonseed flour for the production fermentor (Table 8). Production yields on this medium were in the region of 5×10^9 spores/ml. Drake and Smythe (1963) described a medium comprising cornstarch, cornsteep liquor, casein, sucrose, and yeast extract, which yielded 1×10^{10} spores ml. To stimulate growth and to maximize sporulation and crystal formation in the stationary phase, aeration of the culture is essential and should be maintained at 0.3 vvm; dissolved oxygen should not fall below 20%. Fermentation generally takes from 36 to 48 hr to maximize crystal formation. Monitoring of the process is carried out by total counts of the vegetative cells and spores. Spore-crystal ratios are monitored during the final stages of the fermentation and with *B. thuringiensis* subsp. *kurstaki* give a value of 0.5 at the end of the process. Optimum toxin production is based on the crystal protein concentration or on the number of crystals observed microscopically in a representative number of spores. More direct assays have been based primarily on bioassays performed by adding dilutions of the endotoxin preparation to an artificial diet and feeding susceptible larvae (Dulmage et al., 1971; Moraes and Chaib, 1978). Bioassays generally

Table 8 Media for the Production of δ-Endotoxin from *B. thuringiensis*

Component	Amount (%)	
Soy flour	2-6	
Molasses	0-4	
KH_2PO_4	0.5	
K_2HPO_4	0.5	
$MgSO_4 \cdot 7H_2O$	0.005	
$MnSO_4 \cdot 4H_2O$	0.003	
$FeSO_4 \cdot 7H_2O$	0.001	
$CaCl_2$	0.005	
$Na(NH_4)PO_4 \cdot 4H_2O$	0.15	
(Holmberg et al., 1980)		
Fishmeal	2	
Starch	1	
$CaCO_3$	0.1	
Niacin	0.001	
Mineral salts		
(Pendleton, 1969)		
Beet molasses	1.86	1.0[a]
Cornsteep solids	1.7	0.85[a]
Cottonseed flour	1.4	—
$CaCO_3$	0.1	0.1[a]
(Megna, 1963)		
Cornstarch	6.8	
Cornsteep liquor	4.7	
Casein	1.95	
Sucrose	0.64	
Yeast extract	0.6	
Phosphate buffer	0.6	
(Drake and Smythe, 1963)		

[a]Media for seed stages.

take from 4 to 7 days and require considerable labor in maintaining, rearing, and testing suitable insect larvae. Immunoelectrophoretic assays (Winkler et al., 1971) reduced the assay period to 48 hr, but the time to carry out the assay was still longer than the duration of the fermentation. Andrews et al. (1980) developed a rocket immunoelectrophoresis assay that gave results in 4 hr. The assay was considered essentially qualitative and required a considerable amount of antigen (10 mg/liter). More recently, enzyme-linked immunosorbent assay (ELISA) techniques have been devised that show very close correlation with results from bioassays and give reliable results within 4 hr (Smith and Ulrich, 1983).

3. Processing and Formulation

At the end of crystal formation in the fermentor, the spores and crystals are washed and concentrated by centrifugation. Zamola et al. (1981) discussed the effects of continuous centrifugation for the separation of crystals and spores from the medium on the insecticidal activity. Studies using a solid bowl disk centrifuge and an intermittent discharge centrifuge indicated that the insecticidal activity of the cell paste was dependent not only on the course of the fermentation but on the flow rate through the centrifuge. After washing of the paste it may be filtered to produce a moist filter cake, which is then dried in a vacuum oven (Pendleton, 1969) or spray dried to produce the primary powder. After drying the preparation is formulated to ensure sufficient dispersion to give an acceptable dosage of crystals when applied in the field. It is generally prepared as a wettable powder suspended in water with an inert hydrophilic carrier, such as bentonite. Other additives to aid adhesion or dispersion or to act as attractants may be added prior to spraying.

4. Development of Recombinant Strains for Endotoxin Production

The first reports of plasmids in *B. thuringiensis* (Debabov et al., 1977; Zakharyan et al., 1979) were followed by intense investigation to determine their role in δ-endotoxin production. González et al. (1981) implicated specific plasmids in the production of δ-endotoxin by a number of *B. thuringiensis* strains. The high-frequency transfer of plasmids carrying the structural gene for δ-endotoxin production to strains unable to produce crystals produced Cry^+ transformants (González et al., 1982).

The first reports of molecular cloning of a toxin gene from *B. thuringiensis* were by Schnepf and Whitely (1981) and Whitely et al. (1984). Digests of six plasmids isolated from a commercial strain of *B. thuringiensis* were cloned into the *Bam* H1 site of plasmid pBR322 and the plasmids transformed into *E. coli* HB101. The transformants were selected by ampicillin resistance and examined by radioimmunoassay using partially purified antibodies raised against crystal proteins. One colony was found to cross-react and carried a hybrid plasmid with a 12 kb insert from one of the larger plasmids. Southern blot analysis demonstrated that the sequence hybridized with both a 47 and 30 Md plasmid from the donor strain. The protein produced by the recombinant strain was toxic to larvae of *Manduca sexta*.

The k-73 type insecticidal protein gene from a 47 Md plasmid from *B. thuringiensis* subsp. *kurstaki* was cloned into *E. coli* pBR322. Subsequent bioassays of the toxin produced by *E. coli* indicated specific activities equivalent to those produced by the parent *B. thuringiensis* strain (McLinden et al., 1985). Herrnstadt et al. (1986) similarly cloned a 5.8 kb fragment from *B. thuringiensis* supsp. *san diego* into *E. coli* using pBR322 as cloning vector. Purified toxin from the recombinant organism was toxic to the larvae for elm leaf beetle.

The production of endotoxin by recombinant microorganisms does not yet appear to have replaced the more conventional production processes. There are few reports of fermentation development of recombinant microorganisms for toxin production, but the technology is now available for their exploitation. Recombinant strains offer the potential of enhancing production yields and producing modified toxins that may be more specific or may have an extended range of insecticidal activity.

C. Production of Glucose Isomerase

1. Strain Isolation and Medium Development

The isomerization of glucose to fructose has considerable importance in the starch-processing industry since it enables the production of a product sweeter than sucrose from less expensive starch hydrolysates. There is no apparent reason why cells should require an enzyme for the conversion of glucose to fructose, and in fact glucose isomerases are more accurately described as D-xylose isomerases that have D-glucose isomerase as a side or secondary activity. Glucose isomerase (D-xylose ketol-isomerase, EC 5.3.1.5), which catalyzes the reversible isomerization of D-glucose to D-fructose, has been isolated from a number of different species of microorganisms. The most widely used sources for commercial production are *Streptomyces* (Takahashi et al., 1969) and *B. coagulans* (Yoshimura et al., 1966; Suekane et al., 1974; Diera, 1977).

Yoshimura et al. (1966) described the isolation of over 300 strains from soil and manure on a medium comprising D-xylose, 1%;

yeast extract, 0.1%; K_2HPO_4, 0.05%; $MgSO_4 \cdot 7H_2O$, 0.025%; $MnSO_4 \cdot 7H_2O$, 0.001%; and $CaCO_3$, 0.2%; incubations were carried out for 24 hr at 45°C. Following three successive transfers to fresh medium, the isolates were plated on the medium with 2% agar and examined for glucose isomerase activity. A strain of *B. coagulans* (HN-68) produced D-xylose and D-glucose isomerizing activity when grown on D-xylose as the main carbon source. The production medium comprised D-xylose, 2%; yeast extract, 0.3%; K_2HPO_4, 0.01%; $MgSO_4 \cdot 7H_2O$, 0.05%; $MnSO_4 \cdot 7H_2O$, 0.005%; NH_4Cl, 0.3%; and $CaCO_3$, 0.2%. Maximum enzyme production occurred after 14-16 hr incubation at 40°C and pH 7.0. Glucose isomerase activity was induced in cells grown on D-glucose following exposure to D-xylose.

Outtrup (1976) described a rapid method for the isolation of organisms producing a heat-stable glucose isomerase with a low optimum pH. The procedure involved the plating of soil samples on media incorporating ammonia as the sole nitrogen source and xylose as the sole carbon source and incubating at 60-65°C to ensure only thermophilic strains were isolated. The isolation medium comprised $(NH_4)_2SO_4$, 0.03%; K_2HPO_4, 0.01%; $MgSO_4 \cdot 7H_2O$, 0.005%; KCl 0.005%; $FeSO_4$, 0.001%; xylose, 0.5%; and agar, 2%. Colonies were subcultured to complex medium with xylose as the major carbon source and incubated at 60°C before testing for the presence of glucose isomerase. A total of 18 strains of *Bacillus* were isolated, giving conversion efficiencies of 35-51%. Constitutive mutants, isolated following treatment with *N*-methyl-*N'*-nitro-*N*-nitrosoguanidine, exhibited glucose isomerase activity when grown on a yeast extract medium without xylose or glucose.

Suekane et al. (1974) described the isolation of strains of *B. stearothermophilus* from soil using a medium containing xylose, 0.5%; soluble starch, 0.5%; peptone, 0.5%; cornsteep liquor, 1%; meat extract, 0.5%; yeast extract, 0.25%; NaCl, 0.05%; $MgSO_4 \cdot 7H_2O$, 0.05%; $CoCl_2 \cdot 6H_2O$, 0.024%; and agar, 2%. The medium was prepared at pH 7.0 and inoculated plates incubated for a few days at 60°C. A total of 17 strains with glucose isomerase activity were isolated and identified as strains of *B. stearothermophilus*.

2. Fermentation, Development, and Production

Diers (1976, 1977) reported batch and continuous culture studies using the constitutive mutant strain (NRRL 5650) developed by Outtrup (1976). Glucose isomerase production did not require xylose for induction; the strain grew between 37-65°C with an optimum of 55-60°C and at pH 4.5-8.5. For batch culture a medium based on cornsteep liquor was devised comprising cornsteep liquor (50% dry weight), 3%; K_2HPO_4, 0.015%; $(NH_4)_2SO_4$, 0.5%; glucose, 0.2%; $MgSO_4 \cdot 7H_2O$, 0.01%; $MnSO_4 \cdot H_2O$, 0.01%; and pluronic H-61, 0.03% pH 6.8 with 30% NaOH. In a 10 liter fermentor, air was sparged into

Table 9 Effect of Nutrient Limitation on the Yield of Glucose Isomerase and Productivity from *B. coagulans*

Dilution rate D	Limiting factor	Yield of cell mass (mg dry weight bacteria per mg glucose)	Enzyme yield glucose isomerase (units/mg dry weight bacteria)	Productivity glucose isomerase (units/mg glucose hr)
0.12	O_2	0.17	0.98	2.00×10^{-2}
0.12	C	0.29	0.38	1.16×10^{-2}
0.13	N	0.09	0.06	0.07×10^{-2}
0.11	P	0.09	0.17	0.17×10^{-2}
0.10	Mg	0.175	0.11	0.19×10^{-2}

Source: From Diers (1977).

the culture at 1 vvm, the stirrer speed was maintained at 620 rpm, and the temperature at 50°C. The pH began to rise 13 hr after inoculation, batch growth ended as the oxygen tension reached zero. When the pH reached 6.5, 40% glucose was fed to the vessel at a rate of 2.1 g/liter per hr, enabling the pH to be maintained constant for the next 8 hr. The dissolved oxygen concentration increased and growth terminated. The culture gave a cell yield of 4.2 g/liter and an enzyme yield of 8.8 units of glucose isomerase per milliliter of culture.

Diers (1977) examined the productivity of glucose isomerase under growth-limiting conditions using O_2, C, N, P, and Mg as growth-limiting nutrients. Studies were carried out in a 1 liter fermentor with a working volume of 300-350 ml. The stirrer speed was varied from 250-500 rpm and pH maintained at 7.0, ±0.2°C. The dilution rate was maintained at approximately 0.1 hr^{-1} when examining each nutrient limitation. The dissolved oxygen tension was maintained at zero when O_2 was growth limiting and in other studies at a value higher than 20 mmHg. The medium used for the oxygen limitation studies comprised $(NH_4)_2SO_4$, 2.5 g/liter; K_2HPO_4, 1.0 g/liter; $NaH_2PO_4 \cdot 2H_2O$, 1.0 g/liter; NaCl, 0.25 g/liter; $MgSO_4 \cdot 7H_2O$, 0.1 g/liter; $FeCl_3 \cdot 6H_2O$, 0.02 g/liter; $MnSO_4 \cdot H_2O$, 0.02 g/liter; $ZnSO_4$, 0.0025 g/liter; $CaCl_2$, 0.0075 g/liter; thiamine

Table 10 Medium Constituents and Growth Conditions for the Production of Glucose Isomerase from *Bacillus* Species

Species	C source	N source	Minerals	pH	Temperature	Time	Yield (U/liter of culture)	Reference
B. coagulans HN-68	Xylose	Yeast extract + NH_4Cl	$MgSO_4 \cdot 7H_2O$ $MnSO_4 \cdot 4H_2O$ $CaCO_3$	—	40	14-16	—	Yoshimura, 1966
B. coagulans NRRL-5650	Glucose	CSL $(NH_4)_2SO_4$	$MgSO_4 \cdot 7H_2O$ $MnSO_4 \cdot 4H_2O$ K_2HPO_4	6.8	50	13	8800	Diers, 1977
B. coagulans NRRL-5649-66	Xylose	Peptone	NaCl	—	60	—	—	Outtrup, 1974
B. stearothermophilus	Xylose	Peptone CSL Yeast extract Meat extract	NaCl $MgSO_4 \cdot 7H_2O$ $CoCl \cdot 6H_2O$	—	55	—	3890-12,900	Suekane et al., 1974
Bacillus sp. NRRL B-5350 5351	Xylose Glucose	Peptone Casein	—	—	38	17	—	Aunstrup, 1972

HCl, 0.001 g/liter; glucose, 8 g/liter; EDTA, 0.083 g/liter; and Pluoronic H-61, 0.083 g/liter. In N, P, and Mg limitation studies the medium was modified accordingly, that is $(NH_4)_2SO_4$, 0.4 g/liter; K_2HPO_4, 0.021 g/liter; and $MgSO_4 \cdot 7H_2O$, 0.04 g/liter. The results (Table 9) indicated maximum enzyme yield and productivity under conditions of oxygen limitation.

A summary of medium constituents and growth conditions for the production of glucose isomerase from various strains of *Bacillus* is given in Table 10. Little information is available on the fermentation parameters used for large-scale production. Since the enzyme is intracellular, it can be recovered either following lysis of the organism or by harvesting the cells with the enzyme retained within the cells. Originally the enzyme was utilized in solution, but more recently the use of immobilized enzyme preparations or immobilized cells has permitted much shorter processing times and considerably improved the economics of the process. Much of the current production of high fructose and glucose syrups is based on the use of Novo's *B. coagulans* whole-cell preparation. When the cells are recovered with the intracellular enzyme, considerable loss of enzyme is observed following autolysis of the cells. To avoid the loss of enzyme and to extend the use of cells in fixed-bed reactors, the enzyme is bound to the intracellular matrix by heat fixation or chemical fixation. Following fermentation and immobilization the cells may be separated from the fermentation broth by centrifugation (Antrim et al., 1979). The Novo process for the immobilization of the enzyme involves cross-linking of the harvested cells with glutaraldehyde, breaking into smaller particles, and drying to improve the mechanical strength of the pellets. Alternative procedures include freezing the cross-linked precipitate prior to drying, allowing the centrifuged cells to autolyze before cross-linking, and then flocculating and drying or disrupting the cells and then flocculating, and cross-linking the homogenate. The immobilized enzyme is then used in column reactors at about 60°C to produce syrups containing approximately 42-43% (w/w) fructose (Amotz et al., 1976; Hemmingsen, 1979; Zittan et al., 1975).

VI. CONCLUSIONS

Bacillus occupies a dominant position in modern biotechnology. These bacteria have been and still are major sources of traditional products, such as enzymes and insect toxins, that are manufactured using traditional practices. On the other hand, *B. subtilis* is the most completely understood of all gram-positive bacteria, has a well-characterized genetic background, and retains popularity as a model organism for genetic and biochemical study. It seems that as we complete the first decade of "modern biotechnology" the industrial side of *Bacillus* is finally benefiting from the academic interests. Most of the genes for the important commercial products have been cloned in *B. subtilis* (but to a lesser extent other bacilli), elucidation of the regulation of gene expression is advancing rapidly, and gene transfer between most of the commercially important strains is possible. Perhaps the second decade of modern biotechnology will witness the marriage of industrial and academic research on *Bacillus*, an exciting prospect.

ACKNOWLEDGMENT

The authors would like to thank Mrs. Kate Oultram for typing the manuscript and for her patience and help during its preparation.

REFERENCES

Aiba, S., Kitai, K., and Imanaka, T. (1983). Cloning and expression of thermostable α-amylase gene from *B. stearothermophilus* in *B. stearothermophilus* and *B. subtilis*. *Appl. Environ. Microbiol. 46*: 1059-1065.

Aiba, S., Monden, Y., Ohnishi, M., Koizumi, J., and Ming, Z. (1986). Stabilisation in *B. stearothermophilus* of a recombinant plasmid carrying the homologous α-amylase gene. *J. Chem. Technol. Biotechnol. 36*: 599-609.

Akiba, Y., and Katoh, K. (1986). Microbial ecology of *Bacillus thuringiensis*. V. Selective medium for *B. thuringiensis* vegetative cells. *Appl. Ent. Zool. 21*: 210-215.

Amotz, S., Nielson, T. K., and Thiesen, N. O. (1976). Immobilisation of *Bacillus coagulans*. United States Patent No. 3,980,521.

Andrews, R. E., Ilandolo, J. J., Campbell, B. S., Davidson, L. I., and Bulla, L. A. (1980). Rocket immunoelectrophoresis of the entomocidal parasporal crystal of *Bacillus thuringiensis* subsp. *kurstaki*. *Appl. Environ. Microbiol. 40*: 897-900.

Antrim, R. L., Colilla, W., and Schnyder, B. J. (1979). Glucose isomerase production of high fructose syrups. *Appl. Biochem. Bioeng. 2*: 97-155.

Aronson, A. I., Beckman, W., and Dunn, P. (1986). *Bacillus thuringiensis* and related insect pathogens. *Microbiol. Rev. 50*: 1-24.

Aunstrup, K. (1979). Production, isolation and economics of extracellular enzymes. *Appl. Biochem. Bioeng. 2*: 27-69.

Band, L., and Henner, D. J. (1984). *Bacillus subtilis* requires a "stringent" Shine-Dalgarno region for gene expression. *DNA 3*: 17-21.

Bernlohr, R. W., Maddox, M. K., and Goldberg, N. (1974). Cyclic, guanosine-3',5'-monophosphate in *Escherichia coli* and *Bacillus licheniformis*. *J. Biol. Chem. 249*: 4329-4331.

Boidin, A., and Effront, J. (1917). United States Patent No. 1,227,374.

Botsford, J. L. (1981). Cyclic nucleotides in prokaryotes. *Microbiol. Rev. 45*: 620-642.

Bulla, L. A., Jr., Faust, R. M., Andrews, R., and Goodman, N. (1985). Insecticidal bacilli. In *Molecular Biology of the Bacilli*, vol. 2, D. Dubnau (Ed.). Academic Press, Orlando, Florida, pp. 186-210.

Buonocore, V., Caporale, C., De Rosa, M., and Gambacorta, A. (1976). Stable, inducible thermophilic α-amylase from *Bacillus acidocaldarius*. *J. Bacteriol. 128*: 515-521.

Burbidge, E., and Collier, B. (1968). Production of bacterial amylases. *Proc. Biochem.* November, pp. 53-56.

Chandra, A. K., Medda, S., and Bhadra, A. K. (1980). Production of extracellular thermostable α-amylase by *Bacillus licheniformis*. *J. Ferment. Technol. 58*: 1-10.

Chang, S., Gray, O., Ho, D., Kroger, J., Chang, S.-Y., McLaughlin, J., and Mark, D. (1982). Expression of eukaryotic genes in *B. subtilis* using signals of *penP*. In *Molecular Cloning and Gene Regulation in Bacilli*, A. T. Ganesan, J. Chang, and J. A. Hoch (Eds.). Academic Press, Orlando, Florida, pp. 159-169.

Chang, S., Ho, D., Gray, O., Chang, S.-Y., and McLaughlin, J. (1983). Functional expression of human interferon genes and construction of a partition-proficient plasmid-vector in *B. subtilis*. In *Genetics of Industrial Microorganisms, 1982*, Y. Ikoda and T. Reppu (Eds.). Kodansha Ltd., Tokyo, pp. 227-231.

Clark, B. D., Perlak, F. J., Chu, C.-Y., and Dean, D. H. (1984). The *Bacillus thuringiensis* genetic system. In *Comparative Pathobiology*, vol. 7, T. C. Cheng (Ed.). Plenum Press, New York, pp. 155-171.

Coleman, G., and Elliott, W. H. (1962). Studies on α-amylase formation by *Bacillus subtilis*. *Biochem. J. 83*: 256-263.

Cooper, D. G., and Zajic, J. E. (1980). Surface active compounds from microorganisms. *Adv. Appl. Microbiol. 26*: 229-256.

Corfield, V. A., Read, S. J., Bodmer, J., and Thomson, J. A. (1984). A modified protoplast-regeneration protocol facilitating the detection of cloned exoenzyme genes in *Bacillus subtilis*. *Gene 30*: 17-22.

Davis, P. E., Cohen, D. L., and Whitaker, A. (1980). The production of α-amylase in batch and chemostat culture by *Bacillus stearothermophilus*, *Antonie Van Leeuwenhoek 46*: 391-398.

Deacon, J. W. (1983). *Microbial Control of Plant Pests and Disease.* Van Nostrand-Reinhold, Wokingham, England.

Dean, D. H. (1985). Biochemical genetics of the bacterial insect control agent *Bacillus thuringiensis* and prospects for genetic engineering. *Biotechnol. Genet. Eng. Rev. 2*: 341-363.

Debabov, V. G. (1982). The industrial use of bacilli. In *The Molecular Biology of the Bacilli*, vol. 2, D. Dubnau (Ed.). Academic Press, London, pp. 331-370.

Debabov, V. G., Azizbekyan, R. R., Khlebalina, O. I., D'yachenko, V. V., Gulushka, F. P., and Belykh, R. A. (1977). Isolation and preliminary characterization of extra chromosomal elements of *Bacillus thuringiensis* DNA. *Genetika 13*: 495-501.

Delafield, P. P., Somerville, H. J., and Rittenburg, S. C. (1968). Immunological homology between crystal and spore protein of *Bacillus thuringiensis*. *J. Bacteriol. 96*: 713-720.

Demain, A. L. (1978). Production of nucleotides by microorganisms. In *Economic Microbiology*, vol. 2, *Primary Products of Metabolism*, A. H. Rose (Ed.). Academic Press, London, pp. 178-208.

Demain, A. L., Kennel, Y. H., and Aharonowitz, Y. (1979). Carbon catabolite repression of secondary metabolism. *Symp. Soc. Gen. Microbiol. 29*: 163-186.

Dharmsthiti, S. C., Pantuwatana, S., and Bhumiratana, A. (1985). Production of *Bacillus thuringiensis* subsp. *israelensis* and *Bacillus sphaericus* strain 1593 on media using a byproduct from a monosodium glutamate factory. *J. Invert. Pathol. 46*: 231-238.

Diers, I. (1976). Glucose isomerase in *Bacillus coagulans*. In *Continuous Culture*, vol. 6, A. C. R. Dean, D. C. Ellwood, C. G. T. Evans, and J. Melling (Eds.). Ellis Horwood, Chichester, pp. 208-224.

Diers, I. V. (1977). Method for producing glucose isomerase. United States Patent No. 4,042,460.

Doi, R. H. (1985). Genetic engineering in *Bacillus subtilis*. *Genet. Eng. Biotechnol. Rev. 2*: 121-156.

Doi, R. H., and Wang, L.-F. (1986). Multiple procaryotic ribonucleic acid polymerase sigma factors. *Microbiol. Rev. 50*: 227-243.

Doi, R. H., Gitt, M., Wang, L.-F., Price, C. W., and Kawamura, F. (1985). Major sigma factor, sigma-43, of *Bacillus subtilis* RNA polymerase and interacting *spo0* products are implicated in catabolite control of sporulation. In *Molecular Biology of Microbial Differentiation*, J. A. Hoch and P. Setlow (Eds.). American Society for Microbiology, Washington, D.C., pp. 157-161.

Drake, B. B., and Smythe, C. V. (1963). U.S. Patent No. 3,086,865.

Dulmage, H. T. (1970). Production of the spore-delta-endotoxin complex by variants of *Bacillus thuringiensis* in two fermentation media. *J. Invert. Pathol. 16*: 385-389.

Dulmage, H. T. (1971). Production of delta-endotoxin by eighteen isolates of *Bacillus thuringiensis*, serotype 3, in 3 fermentation media. *J. Invert. Pathol. 18*: 353-358.

Dulmage, H. T., Boening, O. P., Rehnborg, C. S., and Hansen, G. D. (1971). A proposed standard bioassay for formulations of *Bacillus thuringiensis* based on the international unit. *J. Invert. Pathol. 18*:240-245.

Emanuilova, E. I., and Toda, K. (1984). α-Amylase production in batch and continuous cultures by *Bacillus caldolyticus*. *Appl. Microbiol. Biotechnol. 19*: 301-305.

Enei, H., and Hirose, Y. (1984). Recent research on the development of microbial strains for amino acid production. *Genet. Eng. Biotechnol. Rev. 2*: 100-120.

Enei, H., Shibai, H., and Hirose, Y. (1985). 5'-Guanosine monophosphate. In *Comprehensive Biotechnology*, vol. 3, H. W. Blanch, S. Drew, and D. I. C. Wang (Eds.). Pergamon Press, Oxford, pp. 653-658.

Esteban, R., Nebreda, A. R., and Villa, T. G. (1984). Synthesis and regulation of *Bacillus circulans* WL-12 1,3-β-D-glucanases. *J. Gen. Microbiol. 130*: 2483-2487.

Eveleigh, D. E., and Montenencourt, B. S. (1979). Increasing yields of extracellular enzymes. *Adv. Appl. Microbiol. 25*: 57-74.

Fahnestock, S. R., and Fisher, K. E. (1986). Expression of the staphylococcal protein A gene in *Bacillus subtilis* utilizing the promoter from a *Bacillus amyloliquefaciens* α-amylase gene. *J. Bacteriol. 165*: 796-804.

Fencl, Z., and Pazlarova, J. (1982). Production of extracellular enzymes in continuous culture. *Folia Microbiol. 27*: 340-349.

Fencl, Z., Řičica, J., and Kodešová, J. (1972). The use of the multistage chemostat for microbial product formation. *J. Appl. Chem. Biotechnol. 22*: 405-416.

Fencl, Z., Pazlarova, J., Plazek, I., Egorova, L. A., and Loginova, L. G. (1979). Experimental verification of results obtained by mathematical modelling of the production of non α-amylase producing *Bacillus subtilis* mutants during continuous cultivation. *Mikrobiologiya 48:* 93-98.

Fisher, S. H., Rosenkrantz, M. S., and Sonenshein, A. L. (1984). Glutamine synthetase gene of *Bacillus subtilis*. *Gene 32*: 427- 438.

Fouet, A., Aubert, E., Arnaud, M., Klier, A., Rapaport, G., and Le Coq, D. (1984). The sucrose system as a model of genetic regulation in *Bacillus subtilis*. In *Genetics and Biotechnology of Bacilli*, A. T. Ganesan and J. A. Hoch (Eds.). Academic Press, Orlando, Florida, pp. 113-128.

Frangou-Lazaridis, M., and Seddon, B. (1985). Effect of gramicidins on the transcription system of the producer *Bacillus brevis* Nagaro. *J. Gen. Microbiol. 131*: 437-449.

Fukumoto, J., Yamamoto, T., Tsuru, D., and Ichikawa, K. (1957). Some problems on bacterial amylase and proteinase production. In *Proceedings of the International Symposium on Enzyme Chemistry*, Pergamon Press, London, pp. 479-482.

Ghosh, S. B., and Chandra, A. K. (1984). Nutritional requirements and cultural characteristics of *Bacillus apiarius* CBML-152 for the production of thermostable α-amylase. *Zentralbl. Mikrobiol. 139*: 293-304.

Goldberg, L. J., and Margalit, J. (1977). A bacterial spore demonstrating rapid larvicidal activity against *Anopheles sergentii, Uranotaenia unguiculata, Culex univatattus, Aedes aegypti* and *Culex pipiens*. *Mosquito News 37*: 355-358.

González, J. M., Dulmage, H. T., and Carlton, B. C. (1981). Correlation between specific plasmids and endotoxin production in *Bacillus thuringiensis*. *Plasmid 5*: 351-365.

González, J. M., Brown, B. J., and Carlton, B. C. (1982). Transfer of *Bacillus thuringiensis* plasmids coding for δ-endotoxin among strains of *B. thuringiensis* and *B. cereus*. *Proc. Natl. Acad. Sci. USA 79*: 6951-6955.

Grandi, G., Mottes, M., and Sgaramella, V. (1981). Specific pattern of instability of *Escherichia coli hisG* gene cloned in *Bacillus subtilis* via the *Staphylococcus aureus* plasmid pC194. *Plasmid 6*: 99-111.

Gray, G. L., Mainzer, S. E., Rey, M. W., Launsa, M. H., Kindle, K. L., Carmona, C., and Requadt, C. (1986). Structural genes encoding the thermophilic α-amylases of *Bacillus stearothermophilus* and *Bacillus licheniformis*. *J. Bacteriol. 166*: 635-643.

Greasham, R., and Inamine, E. (1986). Nutritional improvement of processes. In *Industrial Microbiology and Biotechnology*, A. L. Demain and N. A. Solomon (Eds.). American Society for Microbiology, Washington, D.C., pp. 41-48.

Griffin, P. J., and Fogarty, W. M. (1973). Production of an amylolytic enzyme by *Bacillus polymyxa* in batch cultures. *J. Appl. Chem. Biotechnol. 23*: 301-308.

Grueninger, H., Sonnleitner, B., and Fiechter, A. (1984). Bacterial diversity in thermophilic aerobic sludge. *Appl. Microbiol. Biotechnol. 19*: 414-421.

Haavik, H. I. (1974). Studies on the formation of bacitracin by *Bacillus licheniformis*. *J. Gen. Microbiol. 84*: 321-326.

Hager, P. W., and Rabinowitz, J. C. (1985). Translational specificity in *Bacillus subtilis*. In *The Molecular Biology of the Bacilli*, vol. II, D. A. Dubnau (Ed.). Academic Press, Orlando, Florida, pp. 1-32.

Hahn, J., and Dubnau, D. (1985). Analysis of plasmid deletional instability in *Bacillus subtilis*. *J. Bacteriol. 162*: 1014-1023.

Heineken, F. G., and O'Conner, R. J. (1972). Continuous culture studies on the biosynthesis of alkaline protease, neutral protease and α-amylase by *Bacillus subtilis* NRRL-B3411. *J. Gen. Microbiol. 73*: 35-44.

Hemmingsen, S. H. (1979). Development of an immobilised glucose isomerase for industrial application. *Appl. Biochem. Bioeng. 2:* 157-183.

Henner, D. J., Young, M., Band, L., and Wells, J. A. (1985). Expression of cloned protease genes in *Bacillus subtilis*. In *Molecular Biology of Microbial Differentiation*, J. A. Hoch and P. Setlow (Eds.). American Society for Microbiology, Washington, D.C., pp. 95-103.

Herrnstadt, C., Scares, G. G., Wilcox, E. R., and Edwards, D. L. (1986). A new strain of *Bacillus thuringiensis* with activity against Coleopteran insects. *Biotechnology 4*: 305-308.

Holmberg, A., Sievanen, R., and Carlberg, G. (1980). Fermentation of *Bacillus thuringiensis* for exotoxin production: Process analysis study. *Biotechnol. Bioeng. 22*: 1707-1724.

Horwath, R. O. (1984). Process for producing thermally stable alpha-amylase. United States Patent No. 4,473,645.

Ingle, M. B., and Boyer, E. W. (1976). Production of industrial enzymes by *Bacillus* species. In *Microbiology 1976*, D. Schlessinger (Ed.). American Society for Microbiology, Washington, D.C., pp. 420-426.

Ingle, M. B., and Ericksson, R. J. (1978). Bacterial α-amylases. *Adv. Appl. Microbiol. 24*: 257-278.

Joyet, P., Guerineau, M., and Heslot, H. (1984). Cloning of a thermostable α-amylase gene from *Bacillus licheniformis* and its expression in *Escherichia coli* and *Bacillus subtilis*. *FEMS Microbiol. Lett. 21:* 353-358.

Katz, E., and Demain, A. L. (1977). The peptide antibiotics of *Bacillus*: Chemistry, biogenesis and possible functions. *Bacteriol. Rev. 41*: 449-474.

Kindle, K. L. (1983). Characteristics and production of thermostable α-amylase. *Appl. Biochem. Biotechnol. 8*: 153-170.

Kinoshita, S., and Nakayama, K. (1978). Amino acids. In *Economic Microbiology*, vol. 2, *Primary Products of Metabolism*, A. H. Rose (Ed.). Academic Press, London, pp. 210-261.

Kleinkauf, H., and von Dohren, H. (1981). Nucleic acid independent synthesis of peptides. *Curr. Top. Microbiol. Immunol. 91*: 129-178.

Kleinkauf, H., and von Dohren, H. (1983). Non-ribosomal peptide formation on multifunctional proteins. *Trends Biochem. Sci. 8*: 281-283.

Klier, A. F., and Lecadet, M. M. (1976). Arguments based on hybridization-competition experiments in favour of the in vitro synthesis of sporulation-specific RNAs by the RNA polymerase of *B. thuringiensis*. *Biochem. Biophys. Res. Commun. 73*: 263-270.

Klier, A., Parsot, C., and Rapoport, G. (1983). In vitro transcription of the cloned chromosomal gene for δ-endotoxin from *Bacillus thuringiensis*. *Nucleic Acids Res. 11*: 3975-3987.

Krause, M., Marakiel, M. A., von Dohren, H., and Kleinkauf, H. (1985). Molecular cloning of an ornithine-activating fragment of the gramicidin S synthetase 2 gene from *Bacillus brevis* and its expression in *Escherichia coli*. *J. Bacteriol. 162*: 1120-1125.

Krieg, A., and Miltenburger, H. G. (1984). Bioinsecticides. 1. *Bacillus thuringiensis*. *Adv. Biotechnol. Proc. 3*: 273-290.

Krishnan, T., and Chandra, A. K. (1982). Effect of oilseed cakes on α-amylase production by *Bacillus licheniformis* CUMC305. *Appl. Environ. Microbiol. 44*: 270-274.

Krych, V. A., Johnson, J. L., and Yousten, A. A. (1980). Deoxyribonucleic acid homologies amongst strains of *Bacillus sphaericus*. *Int. J. Syst. Bacteriol. 30*: 476-484.

Kuroda, H. I., Shimotsu, H., Henner, D. J., and Yanofsky, C. (1986). Regulatory elements common to the *Bacillus pumilus* and *Bacillus subtilis trp* operons. *J. Bacteriol. 167*: 792-798.

Lehtovaara, P., Ulmanen, I., and Palva, I. (1984). In vitro transcription initiation and termination sites of an α-amylase gene from *Bacillus amyloliquefaciens* cloned in *Bacillus subtilis*. *Gene 30*: 11-16.

Lopez, P., Espinosa, M., Greenburg, B., and Lacks, S. A. (1984). Generation of deletions in pneumococcal *mal* genes cloned in *Bacillus subtilis*. *Proc. Natl. Acad. Sci. USA 81*: 5189-5193.

Losick, R., and Pero, J. (1981). Cascades of sigma factors. *Cell 25*: 582-584.

Losick, R., and Youngman, P. (1984). Endospore formation in *Bacillus*. In *Microbial Development*, R. Losick and L. Shapiro (Eds.). American Society for Microbiology, Washington, D.C., pp. 63-88.

Lundström, K. (1984). Expression of the vesicular stomatitis virus membrane glycoprotein gene in *Bacillus subtilis*. *FEMS Microbiol. Lett. 23*: 65-70.

Lundström, K., Palva, I., Karriainen, L., Garoff, H., Sarvas, M., and Peterson, R. F. (1985). Secretion of Semliki Forest virus membrane glycoprotein E1 from *Bacillus subtilis*. *Virus Res. 2*: 69-83.

Madsen, G. B., Norman, B. E., and Slott, S. (1973). A new heat-stable bacterial amylase and its use in high-temperature starch liquefaction. *Die Stärke 25*: 304-308.

Majumdar, S., and Bose, S. K. (1985). Derepression of sporulation and synthesis of mycobacillin and dipicolinic acid by guanosine 3':5'-cyclic monophosphate under conditions of glucose repression in *Bacillus subtilis*. *J. Gen. Microbiol. 131*: 2783-2788.

Malik, V. P. (1979). Genetics of applied microbiology. *Adv. Genet. 20*: 38-126.

Malik, V. S. (1982). Genetics and biochemistry of secondary metabolism. *Adv. Appl. Microbiol. 28*: 28-116.

Marahiel, M. A., Danders, W., Kruase, M., and Kleinkauf, H. (1979). Biological role of gramicidin S in spore formation. *Eur. J. Biochem. 99*: 49-55.

McLinden, J. H., Sabourn, J. R., Clark, B. D., Gensler, D. R., Workman, W. E., and Dean, D. H. (1985). Cloning and expression of an insecticidal k-73 type crystal protein gene from *Bacillus thuringiensis* var *kurstaki* into *E. coli*. *Appl. Environ. Microbiol. 50*: 623-628.

Meers, J. L. (1972). The regulation of α-amylase production in *Bacillus licheniformis*. *Antonie van Leeuwenhoek 38*: 585-590.

Megna, J. C. (1963). United States Patent No. 3,076,922.

Mielenz, J. R., and Mickel, S. (1985). Process for cloning the gene coding for a thermostable alpha-amylase into *E. coli* and *B. subtilis*. United States Patent No. 4,493,893.

Miller, T. L., and Churchill, B. W. (1986). Substrates for large scale fermentations. In *Industrial Microbiology and Biotechnology*, A. L. Demain and N. A. Solomon (Eds.). American Society for Microbiology, Washington, D.C., pp. 122-136.

Millet, J. (1970). Characterization of proteinases excreted by *Bacillus subtilis* Marburg strain during sporulation. *J. Appl. Bacteriol. 33*: 207-219.

Monod, M., Denoya, C., and Dubnau, D. (1986). Sequences and properties of pIM13, a macrolide-lincosamide-streptogramin B resistance plasmid from *Bacillus subtilis*. *J. Bacteriol. 167*: 138-147.

Moraes, O., and Chaib, M. A. (1978). Bioassay for microbial insecticide. *Proc. Biochem.* October: 23-24.

Moran, C. P., Jr., Lang, N., Legrice, S. F. J., Lee, G., Stephens, M., Sonenshein, A. L., Pero, J., and Losick, R. (1982). Nucleotide sequences that signal the initiation of transcription and translation in *Bacillus subtilis*. *Mol. Gen. Genet. 186*: 339-346.

Murphy, N., McConnel, D. J., and Cantwell, B. A. (1984). The DNA sequence of the gene and genetic control sites for the excreted *B. subtilis* enzyme β-glucanase. *Nucleic Acids Res. 12*: 5355-5367.

Murray, C. L., and Rabinowitz, J. C. (1982). Nucleotide sequences of transcription and translation initiation regions in *Bacillus* phage ϕ29 early genes. *J. Biol. Chem. 257*: 1053-1062.

Nam, D. H., and Ryu, D. Y. D. (1985). Relationship between butirosin biosynthesis and sporulation in *Bacillus circulans*. *Antimicrob. Agents Chemother. 27*: 798-801.

Nicholson, W. L., and Chambliss, G. H. (1986). Molecular cloning of cis-acting regulatory alleles of the *Bacillus subtilis amyR* region by using gene conversion transformation. *J. Bacteriol. 165*: 663-670.

Nickerson, K. W., and Bulla, L. A., Jr. (1974). Physiology of spore-forming bacteria associated with insects: Minimal nutritional requirements for growth, sporulation and parasporal crystal formation in *Bacillus thuringiensis*. *Appl. Microbiol. 28*: 124-128.

Nugent, M. E., Primrose, S. B., and Tacon, W. C. A. (1983). The stability of recombinant DNA. *Dev. Ind. Microbiol. 24*: 271-285.

Nybergh, P. M. A. (1984). Production of α-amylase by *Bacillus subtilis* containing a cloned α-amylase gene. In Proc. 3rd Eur. Cong. Biotechnol. Munich, vol. 3, Verlag Chemie, Weinheim, pp. 409-413.

Obeta, J. A. N., and Okafor, N. (1984). Medium for the production of primary powder of *Bacillus thuringiensis* supsp. *israelensis*. *Appl. Environ. Microbiol. 47*: 863-867.

O'Donnell, A. G., Norris, J. R., Berkeley, R. C. W., Claus, D., Kaneko, T., Logan, N. A., and Nozaki, R. (1980). Characterization of *Bacillus subtilis*, *Bacillus pumilus*, *Bacillus licheniformis* and *Bacillus amyloliquefaciens* by pyrolysis gas-liquid chromatography and by DNA-DNA hybridization; biochemical tests and API systems. *Int. J. Syst. Bacteriol. 30*: 448-459.

Ohmura, K., Nakamura, K., Yamazaki, H., Shiroza, T., Yamane, K., Jigami, Y., Tanaka, H., Yoda, K., Yamasaki, M., and Tamura, G. (1984). Length and structural effect of signal peptides derived from *Bacillus subtilis* α-amylase on secretion of *Escherichia coli* β-lactamase in *B. subtilis* cells. *Nucleic Acids Res. 12*: 5307-5319.

Ortlepp, S. A., Ollington, J. F., and McConnell, D. J. (1983). Molecular cloning in *Bacillus subtilis* of a *Bacillus licheniformis* gene encoding a thermostable alpha-amylase gene. *Gene 23*: 267-276.

Ostroff, G. R., and Pene, J. J. (1984). Parameters affecting the stability of recombinant bifunctional plasmid transfer from *Escherichia coli* to *Bacillus subtilis*. In *Genetics and Biotechnology of Bacilli*, A. T. Ganesan and J. A. Hoch (Eds.). Academic Press, Orlando, Florida, pp. 307-322.

Outtrup, H. (1976). Production of glucose isomerase by *Bacillus coagulans*. United States Patent No. 3,979,261.

Outtrup, H., Andresen, O., and Aunstrup, K. (1972). British Patent No. 1,296,839.

Palva, I. (1982). Molecular cloning of α-amylase gene from *Bacillus amyloliquefaciens* and its expression in *B. subtilis*. *Gene 19*: 81-87.

Palva, I., Sarvas, M., Lehtovaara, P., Sibakov, M., and Kääriainen, L. (1982). Secretion of *Escherichia coli* β-lactamase from *Bacillus subtilis* by the aid of α-amylase signal sequences. *Proc. Natl. Acad. Sci. USA 79*: 5582-5586.

Palva, I., Lehtovaara, P., Kääriainen, L., Sibakov, M., Cantell, K., Schein, C. H., Kashiwagi, K., and Weisman, C. (1983). Secretion of interferon by *Bacillus subtilis*. *Gene 22*: 229-235.

Pendleton, I. R. (1969). Insecticides of crystal forming bacteria. *Proc. Biochem.* December: 29-32.

Pfeuller, S. L., and Elliott, W. H. (1969). The extracellular α-amylase of *B. stearothermophilus*. *J. Biol. Chem. 244*: 48-54.

Piggot, P. J. (1985). Sporulation of *Bacillus subtilis*. In *Molecular Biology of the Bacilli*, vol. 1, D. Dubnau (Ed.). Academic Press, Orlando, Florida, pp. 74-109.

Piggott, R. P., Rossiter, A., Ortlepp, S. A., Pembroke, J. R., and Ollington, J. F. (1984). Cloning in *Bacillus subtilis* of an extremely thermostable alpha amylase: Comparison with other heatstable alpha amylases. *Biochem. Biophys. Res. Commun. 122*: 175-183.

Pinches, A., Louw, M. E., and Watson, T. G. (1985). Growth, plasmid stability and α-amylase production in batch fermentation using a recombinant *Bacillus subtilis* strain. *Biotechnol. Lett. 7*: 621-626.

Postma, P. W. (1986). Catabolite repression and related processes. *Symp. Soc. Gen. Microbiol. 39*: 21-49.

Postma, P. W., and Nelson, S. O. (1985). Regulation of carbohydrate metabolism by the bacterial PEP:sugar phosphotransferase system. In *Gene Manipulation and Expression*, R. E. Glass and J. Spizek (Eds.). Croom-Helm, London, pp. 339-355.

Priest, F. G. (1977). Extracellular enzyme synthesis in the genus *Bacillus*. *Bacteriol. Rev. 41*: 711-753.

Priest, F. G. (1981). DNA homology in the genus *Bacillus*. In *The Aerobic Endospore-Forming Bacteria; Classification and Identification*, R. C. W. Berkeley and M. Goodfellow (Eds.). Academic Press, London, pp. 33-57.

Priest, F. G. (1983). Enzyme synthesis: Regulation and process of secretion by microorganisms. In *Microbial Enzymes and Biotechnology*, W. M. Fogarty (Ed.). Applied Science, London, pp. 319-366.

Priest, F. G. (1989). Products from Bacilli. In *Handbooks of Biotechnology*, vol. 2, *Bacillus*, C. R. Hardwood (Ed.). Plenum Press, New York, in press.

Priest, F. G., and Thirunavukkarasu, M. (1985). Synthesis and localization of α-amylase and α-glucosidase in *Bacillus licheniformis* grown in batch and continuous culture. *J. Appl. Bacteriol. 58*: 371-390.

Priest, F. G., Goodfellow, M., Shute, L. A., and Berkeley, R. C. W. (1987). *Bacillus amyloliquefaciens* sp. nov. nom. rev. *Int. J. Syst. Bacteriol. 37*: 69-71.

Ristow, H., Pschorn, W., Hansen, J., and Winkel, U. (1979). Induction of sporulation in *Bacillus brevis* by peptide antibiotics. *Nature 280*: 165-166.

Saunders, C. W., Banner, C. D. B., Fahnestock, S. R., Lindberg, M., Mirot, M. S., Rhodes, C. S., Rudolph, C. F., Schmidt, B. J., Thompson, L. D., Uhlen, M., and Giujer, M. S. (1984). Use of chromosomal integration for the establishment of heterologous genes in *Bacillus subtilis*. In *Protein Transport and Secretion*, D. L. Oxender (Ed.). Alan R. Liss, New York, pp. 329-379.

Schaeffer, P. (1969). Sporulation and the production of antibiotics, exoenzymes and exotoxins. *Bacteriol. Rev. 33*: 48-71.

Schein, C. H., Kashiwagi, K., Fujisawa, A., and Weissman, C. (1986). Secretion of mature IFN-2 and accumulation of uncleaved precursor by *Bacillus subtilis* transformed with a hybrid α-amylase signal sequence-IFN-2 gene. *Bio/Technology 4*: 719-725.

Schnepf, H. E., and Whiteley, A. R. (1981). Cloning and expression of the *Bacillus thuringiensis* crystal protein gene in *E. coli*. *Proc. Natl. Acad. Sci. USA 78:* 2893-2897.

Setlow, P. (1973). Inability to detect cyclic AMP in vegetative or sporulating cells or dormant spores of *Bacillus megaterium*. *Biochem. Biophys. Res. Commun. 52*: 365-372.

Sherratt, D. (1985). Control of plasmid maintenance. *Symp. Soc. Gen. Microbiol. 39*: 239-250.

Shimotsu, H., and Henner, D. J. (1986). Modulation of *Bacillus subtilis* levansucrase gene expression by sucrose and regulation of the steady-state mRNA level by *sacU* and *sacQ* genes. *J. Bacteriol. 168*: 365-373.

Shiroza, T., Nakazawa, K., Tashiro, N., Yamane, K., Yanagi, K., Yamasaki, M., Tamura, G., Saito, H., Kawade, Y., and Taniguchi, T. (1985). Synthesis and secretion of biologically active mouse interferon using a *Bacillus subtilis* α-amylase secretion vector. *Gene 34*: 1-8.

Shivakumar, A., Gundling, G. J., Benson, T. A., Casuto, D., Miller, M. F., and Spear, B. A. (1986). Vegetative expression of the δ-endotoxin genes of *Bacillus thuringiensis* subsp. *kurstaki* in *Bacillus subtilis*. *J. Bacteriol. 166*: 194-204.

Sidler, W., and Zuber, H. (1977). The production of extracellular thermostable neutral proteinase and α-amylase by *Bacillus stearothermophilus*. *Eur. J. Appl. Microbiol. 4*: 255-266.

Sikyta, B., and Fencl, Z. (1976). Continuous production of enzymes In *Continuous Culture, Applications and New Fields*, vol. 6, A. C. R. Dean, D. C. Ellwood, C. G. T. Evans, and J. Melling (Eds.). Ellis Horwood, Chichester, pp. 158-193.

Smith, R. A. (1982). Effect of strain and medium variation on mosquito toxin production by *Bacillus thuringiensis* var. *israelensis*. *Can. J. Microbiol. 28*: 1089-1096.

Smith, R. A., and Ulrich, J. T. (1983). Enzyme-linked immunosorbent assay for quantitative detection of *Bacillus thuringiensis* crystal protein. *Appl. Environ. Microbiol. 45*(2): 586-590.

Smythe, C. V., Drake, B. B., and Neubeek, C. E. (1950). United States Patent No. 2,530,210.

Sonenshein, A. L. (1985). Recent progress in metabolic regulation of sporulation. In *Molecular Biology of Microbial Differentiation*, J. A. Hoch and P. Setlow (Eds.). American Society for Microbiology, Washington, D.C., pp. 185-193.

Srivastava, R. A. K., and Baruah, J. N. (1986). Culture conditions for production of thermostable amylase by *Bacillus stearothermophilus*. *Appl. Environ. Microbiol. 52*: 168-184.

Srivastava, R. A. K.., and Mathur, S. N. (1984). Regulation of of amylase bio-synthesis in growing and non-growing cells of *Bacillus stearothermophilus*. *J. Appl. Bacteriol. 57*: 147-151.

Stackebrandt, E., and Woese, C. R. (1981). The evolution of prokaryotes. *Symp. Soc. Gen. Microbiol. 32*: 1-32.

Stalhl, M., and Ferrari, E. (1984). Replacement of the *Bacillus subtilis* subtilisin structural gene with an in vitro derived deletion mutation. *J. Bacteriol. 158*: 411-418.

Steinmetz, M., LeCoc, D., Aymerich, S., Gonzy Treboul, G., and Gay, P. (1985). The DNA sequence of the gene for the secreted *Bacillus subtilis* enzyme levansucrase and its genetic control sites. *Mol. Gen. Genet. 200*: 220-228.

Stewart, G. C., and Bott, K. F. (1983). DNA sequence of the tandem ribosomal RNA promoter for *B. subtilis* operon *rrnB*. *Nucleic Acids Res. 11*: 6289-6300.

Suekane, M., Kanno, S., Kanno, M., and Shiro, H. (1974). Thermophilic glucose isomerase enzyme preparation. United States Patent No. 3,826,714.

Sun, D., and Takahashi, I. (1984). A catabolite resistance mutation is localised in the *rpo* operon of *Bacillus subtilis*. *Can. J. Microbiol. 30*: 423-429.

Sun, D., and Takahashi, I. (1986). Suppressor mutations for *crs* mutants of *Bacillus subtilis*. *Can. J. Microbiol. 30*: 429-435.

Takahashi, I. (1979). Catabolite repression-resistant mutants of *Bacillus subtilis*. *Can. J. Microbiol. 25*: 1283-1287.

Takahashi, I., Kosugi, Y., and Kanbayashi, A. (1969). Streptomyces glucose isomerase. In *Fermentation Advances*, D. Perlman (Ed.). Academic Press, New York, pp. 561-570.

Tanaka, H., Kobayashi, Y., Itukura, K., and Kamimiya, S. (1981). Concerted induction of cell wall lytic enzymes in *Bacillus circulans* WL-12. *Adv. Biotechnol. 4*: 593-598.

Thirunavukkarasu, M., and Priest, F. G. (1980). Regulation of α-amylase synthesis in *Bacillus licheniformis* NCIB 6346. *FEMS Microbiol. Lett. 7*: 315-318.

Uhlen, M., Flock, J.-I., and Philipson, L. (1981). *RecE* independent deletions of recombinant plasmids in *Bacillus subtilis*. *Plasmid 5*: 161-169.

Ullman, A., and Danchin, A. (1983). Role of cyclic AMP in bacteria. *Adv. Cyclic Nucleotide Res. 15*: 2-53.

Ulmanen, I., Lundström, K., Lehtovaara, K., Sarvus, P., Ruohoene M., and Palva, I. (1985). Transcription and translation of foreign genes in *Bacillus subtilis* by the aid of a secretion vector. *J. Bacteriol. 162*: 176-182.

Underkofler, L. A. (1976). Microbial Enzymes. In *Industrial Microbiology*, B. M. Miller and W. Litsky (Eds.). McGraw-Hill, New York, pp. 128-146.

Vandamme, E. J. (1982). Properties, biogenesis and fermentation of the cyclic decapeptide antibiotic gramicidin S. *Top. Enzyme Ferment. Technol. 5*: 187-261.

Vasantha, N., and Thompson, L. D. (1986). Secretion of a heterologous protein from *Bacillus subtilis* with the aid of protease signal sequences. *J. Bacteriol. 165*: 837-842.

Vehmaanperä, J. O., and Korhola, M. P. (1986). Stability of the recombinant plasmid carrying the *Bacillus amyloliquefaciens* α-amylase gene in *B. subtilis*. *Appl. Microbiol. Biotechnol. 23*: 456-461.

Wang, P.-Z., and Doi, R. H. (1984). Overlapping promoters transcribed by *Bacillus subtilis* sigma-55 and sigma-37 RNA polymerase holoenzymes during growth and stationary phases. *J. Biol. Chem. 259*: 7519-8625.

Ward, O. P. (1983). Proteinases. In *Microbial Enzymes and Biotechnology*, W. M. Fogarty (Ed.). Applied Science, London, pp. 251-317.

Welker, N. E., and Campbell, L. L. (1963a). Effect of carbon sources on formation of α-amylase by *Bacillus stearothermophilus*. *J. Bacteriol. 86*: 681-686.

Welker, N. E., and Campbell, L. L. (1963b). Induction of α-amylase of *Bacillus stearothermophilus* by maltodextrins. *J. Bacteriol. 86*: 687-691.

Wells, J. A., Ferrari, E., Henner, D. J., Estell, D. A., and Chen, E. Y. (1983). Cloning, sequencing and secretion of *Bacillus amyloliquefaciens* subtilisin in *Bacillus subtilis*. *Nucleic Acids Res. 11*: 7911-7925.

Whitely, H. R., Schnepf, H. E., Kronstad, J. W., and Wong, H. C. (1984). Structural and regulatory analysis of a cloned *Bacillus thuringiensis* crystal protein gene. In *Genetics and Biotechnology of Bacilli,* A. T. Ganesan and J. T. Hoch (Eds.). Academic Press, Orlando, FL, pp. 375-386.

Winkler, V. W., Hansen, G. D., and Yoder, J. (1971). Immunochemical analysis of parasporal crystal digests of *Bacillus thuringiensis* as an index of insecticidal activity. *J. Invert. Pathol. 18*: 378-382.

Wong, H. C., and Chang, S. (1986). Identification of a positive retroregulator that stabilizes mRNAs in bacteria. *Proc. Natl. Acad. Sci. USA 83*: 3233-3237.

Wong, H. C., Schnepf, H. E., and Whiteley, H. R. (1983). Transcriptional and translational start sites for the *Bacillus thuringiensis* crystal protein gene. *J. Biol. Chem. 258*: 1960-1967.

Wong, S.-L., and Doi, R. H. (1986). Determination of the signal cleavage peptidase site in the preprosubtilisin of *Bacillus subtilis. J. Biol. Chem. 261*: 10176-10181.

Wong, S.-L., Price, C. W., Goldfarb, D. S., and Doi, R. H. (1984). The subtilisin *E* gene of *Bacillus subtilis* is transcribed from σ^{37} promoter in vivo. *Proc. Natl. Acad. Sci. USA 81*:1184-1188.

Yamane, K., and Maruo, B. (1980). *B. subtilis* α-amylase genes. In *Molecular Breeding and Genetics of Applied Microorganisms,* K. Sakaguchi and M. Ouishi (Eds.). Academic Press, London, pp. 117-123.

Yamazaki, H., Ohmura, K., Nakayam, A., Takeichi, Y., Otozai, K., Yamasaki, M., Tamura, G., and Yamane, K. (1983). α-Amylase genes (*amyR2, amyE*$^{+}$) from α-amylase-hyperproducing *Bacillus subtilis* strain: Molecular cloning and nucleotide sequences. *J. Bacteriol. 156*: 327-337.

Yang, M. Y., Ferrari, E., and Henner, D. J. (1984). Cloning of the neutral protease gene of *Bacillus subtilis* and the use of the cloned gene to create an in vitro-derived deletion. *J. Bacteriol. 160*: 15-21.

Yansura, D. G., and Henner, D. J. (1984). Use of *Escherichia coli lac* repressor and operator to control gene expression in *Bacillus subtilis. Proc. Natl. Acad. Sci. USA 81*: 439-443.

Yoshimura, S., Danno, G., and Natake, M. (1966). Studies on D-glucose isomerizing activity of D-xylose grown cells from *Bacillus coagulans* strain HN-68. *Agr. Biol. Chem. 30*: 1015-1023.

Young, F. E. (1980). Impact of cloning in *Bacillus subtilis* on fundamental and industrial microbiology. *J. Gen. Microbiol. 119*: 1-15.

Zakharyan, R. A., Israelyan, Y. A., Agabalyan, A. S., Tatevosyan, P. E., Akopyan, S. M., and Afikyan, E. K. (1979). Plasmid DNA from *Bacillus thuringiensis. Mikrobiologiya 48*: 226-229.

Zamola, B., Valles, P., Meli, G., Miccoli, P., and Kajfez, P. (1981). Use of the centrifugal separation technique in manufacturing a bioinsecticide based on *Bacillus thuringiensis*. *Biotechnol. Bioeng.* *23*: 1079-1086.

Zeman, N. W., and McCrea, J. M. (1985). Alpha-amylase production using a recombinant DNA organism. *Cereal Foods World* *30*: 777-780.

Zhang, Q., Tsukagoshi, N., Miyashiro, S., and Udaka, S. (1983). Increased production of α-amylase by *Bacillus amyloliquefaciens* in the presence of glycine. *Appl. Environ. Microbiol.* *46*: 293-295.

Zittan, L. (1975). Sweetzyme—a new immobilised glucose isomerase. *Die Stärke* *27*: 236-241.

3

Coryneform Bacteria

ISAMU SHIIO and SHIGERU NAKAMORI
Ajinomoto Co., Inc., Yokohama, Japan

I. INTRODUCTION

Coryneform bacteria have been used for the industrial production of amino acids, nucleotides, and their derivatives, including glutamic acid, lysine, and inosinic acid. These are gram-positive, nonspore-forming, nonmotile, and rod-shaped facultative anaerobes that grow well at pH 5-9 and at 30-37°C on various sugars and organic acids. Most bacterial strains for glutamate production were isolated directly from natural sources and selected by their productivities; those for production of the other amino acids were derived from glutamate producers by mutation. *Brevibacterium ammoniagenes* used for nucleotide production is different from these amino acid producers and was selected for its weak nucleotide-degrading ability. Genetic techniques used for derivation of the producers and for improvement of productivities are mutation, protoplast fusion, and recombinant DNA techniques. Transduction and transformation using chromosomal DNA as the donor are not successful for practical use in these bacteria, although some bacteriophages were isolated for possible transduction use (Momose et al., 1976).

Large-scale submerged cultivation techniques are applied in the industrial processes. Details including the structure of a typical commercial fermentor and methods for the maintenance of pure culture conditions, automatic control of pH and temperature, aeration, and agitation are described by Hirose and Okada (1979). More than 99% of the world supply of amino acids by fermentation processes is accounted for by L-glutamic acid and L-lysine. The

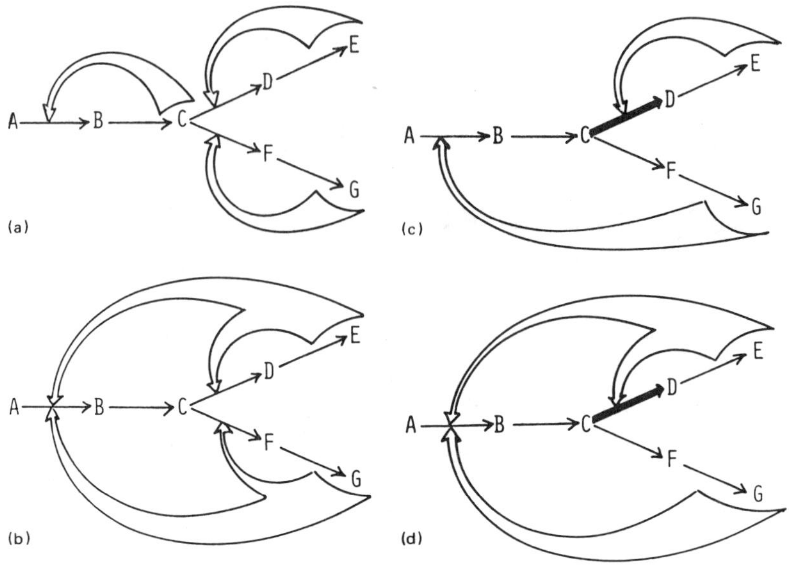

Figure 1 Regulatory patterns in branched biosynthetic pathways: (a) sequential control, (b) multiple end product control, (c) single end product control with preferential synthesis, (d) mixed pattern of b and c. (From Shiio, 1982.)

annual production was 340,000 as monosodium salt and 34,000 tons as HCl salt, respectively, in 1980 (Akashi, 1982).

A. Regulation of Biosynthesis and Overproduction

Biosynthesis of amino acids and nucleotides is regulated through feedback control by these metabolic end products. Then, for overproduction, it is essential to establish genetically or physiologically some conditions in which the control does not operate. Two general feedback control patterns are theoretically possible in branched complex pathways for the biosynthesis of these primary metabolites. The sequential control pattern as illustrated in Fig. 1a consists of feedback control by end products in their specific branches and by the last common intermediate in the common pathway. The pattern in Fig. 1b is characterized by multiple end product control in the common pathway. However, experimental data are often not consistent with these patterns. Two types of deviation from pattern b were observed, as summarized in Fig. 1c: lack of feedback control at one of the specific branches and control by one or some but not all of the end products on the common pathway. Control by an end product G does not disturb the synthesis of another end product E, if E is synthesized in preference to G under conditions in which control in the specific branches does not operate. In other words, preferential synthesis of E is one of the essential elements for this pattern. Feedback control in the E-specific branch not only prevents overproduction of E but also switches the metabolic flow into G. Figure 1d is a combination of patterns b and c. This pattern, in which control in the common pathway by G effectively operates only in the presence of excess E, is much safer than pattern C (Shiio, 1978, 1982). Remarkable preferential synthesis in branched pathways has often been observed, for example, with aromatic biosynthesis by Davis (1952). Five general principles for overproduction were found, which established conditions in which controls did not operate: (1) genetic removal of feedback control, (2) alteration of preferential or balanced synthesis, (3) elevation of intracellular concentration of regulatory enzyme substrate that competed with feedback effectors, (4) limitation of effector(s) in the multiple end product control, and (5) decrease in the intracellular concentration of the end product itself (Shiio, 1982). Mutants insensitive to feedback control may be selected by resistance to analogs of the feedback effectors. The concentration of substrate for a regulatory enzyme may be elevated by removal of feedback control on its synthesis, if such control is present. When the product is an intermediate but not a metabolic end product itself, feedback effectors may be limited simply by blocking the pathway downstream of the product and adding limited amounts of the effectors.

B. Promotion of Overproduction

Removal of feedback control or prevention of its operation does not necessarily provide an amount of product large enough for industrial purposes. Five general methods for increasing the yield are known: (1) combination of methods for overproduction, (2) increase of enzyme activities specific for biosynthesis, (3) increase in supply of biosynthetic precursors, (4) decrease or removal of enzyme activities involved in degradation of the product or its precursor, and (5) decrease in cellular transport activity for the product (Shiio, 1982, 1983). These methods alone, except method 1, do not cause a marked overproduction but promote it. The coryneform bacteria were found not to degrade amino acids once produced under the production conditions, although they were able to utilize some of them as carbon and/or nitrogen sources (see also Sec. II.A).

II. ACIDIC AMINO ACIDS

A. Glutamic Acid from Sugars

Various coryneform bacteria that produce large amounts of L-glutamate were isolated from natural sources. The representative species are *Corynebacterium glutamicum* (Kinoshita et al., 1958), *Brevibacterium lactofermentum*, *Brevibacterium flavum* (Okumura et al., 1962), and *Brevibacterium thiogenitalis* (Kanzaki et al., 1967), among which the first was designated *Micrococcus glutamicus* when isolated. *B. lactofermentum* produced about 50 g/liter of glutamic acid when cultured in a medium containing 100 g/liter of sugar for 28 hr. The concentration of glutamate produced was further increased by feeding the sugar during cultivation (Hirose and Okada, 1979). These strains required biotin for growth and produced glutamate specifically under biotin limitation (Tanaka et al., 1960). The limitation was found to make the cells permeable to glutamate through a limited supply of fatty acids, constituents of the cell membrane. The addition of penicillin or limitation of oleic acid, which replaced biotin for growth, also caused overproduction by permeabilizing the cells (Shiio et al., 1962b, 1963; Ôtsuka and Shiio, 1968). Conditions for glutamate production under excess biotin, which was often found in natural sugar sources, were required for its industrial application in addition to the penicillin method (Somerson and Phillips, 1961) as just described. Based on the mechanism found, the effects of various surface-active agents, fatty acids, and their derivatives were examined. Among them, Tween 60 (polyoxyethylene sorbitan monostearate) was effective on production under excess biotin by permeabilizing the cells. The effect was specific to its fatty acid constituents and, therefore, not due to that as a surface-active agent but probably as an antimetabolite of fatty acids that inhibited the formation of lipids in the cell membrane. Tween 60 had no effect on production with resting cells (Shiio et al., 1963). Glutamate as well as other primary metabolites is synthesized under feedback control. In *B. flavum*, phosphoenolpyruvate (PEP) carboxylase, which catalyzes the formation of oxalacetate, a common intermediate for glutamate and aspartate biosynthesis, is repressed by glutamate or aspartate and inhibited by aspartate or more strongly by aspartate plus 2-oxoglutarate, the direct precursor of glutamate. Glutamate dehydrogenase, which catalyzes the final reaction for glutamate biosynthesis, is inhibited by glutamate (Shiio and Ozaki, 1970; Shiio and Ujigawa, 1978; Mori and Shiio, 1985). Therefore, glutamate was overproduced only under special conditions in which it was rapidly excreted to outside the cells and its intracellular concentration would be maintained at a lower level than that required for operation of feedback control. Moreover, aspartate, another feedback effector, was also found to be excreted under these conditions (Shiio et al., 1962a). The intracellular levels of glutamate and aspartate may also depend on cellular transport ability. *B. flavum* as well as other bacteria has the ability to transport and concentrate these amino acids in the cells. Mutants with impaired transport activities produced glutamate about 10% more than did the parent. These mutants were selected by the inability to utilize glutamate as carbon and nitrogen sources (Shiio et al., 1982b; Mori and Shiio, 1983). Glutamate, once produced, did not decrease during prolonged culture periods. Furthermore, the cells did not consume glutamate in a buffer solution, whereas those grown on glutamate did (Shiio et al., 1982b).

B. Glutamic Acid from Other Carbon Sources

The possibility of using acetic acid, a petrochemical product, as an industrial carbon source for production was suggested by the result that *B. flavum* grown on glucose produced glutamate from acetate in a buffer, with yield comparable to that from glucose. In fact, it was able to grow on acetic acid at a concentration lower than 5% and to produce glutamate. Feeding of acetic acid was useful not only for charging the additional carbon source but also for maintaining pH during cultivation. Production from acetate also depended on the amount of biotin added, but the optimal amount was one-tenth to one-fifth as much as that for glucose as carbon source (Tsunoda et al., 1961a,b). This is probably due to more effective fatty acid synthesis from acetate than from glucose. Kanzaki et al. (1972) reported that a mutant of *B. thiogenitalis* that did not respond to biotin but to oleic acid produced 66 g/liter of glutamic acid from acetate with a 50% yield under limitation of oleic acid.

These strains were selected by their productivities from glucose and therefore may have an excellent machinery for glucose but probably not for acetate. An enhancement of metabolic abilities specific to acetate was possible to promote production from acetate. Among enzymes specific to acetate utilization, phosphate acetyltransferase activity showed a linear relation to the glutamate productivity when revertants from a *B. flavum* mutant lacking the enzyme were cultured. A revertant having 1.5-fold higher enzyme activity than the original wild strain produced 45 g/liter of glutamic acid with 55% yield, 1.5-fold that of the original strain. Moreover, the strain produced 100 g/liter with 44% yield with continuous feeding of acetic acid (Shiio et al., 1969; Shiio, 1982).

Glutamate production from petroleum hydrocarbons was examined. Among various stock microbial cultures tested, *Corynebacterium hydrocarboclastus* and *Corynebacterium oleophilus* were the best species when kerosene was used. These strains did not require biotin, but production was promoted more than 10-fold with penicillin.

Addition of Tween 60 was not effective. *C. hydrocarboclastus* R-7, which was isolated from a natural source and selected by its ability to utilize n-alkanes, produced 20 g/liter of glutamic acid with 45% yield based on carbon source consumed when cultured in a medium containing 8% (v/v) *n*-hexadecane for 40 hr (Ôtsuka et al., 1964; Shiio and Uchio, 1969). Furthermore, a penicillin-resistant mutant derived from strain R-7 maintained its ability to utilize n-alkanes during prolonged culture in contrast to the parent and produced 84 g/liter of glutamic acid in 96 hr with feeding of carbon sources (Kobayashi et al., 1971). Kikuchi et al. (1972) reported that a glycerol auxotroph of *Corynebacterium alkanolyticum* produced 72 g/liter of glutamic acid when cultured for 80 hr in a medium containing 10% (w/v) n-paraffins under the limitation of glycerol, which made the cells permeable to glutamate.

C. Aspartic Acid

B. flavum wild strain produced a small amount of L-aspartate in addition to glutamate. Production depended on limitation of biotin, which made the cell permeable to aspartate as well as glutamate (Shiio et al., 1962a). The production of aspartate lower than that of glutamate reflects a preferential synthesis of the latter at the branch point of the biosynthetic pathway. This is probably due to a much stronger affinity for the common intermediate, oxalacetate, of citrate synthase on the glutamate side, than that of aspartate aminotransferase on the aspartate side of the branch point (Shiio et al., 1982a). In fact, mutant 15-8, whose citrate synthase had lower activity and much weaker affinity for oxalacetate, produced 10 g/liter of aspartic acid with 30% yield, 10-fold more than did the original strain. The mutant had a PEP carboxylase less sensitive to feedback inhibition, which might also contribute to the increased production. This mutant was isolated as a revertant from a citrate synthase-lacking mutant of *B. flavum*. Moreover, a pyruvate kinase-lacking mutant derived from strain 15-8 produced 22.6 g/liter of aspartic acid, more than twofold greater than the parent, when cultured in a medium containing 100 g/liter of glucose. The mutant also produced 6 g/liter of glutamic acid, whereas the original wild strain produced 1 and 43 g/liter of aspartic and glutamic acids, respectively. Thus, the ratio of aspartic to glutamic acid produced increased from 0.023 to 3.8. However, the total amount of the two amino acids produced was reduced, probably owing to an elevated concentration of aspartate, the feedback inhibitor of the common regulatory enzyme, PEP carboxylase. The optimal concentration of biotin for aspartate production in the mutant was about 3 μg/liter, almost the same as that for glutamate production in the wild strain (Shiio et al., 1982c; Mori and Shiio, 1984).

III. AMINO ACIDS DERIVED FROM ASPARTATE

Lysine, threonine, isoleucine, and methionine are synthesized from aspartate in coryneform bacteria. The first committed enzyme in the common pathway, aspartokinase, is feedback inhibited concertedly by lysine and threonine (Shiio and Miyajima, 1969). This seems to be the only regulatory mechanism for lysine biosynthesis, as no other feedback control was observed with any enzymes in the common and lysine-specific pathways. These results are markedly different from those in enteric bacteria. Another difference may be the presence of N-acetyl intermediates on the specific pathway in contrast to N-succinyl derivatives in enterics. Whereas a mutant of *Escherichia coli* lacking 2-oxoglutarate dehydrogenase, which catalyzes the formation of succinyl-CoA, required succinate or lysine plus methionine for growth, that of *B. flavum* did not (Miyajima et al., 1968; Miyajima and Shiio, 1970a, 1971; Tosaka and Takinami, 1978; Shiio et al., 1982b). Homoserine dehydrogenase, the first-step enzyme in the threonine-, methionine-, and isoleucine-specific pathway is feedback inhibited by threonine and repressed by methionine, in contrast to the two isoenzymes in enterics. The enzyme in *B. flavum* wild strain grown in a minimal medium was at a fully derepressed level and decreased one-third to one-seventh by the addition of methionine (Miyajima and Shiio, 1970b, 1971).

A. Lysine

Three groups of mutants were found to produce large amounts of L-lysine. The first group is that having aspartokinase insensitive to concerted feedback inhibition by lysine plus threonine and was selected by resistance to a lysine analog, *S*-(2-aminoethyl)-L-cysteine (AEC) plus threonine. The growth of *B. flavum* was completely inhibited by AEC only in the presence of threonine and was specifically recovered by lysine. This is consistent with feedback control of lysine biosynthesis by the concerted inhibition of aspartokinase. The feedback-resistant mutants derived from *B. flavum* wild strain produced 15-18 g/liter, maximally 30 g/liter, of L-lysine as its HCl salt when cultured in a medium containing 100 g/liter of glucose for 72 hr. Production occurred during growth and continued even into the stationary stage. It was not inhibited by threonine in contrast to that by the second and third groups of mutants. Cane molasses was a good carbon source without any special additions, in contrast to glutamate production (Sano and Shiio, 1970). Aspartokinase from a representative strain, FAI-30, was not inhibited by lysine plus threonine but was activated by threonine in contrast to the parent enzyme. Although both the enzymes were weakly inhibited by lysine alone, inhibition of the former was reversed by threonine, whereas that of the latter promoted (Shiio et al.,

1970b). Tosaka et al. (1978b) reported a feedback-resistant mutant from *B. lactofermentum* that produced 18 g/liter of lysine. The enzyme of this mutant was not affected by threonine and, therefore, weakly inhibited by lysine plus threonine to a similar extent as by lysine alone, in contrast to that of strain FAI-30.

The second group of mutants producing lysine is that requiring threonine for growth. They produced lysine only when cultured under the limitation of threonine, in which the concerted feedback inhibition of aspartokinase did not operate. Production also depended on enzymes blocked in the mutants. Whereas threonine synthetase-lacking mutants produced little lysine, homoserine kinase-lacking mutants produced both homoserine and lysine. The amounts of lysine produced were 17 g/liter on average and 26 g/liter maximum, whereas those produced by homoserine dehydrogenase-lacking mutants were 36 and 42 g/liter, respectively. The latter mutants require homoserine or methionine in addition to threonine (Sano and Shiio, 1967). A homoserine auxotroph of *C. glutamicum* produced lysine with 29% yield under the limitation of homoserine (Nakayama et al., 1961b).

As described previously, lysine biosynthesis is feedback regulated by lysine on a premise of the preferential synthesis of threonine, probably owing to about a 15-fold higher activity of homoserine dehydrogenase on the threonine side than that of dihydrodipicolinate synthetase on the lysine side of the biosynthetic branch point. Therefore, a decrease in homoserine dehydrogenase activity causes overproduction of lysine. In fact, *B. flavum* wild strain produced 5 g/liter of lysine when the enzyme was repressed to one-third of the normal level by the addition of methionine (Miyajima and Shiio, 1970a, 1970b, 1971). Furthermore, a genetic decrease in the enzyme to one-thirtieth of the normal level gave a maximal lysine production of 25 g/liter, as observed in mutant S-20.

The third group of mutants including this strain is prototrophic. The growth rate of strain S-20 in exponential phase was shown to be almost the same as that of the wild strain in either the absence or presence of homoserine, in contrast to homoserine auxotrophs. In contrast to the wild strain, growth was inhibited by the addition of threonine or methionine and recovered by further addition of methionine or threonine, presumably owing to inhibition or repression of homoserine dehydrogenase, respectively. These sensitive mutants produced lysine mainly during growth, whereas mutants requiring threonine produced lysine during the stationary phase. The sensitive mutants were isolated by two methods: as prototrophic revertants from homoserine auxotrophs and as threonine- or methionine-sensitive mutants from the wild-type strains (Shiio and Sano, 1969; Sano and Shiio, 1971).

A combination of two of these three kinds of mutations increased the yield of overproduction. A threonine-sensitive threonine auxotroph of *B. flavum* produced 33 g/liter of lysine, more than did strains with either single mutation (Shiio and Sano, 1969). An AEC-resistant homoserine and leucine double auxotroph of *C. glutamicum* produced 40 g/liter of lysine, whereas the parent homoserine and leucine double auxotroph made only 35 g/liter (Nakayama and Araki, 1982). Although leucine limitation in auxotrophs did not cause the overproduction, it increased the production in AEC-resistant mutants of *B. lactofermentum* and *C. glutamicum* to 41 and 36 g/liter from 18 g/liter, respectively. It was found to derepress dihydrodipicolinate synthetase specifically (Tosaka et al., 1978a,c).

In *B. flavum*, lysine is synthesized from three intermediates of the central metabolism of sugar, phosphoenolpyruvate, pyruvate, and acetyl-CoA. Enzymes catabolizing these raw materials were examined for the effect of their reduced activities. A reduction in citrate synthase activity, which was involved in acetyl-CoA catabolism, increased lysine production in feedback-resistant mutants to 28 from 15 g/liter but did not affect that in homoserine auxotrophs or methionine-sensitive mutants. On the other hand, the lack of pyruvate kinase, which catalyzed the degradation of PEP, stimulated production in each of the three groups of mutants discussed earlier. A representative feedback-resistant mutant, a methionine-sensitive mutant, and strain No. 22, a homoserine auxotroph, all of which lacked pyruvate kinase and had low citrate synthase activities, produced 43, 41, and 51 g/liter of lysine, respectively. Mutants lacking pyruvate kinase were selected by resistance to AEC plus threonine or by the inability to utilize ribose as a carbon source. Simple mutants with low citrate synthase activities or lacking pyruvate kinase did not produce lysine (Ozaki and Shiio, 1983; Shiio et al., 1984b, 1987). A mutant with low pyruvate dehydrogenase activity, strain KD-11, was isolated as a prototrophic revertant from an acetate auxotroph of strain No. 22 and produced 55 g/liter of lysine with 55% yield. The enzyme from strain KD-11 had a very low affinity for one of the substrates, NAD^+, and therefore would show a low activity under physiological conditions. The parent acetate auxotroph produced little lysine (Shiio, et al., 1984c). A mutant with low pyruvate dehydrogenase activity was also derived from a lysine-producing mutant of *B. lactofermentum* but selected by sensitivity to β-fluoropyruvate, a substrate analog for the enzyme. The parent and the mutant with 27% of the enzyme activity produced 43 and 51 g/liter of lysine, respectively (Tosaka et al., 1985).

B. Threonine

As described earlier, the regulatory pattern of lysine and threonine biosynthesis in *B. flavum* is given by Fig. 1d, where threonine synthesis is regulated by simple feedback inhibition of homoserine dehydrogenase in the specific pathway. Thus, only one group of mutants was found for threonine production. These mutants were selected by resistance to α-amino-β-hydroxyvaleric acid (AHV), an analog of threonine, which inhibited growth of the wild strain as well as its homoserine dehydrogenase activity. Furthermore, the growth inhibition was specifically reversed by threonine. AHV-resistant mutants of *B. flavum* and *C. acetoacidophilum* produced 11 and 6 g/liter of L-threonine at maximum, respectively, when cultured in a medium containing 100 g/liter of glucose for 64 hr. A mutant BB-69, which was derived from the AHV-resistant mutant B-183 of *B. flavum* and selected by resistance to a higher concentration of AHV, produced 14 g/liter. In these mutants, homoserine dehydrogenase was insensitive to feedback inhibition, whereas aspartokinase was normally sensitive to the concerted inhibition by lysine plus threonine. This overproduction in the presence of normal aspartokinase is probably due to a preferential synthesis of threonine, which maintains the intracellular lysine level low enough to prevent operation of its feedback control. Thus, the addition of lysine specifically inhibited the production (Shiio and Nakamori, 1970; Shiio et al., 1970a). The second enzyme in the threonine-specific pathway, homoserine kinase, which was inhibited by threonine but more weakly than was homoserine dehydrogenase, was also normally sensitive to the inhibition in these threonine producers (Miyajima and Shiio, 1972a). Kase and Nakayama (1972) reported that AHV-resistant mutants derived from a methionine auxotroph of *C. glutamicum* produced 3.7 g/liter of threonine at maximum and had similar enzymatic alterations as did *B. flavum* mutants.

Possible increases in the production yield by genetic or physiological removal of feedback inhibition of aspartokinase and of repression of homoserine dehydrogenase were examined, as the production by strain BB-69 was inhibited by the addition of lysine or methionine. Thus, lysine auxotrophs from AHV-resistant mutant B-2 whose homoserine dehydrogenase was partially desensitized to feedback inhibition produced 10 g/liter of threonine at maximum, 50% more than did the parent, under limitation of lysine, whereas those from strain BB-69 did not produce more than the parent. This indicates a positive effect only when feedback control in the key enzyme is incompletely removed. Methionine auxotrophs from either strain did not produce more than the parents, except that strain BBM-21, derived from strain BB-69, produced 18 g/liter of threonine but little homoserine, whereas the parent produced 2-3 g/liter of homoserine in addition to threonine (Nakamori and Shiio, 1972). AHV-resistant mutants were derived from lysine-producing AEC-resistant mutants of *B. flavum*, the aspartokinases of the latter being insensitive to feedback inhibition. A representative strain FAB-44 derived from strain FAI-30 (see above), whose homoserine dehydrogenase was insensitive to feedback inhibition, produced 15 g/liter of threonine, 50% more than the simple AHV-resistant mutant B-183 but only slightly more than strain BB-69 (see above), which was resistant to a higher concentration of AHV. Therefore, the increase may not be due to the removal of concerted feedback inhibition of aspartokinase but to the differing effects of threonine on the aspartokinases of these threonine producers; the enzymes in strains B-183, BB-69, and FAI-30 are inhibited by a saturating concentration of threonine, 45, 20, and 0%, respectively. Strain FAB-44 did not produce lysine but did produce threonine, indicating a preferential synthesis of threonine (Nakamori and Shiio, 1973; Shiio, 1978). On the other hand, Kase and Nakayama (1972) reported that removal of feedback inhibition of aspartokinase in a AHV-resistant methionine auxotroph of *C. glutamicum* increased threonine production, but only to 9.5 from 2.3 g/liter in the parent. The production further increased to 14 g/liter when cultured for 96 hr under optimal conditions. Karasawa et al. (1986) reported that the introduction of lysine auxotrophy by protoplast fusion increased threonine production to 18 g/liter. The parent, an AEC- and AHV-resistant isoleucine and leucine double auxotroph of *B. lactofermentum*, produced 6 g/liter of lysine in addition to 13 g/liter of threonine.

C. Isoleucine

Isoleucine biosynthesis in *B. flavum* is regulated by feedback inhibition of threonine deaminase, the first-step enzyme of the specific pathway. The inhibition is competitively reversed by threonine, and the enzyme activity has a sigmoidal relation to the substrate threonine concentration (Miyajima and Shiio, 1972b). Therefore, elevated concentrations of threonine in AHV-resistant threonine producers caused overproduction of L-isoleucine but only slightly. Some AHV-resistant mutants were found to produce 10-15 g/liter of isoleucine, rather than threonine, when selected by resistance to lower concentrations of the analog than those used for selecting threonine producers. These isoleucine producers had homoserine dehydrogenases insensitive to feedback inhibition but normally sensitive threonine deaminases. Thus, this is an example of principle 3 for overproduction as described in Sec. I.A. To increase the production yield, a mutant having about twofold elevated threonine deaminase activity was derived from an AHV-resistant isoleucine producer and selected by resistance to *O*-methylthreonine, a substrate analog for the enzyme, which inhibited growth of the parent. Growth inhibition of the parent was reversed specifically by further addition of isoleucine. The mutant and its parent produced 14.5 and 11 g/liter of isoleucine, respectively (Shiio et al., 1973). An ethionine-resistant mutant with threonine deaminase insensitive to feedback inhibition was derived from a threonine-producing AEC- and AHV-resistant mutant of *B. flavum*, whose aspartokinase and homoserine dehydrogenase were insensitive to feedback inhibition. Whereas the parent strain produced 11 and 1.6 g/liter of threonine and isoleucine, respectively (Nakamori and Shiio, 1973), the mutant produced isoleucine alone, 11.3 g/liter. The latter strain produced 20 g/liter when cultured in a medium containing acetic acid as carbon source (Ikeda et al., 1976). An isoleucine-producing mutant of *C. glutamicum* was also derived from a threonine-producing AEC- and AHV-resistant methionine auxotroph and selected stepwise by resistance to thiaisoleucine, ethionine, 4-azaleucine, and α-aminobutyric acid. The mutant produced 9.7 g/liter of isoleucine but little threonine in a molasses medium (Kase and Nakayama, 1977).

IV. AROMATIC AMINO ACIDS

Tryptophan, phenylalanine, and tyrosine are synthesized from erythrose 4-phosphate and phosphoenolpyruvate through chorismate, the branch point intermediate between tryptophan and the other two (Fig. 2). Between phenylalanine and tyrosine, prephenate is the biosynthetic branch point. In contrast to enteric bacteria, coryneform bacteria have pretyrosine as an intermediate in the tyrosine-specific pathway (Fazel and Jensen, 1979). Regulatory mechanisms for aromatic amino acid biosynthesis in *B. flavum* are also different from those in enteric bacteria or bacilli. 3-Deoxy-D-*arabino*-heptulosonate-7-phosphate (DAHP) synthase, the first enzyme of the common pathway, is synergistically inhibited by phenylalanine and tyrosine. The enzyme, as well as dehydroquinate synthase, shikimate dehydrogenase, shikimate kinase, and 5-enolpyruvylshikimate-3-phosphate synthase, is repressed by tyrosine. Chorismate mutase, which catalyzes the conversion of chorismate to prephenate and is a dissociable complex with DAHP synthase as a regulatory subunit, is only partially inhibited by the two end products, 65% at maximum under physiological conditions in which dissociation is negligible. Another subunit of the enzyme is repressed by tyrosine, more strongly than DAHP synthase (Shiio et al., 1974; Shiio and Sugimoto, 1979, 1981a; Sugimoto and Shiio, 1985). Whereas enzymes of the tyrosine-specific pathway are not regulated, prephenate dehydratase of the phenylalanine pathway is feedback inhibited by phenylalanine and recovered competitively by tyrosine,

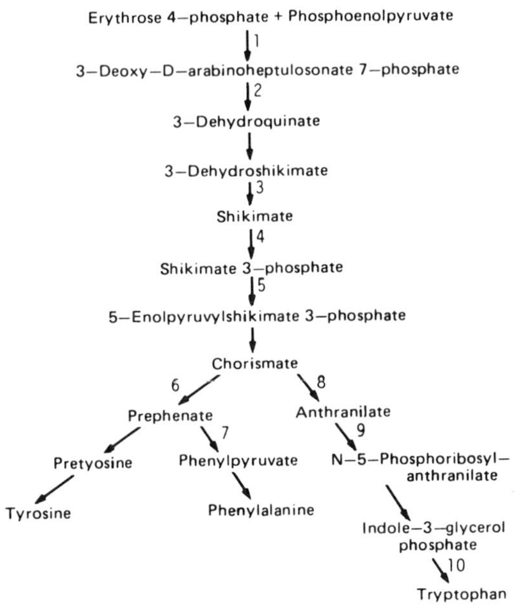

Figure 2 Aromatic amino acid biosynthetic pathway in coryneform bacteria. The reactions are catalyzed by (1) 3-deoxy-D-*arabino*-heptulosonate-7-phosphate synthase; (2) dehydroquinate synthase; (3) shikimate dehydrogenase; (4) shikimate kinase; (5) 5-enolpyruvylshikimate-3-phosphate synthase; (6) chorismate mutase; (7) prephenate dehydratase; (8) anthranilate synthase; (9) anthranilate phosphoribosyltransferase; (10) tryptophan synthase.

causing a balanced synthesis of the two amino acids (Fazel and Jensen, 1979; Shiio and Sugimoto, 1981b; Sugimoto and Shiio, 1974). Anthranilate synthase of the tryptophan pathway is feedback inhibited, and all the enzymes are repressed by tryptophan but not coordinately (Shiio et al., 1972a; Sugimoto and Shiio, 1977). Therefore, the regulatory pattern for phenylalanine and tyrosine biosynthesis is a modification of Fig. 1d, but that for these two amino acids on the one hand and tryptophan on the other hand is as shown in Fig. 1c.

A. Phenylalanine and Tyrosine

Four groups of mutants were found to produce L-phenylalanine and/or L-tyrosine. Most (95%) of the *m*-fluorophenylalanine (MFP)-resistant mutants of *B. flavum* had feedback-resistant DAHP synthase and produced both the amino acids to 2 g/liter each at maximum.

The remaining 5% were those having feedback-resistant prephenate dehydratases and produced phenylalanine to 2 g/liter at maximum when cultured in a medium containing 100 g/liter of glucose for 72 hr. Production of phenylalanine and tyrosine by the former mutants was not inhibited by phenylalanine or tyrosine, whereas production of phenylalanine by the latter mutants was inhibited by tyrosine. This is consistent with the presence of normal DAHP synthase in the latter mutants, whose regulation may not operate under the production conditions owing to a tyrosine limitation caused by the preferential synthesis of phenylalanine. Growth inhibition by MFP was reversed by phenylalanine but not by tyrosine. MFP also inhibited DAHP synthase, synergistically with tyrosine, as well as prephenate dehydratase, whereas *p*-fluorophenylalanine (PFP) inhibited the latter enzyme alone (Shiio and Sugimoto, 1976; Shiio et al., 1974; Sugimoto et al., 1973; Sugimoto and Shiio, 1974). Hagino and Nakayama (1973) found that PFP-resistant mutants of *C. glutamicum* produced phenylalanine, 4.5 g/liter at maximum, but not tyrosine. The third and fourth groups of mutants, tyrosine and pehnylalanine auxotrophs of *B. flavum* that lacked pretyrosine dehydrogenase and prephenate dehydratase, respectively, produced phenylalanine and tyrosine, 1.6 and 1.5 g/liter, respectively, under limitation of required substances, conditions in which feedback control of DAHP synthase did not operate. Control of prephenate dehydratase may be overcome competitively by the overproduced substrate (Sugimoto et al., 1973).

A combination of these methods increased the production yield. A PFP-resistant mutant derived from a tyrosine auxotroph of *B. lactofermentum* produced 7.1 g/liter of phenylalanine, whereas the parent produced 4.2 g/liter when cultured in a medium containing 130 g/liter of glucose. Production was elevated to 22 g/liter by the addition of 5-methyltryptophan (5MT) resistance, as observed with the strain No. 123, and further to 25 g/liter by the addition of decoyinine sensitivity. These higher productivities were partially due to the culture medium containing acetate and fumarate in addition to glucose. Strain T-222 produced 18 g/liter in the presence of these organic acids but only 9 g/liter in the absence of fumarate (Goto et al., 1981; Tsuchida et al., 1987a).

On the other hand, a PFP- and *p*-aminophenylalanine (PAP)-resistant tyrosine auxotroph of *C. glutamicum* produced 9.5 g/liter of phenylalanine in a cane molasses medium containing 100 g/liter of sugar. This strain had prephenate dehydratase less sensitive to feedback inhibition, but normal DAHP synthase. Stepwise addition of 3-aminotyrosine, PAP, PFP, and tyrosine hydroxamate resistance to a phenylalanine auxotroph increased tyrosine production to 17.6 from 3 g/liter. The resistant mutant contained feedback-insensitive DAHP synthase (Hagino and Nakayama, 1973, 1974, 1975b).

B. Tryptophan

Only one group of tryptophan-producing mutants was found that was resistant to feedback control. A feedback-resistant strain, No. 833, which was derived from *B. flavum* wild strain by two-step mutation and selected by resistance to lower and higher concentrations of 5-fluorotryptophan (5FT), successively, produced 2.4 g/liter of L-tryptophan in a medium containing 100 g/liter of glucose, whereas its intermediate strain produced 0.4 g/liter. The former had anthranilate synthase and phosphoribosyltransferase less sensitive to feedback inhibition than those of the latter, which in turn were much less sensitive than those of the wild strain. All enzymes of the tryptophan-specific pathway except tryptophan synthase were in elevated levels in these strains, owing to genetic derepression. Removal of feedback inhibition in the presence of normal repression control caused overproduction of anthranilate but not tryptophan. Derepression of any two of anthranilate synthase, phosphoribosyltransferase, or tryptophan synthase was essential for tryptophan production. 5FT repressed all tryptophan enzymes tested and inhibited anthranilate synthase and phosphoribosyltransferase activities as well as the growth of the wild strain, which was recovered by tryptophan. To increase the production yield, a MFP-resistant mutant having feedback-insensitive DAHP synthase and its phenylalanine auxotroph were derived from strain 833 and produced 4.5 and 6.2 g/liter of tryptophan, respectively. 5MT-resistant mutants derived from the wild strain and its phenylalanine and tyrosine double auxotroph produced 0.028 and 1.9 g/liter of tryptophan, respectively. Production in these mutants except MFP-resistant strains was strongly inhibited by the addition of both phenylalanine and tyrosine or tyrosine alone (Shiio et al., 1972b; Shiio and Sugimoto, 1978; Shiio et al., 1975). A phenylalanine and tyrosine double auxotroph lacking chorismate mutase produced about 1 g/liter of anthranilic acid. This is probably explained by the result that feedback inhibition of anthranilate synthase was competitively overcome by a high concentration of its substrate, chorismate, which was accumulated under limitation of tyrosine. Anthranilate produced was not converted to tryptophan by genetic derepression of the tryptophan enzymes, probably owing to the presence of weak but noncompetitive feedback inhibition of phosphoribosyltransferase. Genetic removal of feedback inhibition of this enzyme and anthranilate synthase was simultaneous in any mutants tested (Sugimoto and Shiio, 1983). A 5FT-resistant mutant, No. 187, was derived from a PFP-resistant tyrosine auxotroph, No. 239, of *B. flavum*, owing to its excellent productivity, and produced 8 g/liter of tryptophan but only 0.2 g/liter of phenylalanine in a medium containing 130 g/liter of glucose, whereas the parent produced 17 g/liter of phenylalanine.

Strain 187 had feedback-resistant anthranilate synthase and phosphoribosyltransferase and derepressed levels of the tryptophan enzymes but normally feedback-sensitive enzymes of the common pathway in spite of its MFP resistance. A mutant, A-100, which was derived from strain 187 and selected by resistance to azaserine, a substrate analog for anthranilate synthase, produced 10 g/liter of tryptophan and had twofold elevated levals of the enzyme, partially feedback-resistant DAHP synthase, and increased activities of several enzymes involved in the common pathway. A mutant derived from strain 239 (above) and selected by 5FT resistance in the presence of phenylalanine and tyrosine also produced 10 g/liter of tryptophan with a similar enzymatic alteration as that in strain A-100, except that anthranilate synthase was in a similar level as that in strain 187. Sulfonamides inhibit dihydropteroate synthase competitively with its substrate *p*-aminobenzoate, which is synthesized from chorismate. A mutant, S-225, resistant to sulfaguanidine, a sulfonamide, was derived from strain A-100 and produced 19 g/liter of tryptophan in addition to an increased amount of chorismate. The maximum production yield based on sugar added was 17% with sucrose. The strain had elevated levels of DAHP synthase and dihydroquinate synthase, which were regulated more strongly than were the other common enzymes. The addition of sulfaguanidine resistance to 5MT-resistant mutants of *C. glutamicum* also increased production, 2.2-fold, to 9.7 g/liter at maximum (Shiio et al., 1982d, 1984a; Sugimoto and Shiio, 1985). Hagino and Nakayama (1975a,b) reported that a mutant resistant to various tryptophan analogs, which was derived from a phenylalanine and tyrosine double auxotroph of *C. glutamicum*, produced 4.9 g/liter of tryptophan, and this was increased to 12 g/liter by further addition of resistance to several analogs of phenylalanine and tyrosine. The final strain had feedback-resistant anthranilate synthase but normal DAHP synthase.

V. OTHER AMINO ACIDS

A. Arginine, Ornithine, and Citrulline

Arginine is synthesized from glutamate by eight enzyme reactions via ornithine and citrulline. Arginine represses the formation of these biosynthetic enzymes and inhibits the second enzyme, *N*-acetylglutamokinase, in *C. glutamicum* (Yoshida et al., 1979). Thus, if the arginine supply could be limited, intermediates would accumulate. In fact, a citrulline-requiring mutant of *C. glutamicum* produced 26 g/liter and an arginine-requiring mutant of *B. lactofermentum* produced 40 g/liter of L-ornithine, respectively, under the suboptimal concentrations of these amino acids (Kinoshita et al., 1957; Okumura et al., 1968). Similarly, an arginine auxotroph of *C.*

glutamicum produced 10.7 g/liter of L-citrulline (Nakayama and Hagino, 1966). An arginine-producing mutant, *C. glutamicum* KY-10557, was selected as a D-serine-sensitive, arginine hydroxamate- and 2-thiazolealanine-resistant mutant (Nakayama and Yoshida, 1972). In the mutant, which produced 25 g/liter of L-arginine from 15% glucose, several enzymes were derepressed. *N*-acetylglutamokinase was not inhibited by arginine, and the permeability of arginine increased compared with that the the parent (Yoshida et al., 1979). A 2-thiazolealanine-resistant mutant of *B. flavum* produced more arginine, 34 g/liter, with a 24% yield from glucose (Kubota et al., 1973). *N*-acetylglutamokinase of the mutant was derepressed remarkably (Morioka et al., 1977).

B. Proline

The production of proline was first observed with isoleucine auxotrophs of *B. flavum* under a suboptimal concentration of isoleucine. One of them, mutant No. 14-5, produced 12 g/liter of L-proline from 10% glucose. In the mutant, no alteration was found in proline biosynthetic enzymes. The overproduction was explained as follows. Intracellularly overproduced glutamate and ATP were converted to proline, which permeated the cell membrane more easily than glutamate (Yoshinaga, 1969). A sulfaguanidine-resistant mutant, No. 191, derived from No. 14-5, produced 30-35 g/liter of proline. As sulfaguanidine inhibited purine biosynthesis in microorganisms, the intracellular content of ATP was thus presumed to be raised in the mutant (Tsuchida et al., 1986). Proline producers were also derived from *B. flavum*, *B. lactofermentum*, and *C. glutamicum* as mutants resistant to 3,4-dehydroproline, an analog of proline. The yields were 10-18 g/liter from 10% glucose. A 3,4-dehydroproline-resistant mutant derived from No. 191 produced 40 g/liter of proline from 10% glucose (Nakamori et al., 1982). A histidine auxotroph of *Brevibacterium* sp. and various auxotrophs of *C. glutamicum* also produced 25-31 g/liter of proline (Yamatodani et al., 1967; Araki et al., 1975). Other proline producers were selected from wild-type *C. glutamicum*, and culture conditions were examined (Nakanishi et al., 1973). The production was 44 g/liter, 25% yield, maximally.

C. Valine and Leucine

In wild-type *B. lactofermentum*, acetohydroxyacid synthetase, a common biosynthetic enzyme of valine and leucine, is inhibited by valine and is repressed by valine plus leucine plus isoleucine. The activity and the formation of α-isopropylmalate synthase, a key enzyme for leucine biosynthesis, is inhibited and repressed by leucine, respectively (Tsuchida and Momose, 1975). In the early studies of of amino acid fermentation, wild-type *B. ammoniagenes* had been shown to produce 4.7 g/liter of L-valine from 10% glucose. An isoleucine and leucine double auxotroph of *C. glutamicum* produced 10.3 g/liter of valine (Nakayama et al., 1961a) under limitation of isoleucine and leucine, conditions in which valine biosynthetic enzymes would be derepressed. A more potent valine producer was selected from wild-type *B. lactofermentum* as a mutant resistant against 2-thiazolealanine. The strain produced 31 g/liter, 31% yield from 10% glucose (Tsuchida et al., 1975). Acetohydroxyacid synthase was not inhibited by valine or repressed by the three amino acids (Tsuchida and Momose, 1975). Potent L-leucine producers were also selected as 2-thiazolealanine-resistant mutants from an isoleucine and methionine double auxotroph of *B. lactofermentum*. The best strain produced 28 g/liter from 10% glucose (Tsuchida et al., 1974). α-Isopropylmalate synthase of the mutant was not inhibited or repressed by leucine, respectively (Tsuchida and Momose, 1975).

D. Histidine

Histidine auxotrophs of *B. flavum* were shown to produce large amounts of histidinol under the limitation of histidine supply. Histidine production by microbial conversion of histidinol has been shown (Kubota et al., 1971). Direct fermentation of histidine has been established by the use of histidine analog-resistant mutants of *C. glutamicum* and *B. flavum*. 1,2,4-Triazolealanine-resistant mutants of *C. glutamicum* produced 6-7 g/liter of L-histidine. The yield was improved by sequentially deriving mutants resistant to 6-mercaptopurine, 8-azaguanine, 2-thiouracil, and 5-methyltryptophan. In the final strain, the production rose to 15 g/liter, with the yield 10% from glucose (Araki et al., 1974). Phosphoribosyl-ATP pyrophosphorylase of the mutant became 100-fold resistant to feedback inhibition by histidine, and the enzyme level was twofold higher than that of the parent (Araki and Nakayama, 1974). A 2-thiazolealanine-resistant mutant of *B. flavum* produced 4 g/liter of histidine. The yield was improved up to 10 g/liter by the sequential mutation and selection of sulfaguanidine-, α-amino-β-hydroxyvaleric acid-, and ethionine-resistant mutants (Kamijo et al., 1973).

E. Serine, Alanine, Glutamine, and *N*-Acetylglutamine

At present, the direct fermentation of serine from glucose has not been reported. Serine producers were screened among microorganisms from natural sources in a medium containing 2% glycine. A newly isolated bacterium, *Corynebacterium glycinophilum*, produced 8 g/liter of L-serine, maximally. The bacterium did not produce any serine without glycine. Serine hydroxymethyltransferase, which catalyzed the conversion of glycine to serine, was induced by glycine. Mutants with weak activities of serine dehydratase, which catalyzed the conversion of serine to pyruvate and ammonia, were selected as those that could not grow on a medium containing serine as a sole source of nitrogen. Mutants with weak activities of the enzyme were also selected as various auxotrophs. Among them, the best strain produced 14 g/liter of serine, with a molar yield of 33% from 30 g/liter of glycine (Kubota, 1985). Alanine production was investigated in early studies of amino acid production. Several kinds of coryneform bacteria have been shown to produce 15-40 g/liter of alanine from glucose and pentose materials, such as xylose and arabinose. However, the alanine produced was the DL form in every case (Kitai, 1972).

The first work on L-glutamine production was reported with wild-type *B. flavum* and *B. lactofermentum* (Tsunoda et al., 1961). The productivity has been improved by the use of sulfaguanidine-resistant mutants of *B. flavum* (Tsuchida et al., 1987b). Glutamate production by wild-type *C. glutamicum* was altered to that of glutamine and *N*-acetylglutamine by changes in the culture medium composition (Nakanishi et al., 1977).

VI. NUCLEOTIDES AND RELATED COMPOUNDS

Production of nucleotides and related compounds by biotechnological methods has been reported with various kinds of microorganisms (Hirose et al., 1979). In this section, we refer only to those produced by coryneform bacteria. These include direct fermentation from carbon sources, such as glucose, and microbial conversion of intermediates. In order to obtain production by direct fermentation, unfavorable genetic traits, such as regulation of nucleotide biosynthesis by the end products (feedback inhibition and repression), enzyme activities for the degradation of nucleotides to the corresponding nucleosides and bases, and permeability barriers to nucleotides, were overcome by the selection of original strains and mutants.

A. Purine Nucleotides and Related Compounds

B. ammoniagenes has been successfully applied in the production of purine nucleotides and related compounds. The strain was originally selected by its low degradation activity for nucleotides. An adenine bradytroph of strain KY13102 accumulated 12.8 g/liter of inosine-5'-monophosphate (IMP) as the disodium and 7.5 hydrate form (Furuya et al., 1968). The production of IMP was reduced and converted to that of hypoxanthine by manganese ion at more than the optimum concentration (10 μg/liter). A manganese-insensitive mutant in which production was not affected by manganese ion was selected from strain KY13102. The production of IMP has been further improved by multiple sequential mutagenesis and colony isolation.

The final strain, KY13369, an adenine bradytrophic and guanine auxotrophic strain, produced 27 g/liter of IMP, with a 20% yield from glucose. The improvement was explained by (1) weak IMP degradation activity; (2) release of feedback inhibition by AMP, ADP, ATP, and GMP and repression by adenine on PRPP amidotransferase, the first enzyme for purine biosynthesis, by the limited supply of adenine and guanine in the medium; and (3) improved permeability of IMP into the medium; the latter mutation was selected by manganese insensitivity (Teshiba and Furuya, 1982).

From the same original strain as for the IMP producer, an inosine producer, KY13761, was derived as a 6-mercaptoguanine- and 6-thiopurine-resistant, adenine bradytrophic, and guanine auxotrophic mutant. The production of inosine was further improved by sequential and multiple colony isolation and reached maximally 30 g/liter from 15% sugar. The production was not affected by manganese ion, in contrast to that of IMP. In this mutant, degradative activity for nucleotides to corresponding bases was remarkably lower than that of the parent strain, but the activity for purine nucleosides was not altered (Kotani et al., 1978). From strain KY13102, a potent producer of xanthosine-5'-monophosphate (XMP) has been selected as a guanine auxotrophic and nucleotidase-deficient mutant. This strain produced 5-6 g/liter of XMP from 10% glucose. The production of guanosine-5'-monophosphate (GMP) by microbial conversion of XMP was attained with a mutant of *B. ammoniagenes*, which had a high activity of XMP aminase. The mutant was obtained from a nucleotidase deficient strain as a mutant resistant to the antibiotic, decoynine, which inhibits XMP aminase specifically in microorganisms. Using the mutant cells as a source of enzyme, purified XMP was converted to GMP with a molar yield of 80%, 45 g/liter, in a 22 hr incubation (Fujio et al., 1984). The wild-type strain of *B. ammoniagenes* has been shown to produce ATP (1.57 g/liter), ADP (1.59 g/liter), and AMP (2.16 g/liter) by the addition of 20 g/liter of adenine during the course of fermentation. Small amounts of GTP, GDP, and GMP could be produced by the addition of guanine (Tanaka et al., 1968). The cells, when treated with surface-active agents, converted adenine to ATP with a molar yield of 83% (Fujio and Furuya, 1983).

Cyclic adenosine-3',5'-monophosphate (cAMP) was produced in culture fluids of *Brevibacterium liquefaciens* with the addition of adenine and hypoxanthine. Similar production from adenine was also reported by the cultivation of *Corynebacterium murisepticum* and *Microbacterium* sp. (Furuya et al., 1976). The addition of adenine and nicotinic acid during cultivation of strain *B. ammoniagenes* allowed the formation of 2.3 g/liter of NAD (Nakayama et al., 1968). Immobilized cells of *B. ammoniagenes* formed NADP continuously from ATP plus NAD (Murata et al., 1979; Hayashi et al., 1979). Coenzyme A was synthesized from D-panthothenic acid and cysteine by enzymatic reaction using cells of *B. ammoniagenes* with yields of 3-6 g/liter. The yield was improved to 11.5 g/liter by an increasing pantothenate kinase activity in a mutant resistant to D-oxypantotheine (Shimizu et al., 1984). Small amounts of guanosine polyphosphate (ppGpp) were found in the culture fluids of *B. ammoniagenes* (Sato and Furuya, 1975).

B. Pyrimidine Nucleotides and Related Compounds

An uracil-requiring mutant of *C. glutamicum* and *B. ammoniagenes* produced 14 g/liter of orotic acid from 10% glucose and 5 g/liter from 5% glucose, respectively. The same type of mutant of *Arthrobacter paraffineus* produced 6.6 g/liter of orotic acid and 8 g/liter of orotidine simultaneously in a medium containing 10% of *n*-paraffin (Furuya et al., 1976).

The formation of uridine-5'-monophosphate (UMP) from orotic acid was shown using cell extracts of *C. glutamicum*. Microbial conversion of orotic acid and uracil to UMP was shown with *Brevibacterium insectiphilum* and *B. ammoniagenes* (Ogata, 1976). Cytosine was produced by a mutant of *C. glutamicum*, which was selected for resistance to 2-thiazolealanine, 1,2,4-triazolealanine, and 2-fluoroadenine with a yield of 2.3 g/liter (Kobayashi et al., 1978).

VII. APPLICATION OF GENETIC ENGINEERING TECHNIQUES TO THE PRODUCTION OF AMINO ACIDS

The first successful application of genetic engineering techniques to the production of amino acids was obtained in threonine production with *E. coli* K-12 (Debabov, 1983; Miwa et al., 1983). The results suggested improved amino acid production in coryneform bacteria. For the effective application of these techniques, however, establishment of host-vector systems in these bacteria was necessary. Plasmids were sought as vectors among these bacteria, and various types of plasmids were found and characterized (Kaneko et al., 1979; Miwa et al., 1984; Sandoval et al., 1984). Many of them were cryptic plasmids. Among them, pAM330, found in *B. lactofermentum*, was further analyzed and the entire nucleotide sequence, 4448 bp, was determined (Yamaguchi et al., 1986). A plasmid pCG11, which had a selectable spectinomycin-resistant marker, was found in *C. glutamicum* and was tested for its ability to transform protoplasts of the host strain (Katsumata et al., 1984b). On the other hand, selection markers, such as resistance against chloramphenicol, ampicillin, and kanamycin originating from *E. coli* and *Staphylococcus* plasmids, were found to be expressed in coryneform bacteria. From these results, new shuttle vectors were constructed in various strains by combining cryptic plasmids with those from *E. coli* and other microorganisms (Ozaki et al., 1984; Santamaria et al., 1984; Miwa et al., 1985; Yoshihama et al., 1985). A plasmid vector, pAJ655, with a molecular weight of 10.4 kb was constructed as a hybrid plasmid from pBR325 and pAM330, whereas pAJ1844 with a molecular weight of 9.0 kb was constructed from pBR325 and pHM-1519 found in *C. glutamicum*. Both of these had chloramphenical resistance markers. Self-cloning vectors, such as pAJ228 and pAJ224, for *B. lactofermentum* were also constructed from pAM330 using a trimethoprim resistance marker derived from the chromosomal DNA of a *B. lactofermentum* mutant and a synthetic oligonucleotide polylinker with multiple cloning sites, such as *Bam* HI, *Pst* I, and *Sal* I (Takagi et al., 1985).

A. Threonine and Isoleucine

The first improvement in the production of amino acids with coryneform bacteria was obtained in threonine production by *B. lactofermentum*. A 3.2 kb chromosomal DNA fragment encoding the homoserine dehydrogenase of a threonine-producing mutant, M-15, was inserted into a *Pst* I site of a vector plasmid, pAJ1844. Strain M-15, which produced 15 g/liter of threonine and 12 g/liter of lysine monohydrochloride simultaneously from 10% glucose, was transformed with this hybrid plasmid, pAJ210. In the transformant, AJ12021, the activity of homoserine dehydrogenase was twofold higher than that of M-15, and at the same time, the production of threonine increased up to 25 g/liter but that of lysine monohydrochloride decreased to 1.1 g/liter (Nakamori et al., 1987). Further improvement of the yield was attained by cloning a gene encoding the homoserine kinase of the same strain M-15 by the use of a self-cloning vector pAJ228, which was compatible with pAJ210 and was introduced into a strain AJ12021. In the transformant, the activity of homoserine kinase as well as homoserine dehydrogenase increased remarkably and the production of threonine was 33 g/liter (Morinaga et al., 1987). Improvement in threonine production was also attained in *C. glutamicum* by cloning genes of the wild-type threonine operon of *E. coli* followed by deriving threonine analog-resistant strains. Production by the strain thus obtained was 21 g/liter, but that of the parent was 5.6 g/liter (Katsumata et al., 1984a). Similarly, the genes of the threonine operon of *E. coli* were introduced into an isoleucine-producing mutant of *C. glutamicum*, and isoleucine production increased from 4.6 to 8.4 g/liter. A gene encoding homoserine dehydrogenase of wild-type *C. glutamicum* was self-cloned, and threonine analog-resistant mutants were selected. In the mutant, isoleucine production increased from 4.6 to 11.2 g/liter (Katsumata et al., 1985).

B. Histidine

The *E. coli his G* gene, which codes for phosphoribosyl-ATP pyrophosphorylase, a key enzyme for histidine biosynthesis, was shown to be expressed in a histidine auxotroph of *C. glutamicum*. By deriving 2-thiazolealanine-resistant mutants from a *C. glutamicum* strain that harbored the *his G* gene, plasmids with deregulated *his G* enzyme were obtained. A histidine-producing mutant of *C. glutamicum* was transformed with the plasmid. Histidine production of the obtained transformant was strengthened to 15.3 from 7.6 g/liter in the parent (Mizukami et al., 1984).

C. Tryptophan, Phenylalanine, and Tyrosine

A chromosomal DNA fragment of wild-type *B. lactofermentum*, which complemented a tryptophan synthase-deficient mutant, was inserted into a *Pst* I site of a vector, pAJ1844. A tryptophan-producing mutant was transformed with this hybrid plasmid. In the transformant, tryptophan synthase activity was sevenfold and tryptophan production was threefold higher (4.0 g/liter) than those of the parent (Ishida et al., 1985). By a similar system, DNA encoding 3-dehydroquinate synthase, shikimate kinase, and shikimate dehydrogenase was cloned. Introduction of these genes into tryptophan-producing mutants increased enzyme activities up to 2- to 12-fold over the parent. Tryptophan production increased threefold (4.2 g/liter) higher than the parent (Matsui et al., 1985).

A *Bam* HI-digested fragment of chromosomal DNA of a phenylalanine-producing mutant of *C-glutamicum*, strain K-38, was inserted into a *Bgl* II site of a vector pCG11 with a marker for spectinomycin resistance. The hybrid plasmid thus obtained, which complemented a chorismate mutase deficiency of phenylalanine and a tyrosine double auxotrophic mutant of *C. glutamicum*, was reintroduced into strain K-38. Phenylalanine production by the transformant was 19 g/liter, whereas that by the parent was 11 g/liter (Ozaki et al., 1985). Increased phenylalanine production to 17 g/liter was also reported by introducing the *phe A* gene of *E. coli*, mutagenized for resistance to a phenylalanine analog, into the same phenylalanine-producing mutant (Ikeda et al., 1985). A chromosomal DNA fragment of a phenylalanine-producing mutant of *B. lactofermentum*, which was selected by *m*-fluorophenylalanine resistance and encoded deoxy-D-arabinoheptulosonate-7-phosphate synthase, a key enzyme for aromatic amino acid biosynthesis and insensitive to feedback inhibition, was introduced into a plasmid pAJ1844. A phenylalanine and a phenylalanine and tyrosine producer of *B. lactofermentum* were transformed with the hybrid plasmid. Phenylalanine production of the former increased up to 21 g/liter from that of the parent, 17 g/liter, and production of phenylalanine and tyrosine of the latter increased to 8.9 from 6.2 g/liter and to 9.5 from 7.1 g/liter in the mutant and parent, respectively (Satoh et al., 1985).

VIII. SUMMARY

Strain improvement has played a major role in developing the productivity of these primary metabolite fermentations because of the presence of feedback control, which prevents overproduction in the wild-type strains. As shown by the experimental results described in this chapter, this can be made more effective when a detailed knowledge of the relevant metabolic pathways and their regulation is available and, furthermore, when intermediate strains obtained during the improvement are followed by biochemical analysis of their genetic alteration. Another important development of these fermentations has come from finding and understanding the specific fermentation conditions that make the cells permeable to the products, including acidic amino acids and nucleotides. The remarkable effects of biotin, penicillin, and fatty acid ester surfactants and their mechanisms were first found during these studies on fermentation development.

Some bacterial strains other than coryneform bacteria have also been used for the industrial production of amino acids and nucleotides. In these strains, removal of product degrading ability was essential for better yields. This was, however, not necessary in coryneform bacteria, because of the absence of product degrading ability, at least under the production conditions. This is one of the reasons that coryneform bacteria are used for these purposes. Another favorable characteristic of these bacteria is their strong ability to catabolize industrial carbon sources, including glucose, sucrose, acetate, and hydrocarbons. The same strong abilities would, however, be used for degrading the intermediate metabolites to CO_2 in preference to biosynthesis of primary metabolites at their branch points. Therefore, improvements to the catabolic pathway, as well as to the biosynthetic pathway, are important in improving the productivities, as observed for the lysine and aspartate fermentations.

Genetic engineering techniques were successfully applied to threonine fermentation with *B. lactofermentum*, giving a high production yield that had not previously been obtained by conventional techniques. The new techniques can easily elevate the levels of some biosynthetic enzymes that might be hard to do by conventional methods. On the other hand, removal of feedback control is essential, even for overproduction with the new techniques. Therefore, these techniques complement each other.

REFERENCES

Akashi, T. (1982). Amino acid production and use to improve nutrition of foods and feeds. *Plenary. Lect. CHEMRAWN II*, p. 14.

Araki, K., and Nakayama, K. (1974). Feedback-resistant phosphoribosyl-ATP pyrophosphorylase in L-histidine-producing mutants of *Corynebacterium glutamicum. Agric. Biol. Chem. 38*: 2209-2218.

Araki, K., Kato, F., Arai, Y., and Nakayama, K. (1974). Histidine production by auxotrophic histidine analog-resistant mutants of *Corynebacterium glutamicum. Agric. Biol. Chem. 38*: 189-194.

Araki, K., Takasawa, Y., and Nakajima, J. (1975). Fermentative production of L-proline with auxotrophs of *Corynebacterium glutamicum. Agric. Biol. Chem. 39*: 1193-1200.

Davis, B. D. (1952). Aromatic biosynthesis. IV. Preferential conversion in incompletely blocked mutants, of common precursor of several metabolites. *J. Bacteriol. 64*: 729-748.

Debabov, V. G. (1983). Construction of strains producing L-threonine. *Proc. IVth Int. Sym. Genetics of Industrial Microorganisms*, 1982, Y. Ikeda and T. Beppu (Eds.). Kodansha Ltd., Tokyo, pp. 254-256.

Fazel, A. M., and Jensen, R. A. (1979). Obligatory biosynthesis of L-tyrosine via the pretyrosine branchlet in coryneform bacteria. *J. Bacteriol. 138*: 805-815.

Fujio, T., and Furuya, A. (1983). Production of ATP from adenine by *Brevibacterium ammoniagenes. J. Ferment. Technol. 61*: 261-267.

Fujio, T., Kotani, T., and Furuya, A. (1984). Production of 5'-guanylic acid by the enzymatic conversion of 5'-xanthylic acid. *J. Ferment. Technol. 62*: 131-137.

Furuya, A., Abe, S., and Kinoshita, S. (1968). Production of nucleic acid-related substances by fermentative processes. XIX. Accumulation of 5'-inosinic acid by a mutant of *Brevibacterium ammoniagenes. Appl. Microbiol. 16*: 981-987.

Furuya, A., Ishiyama, J., Sasajima, K., and Kuninaka, A. (1976). Production of other nucleic acid-related substances. In *Microbial Production of Nucleic Acid-Related Substances*, K. Ogata, S. Kinoshita, T. Tsunoda, and K. Aida (Eds.). Halsted Press, New York, pp. 184-205.

Goto, E., Ishihara, M., Sakurai, S., Enei, H., and Takinami, K. (1981). Method for L-phenylalanine production. Japanese Patent Application (Kokai) No. 56-64,793.

Hagino, H., and Nakayama, K. (1973). L-Tyrosine production by analog-resistant mutants derived from a phenylalanine auxotroph of *Corynebacterium glutamicum*. *Agric. Biol. Chem. 37*: 2013-2023.

Hagino, H., and Nakayama, K. (1974). L-Phenylalanine production by analog-resistant mutants of *Corynebacterium glutamicum*. *Agric. Biol. Chem. 38*: 157-161.

Hagino, H., and Nakayama, K. (1975a). L-Tryptophan production by analog-resistant mutants derived from a phenylalanine and tyrosine double auxotroph of *Corynebacterium glutamicum*. *Agric. Biol. Chem. 39*: 343-349.

Hagino, H., and Nakayama, K. (1975b). The biosynthetic control in aromatic amino acid producing mutants of *Corynebacterium glutamicum*. *Agric. Biol. Chem. 39*: 351-361.

Hayashi, T., Tanaka, T., and Kawashima, K. (1979). Immobilization of microbial cells containing NAD kinase. *Biotechnol. Bioeng. 21*: 1019-1030.

Hirose, Y., and Okada, H. (1979). Microbial production of amino acids. In *Microbial Technology*, 2nd ed., vol. 1, H. J. Peppler and D. Perlman (Eds.). Academic Press, New York, pp. 211-244.

Hirose, Y., Enei, H., and Shibai, H. (1979). Nucleosides and Nucleotides. *Annu. Rep. Ferment. Proc. 3*: 253-274.

Ikeda, S., Fujita, I., and Yoshinaga, F. (1976). Screening of L-isoleucine producers among ethionine-resistant mutants of L-threonine producing bacteria. *Agric. Biol. Chem. 40*: 511-516.

Ikeda, M., Ozaki, A., Katsumata, R., and Oka, T. (1985). Molecular breeding of L-phenylalanine producing strains in *Corynebacterium glutamicum* by the cloning of phe A gene from *Escherichia coli*. *Abst. Ann. Meet. (Agric. Chem. Soc. Jpn.)*, p. 30.

Ishida, M., Matsui, K., Miwa, K., Nakamori, S., and Sano, K. (1985). Amplification of tryptophan synthase gene of *Brevibacterium lactofermentum*. *Abst. Ann. Meet. (Soc. Ferment. Technol. Jpn.)*, p. 34.

Kamijyo, H., Mihara, O., and Kubota, K. (1973). L-Histidine production by histidine analog-resistant mutants of glutamate producing bacteria. *Abst. Ann. Meet. (Agric. Chem. Soc. Jpn.)*, p. 112.

Kaneko, H., Tanaka, T., and Sakaguchi, K. (1979). Isolation and characterization of plasmids from *Brevibacterium lactofermentum*. *Agric. Biol. Chem. 43:* 867-868.

Kanzaki, T., Isobe, K., Okazaki, H., Mochizuki, K., and Fukuda, H. (1967). L-Glutamic acid fermentation. Part I. Selection of an oleic acid-requiring mutant and its properties. *Agric. Biol. Chem. 31*: 1307-1313.

Kanzaki, T., Kitano, K., Sumino, Y., and Okazaki, H. (1972). L-Glutamate production with acetic acid by an oleic acid-requiring mutant. Part I. Cultural characteristics. *J. Agric. Chem. Soc. Jpn. 46*: 95-101.

Karasawa, M., Tosaka, O., Ikeda, S., and Yoshii, H. (1986). Application of protoplast fusion to the development of L-threonine and L-lysine producers. *Agric. Biol. Chem. 50*: 339-346.

Kase, H., and Nakayama, K. (1972). Production of L-threonine by analog-resistant mutants. *Agric. Biol. Chem. 36*: 1611-1621.

Kase, H., and Nakayama, K. (1977). L-Isoleucine production by analog-resistant mutants derived from threonine-producing strains of *Corynebacterium glutamicum*. *Agric. Biol. Chem. 41*: 109-116.

Katsumata, R., Hara, M., Mizukami, T., Oka, T., and Furuya, A. (1984a). Molecular breeding of L-threonine producing strains in *Corynebacterium glutamicum*. *Abst. Ann. Meet. (Agric. Chem. Soc. Jpn.)*, p. 248.

Katsumata, R., Ozaki, A., Oka, T., and Furuya, A. (1984b). Protoplast transformation of glutamic acid producing bacteria with plasmid DNA. *J. Bacteriol. 159*: 306-311.

Katsumata, R., Mizukami, T., Hara, M., and Oka, T. (1985). Molecular breeding of L-isoleucine producing strains in *Corynebacterium glutamicum*. *Abst. Ann. Meet. (Agric. Chem. Soc. Jpn.)*, p. 421.

Kikuchi, M., Doi, M., Suzuki, M., and Nakao, Y. (1972). Culture conditions for the production of L-glutamic acid from n-paraffins by glycerol auxotroph GL-21. *Agric. Biol. Chem. 36*: 1141-1146.

Kinoshita, S., Nakayama, K., and Udakd, S. (1957). The fermentative production of L-ornithine. *J. Gen. Appl. Microbiol. 3*: 276-277.

Kinoshita, S., Nakayama, K., and Akita, S. (1958). Taxonomical study of glutamic acid accumulating bacteria, *Micrococcus glutamicus* nov. sp. *Bull. Agric. Chem. Soc. Jpn. 22*: 176-185.

Kitai, A. (1972). Alanine fermentation. In *The Microbial Production of Amino Acids*, K. Yamada, S. Kinoshita, T. Tsunoda, and K. Aida (Eds.). Halsted Press, New York, pp. 325-337.

Kobayashi, K., Ikeda, S., Takinami, K., Hirose, Y., and Shiro, T. (1971). Production of L-glutamic acid from hydrocarbon by penicillin-resistant mutants of *Corynebacterium hydrocarboclastus*. *Agric. Biol. Chem. 35*: 1241-1247.

Kobayashi, S., Araki, K., and Nakayama, K. (1978). Cytosine producing mutants of *Brevibacterium ammoniagenes*. *J. Agric. Chem. Soc. Jpn. 52*: 129-133.

Kotani, Y., Yamaguchi, K., Kato, F., and Furuya, A. (1978). Inosine accumulation by mutants of *Brevibacterium ammoniagenes*. Strain improvement and culture conditions. *Agric. Biol. Chem. 42*: 399-405.

Kubota, K. (1985). Improved production of L-serine by mutants of *Corynebacterium glycinophilum* with less serine dehydratase activity. *Agric. Biol. Chem. 49*: 7-12.

Kubota, K., Shiro, T., and Okumura, S. (1971). Fermentative

production of L-histidine. I. Production of L-histidinol by a histidine auxotroph of *Brevibacterium flavum* and formation of L-histidine from L-histidinol by microorganisms. *J. Gen. Appl. Microbiol. 17*: 1-12.

Kubota, K., Onoda, T., Kamijyo, H., Yoshinaga, F., and Okumura, S. (1973). Microbial production of L-arginine. I. Production of L-arginine by mutants of glutamate producing bacteria. *J. Gen. Appl. Microbiol. 19*: 339-352.

Matsui, K., Miwa, K., and Sano, K. (1988). Cloning of a gene cluster of *aroB, aroE* and *aroL* for aromatic amino acid biosynthesis in *Brevibacterium lactofermentum,* a glutamic acid-producing bacterium. *Agrig. Biol. Chem. 52*: 525-531.

Miwa, K., Tsuchida, Y., Kurahashi, O., Nakamori, S., Sano, K., and Momose, H. (1983). Construction of L-threonine overproducing strains of *Escherichia coli* K-12 using recombinant DNA techniques. *Agric. Biol. Chem. 47*: 2329-2334.

Miwa, K., Matsui, H., Terabe, M., Nakamori, S., Sano, K., and Momose, H. (1984). Cryptic plasmids in glutamic acid-producing bacteria. *Agric. Biol. Chem. 48*: 2901-2903.

Miwa, K., Matsui, K., Terabe, M., Itoh, K., Ishida, M., Takagi, H., Nakamori, S., and Sano, K. (1985). Construction of novel shuttle vectors and a cosmid vector for the glutamate-producing bacteria, *Brevibacterium lactofermentum* and *Corynebacterium glutamicum. Gene 39*: 281-286.

Miyajima, R., and Shiio, I. (1970a). Regulation of aspartate family amino acid biosynthesis in *Brevibacterium flavum.* Part II. The first and second step enzymes of lysine biosynthesis. *Agric. Biol. Chem. 34*: 1275-1282.

Miyajima, R., and Shiio, I. (1970b). Regulation of aspartate family amino acid biosynthesis in *Brevibacterium flavum.* III. Properties of homoserine dehydrogenase. *J. Biochem. 68*: 311-319.

Miyajima, R., and Shiio, I. (1971). Regulation of aspartate family amino acid biosynthesis in *Brevibacterium flavum.* Part IV. Repression of the enzymes in threonine biosynthesis. *Agric. Biol. Chem. 35*: 424-430.

Miyajima, R., and Shiio, I. (1972a). Regulation of aspartate family amino acid biosynthesis in *Brevibacterium flavum.* V. Properties of homoserine kinase. *J. Biochem. 71*: 219-226.

Miyajima, R., and Shiio, I. (1972b). Regulation of aspartate family amino acid biosynthesis in *Brevibacterium flavum.* VI. Effect of isoleucine and valine on threonine dehydratase activity and its formation. *J. Biochem. 71*: 951-960.

Miyajima, R., Ôtsuka, S., and Shiio, I. (1968). Regulation of aspartate family amino acid biosynthesis in *Brevibacterium flavum.* I. Inhibition by amino acids of the enzymes in threonine biosynthesis. *J. Biochem. 63*: 139-148.

Mizukami, T., Katsumatam, R., and Oka, T. (1984). Molecular breeding of L-histidine producers in *Corynebacterium glutamicum*. *Abst. Ann. Meet. (Agric. Chem. Soc. Jpn.)*, p. 249.

Momose, H., Miyashiro, S., and Ooba, M. (1976). On the transducing phages in glutamic acid-producing bacteria. *J. Gen. Appl. Microbiol. 22*: 119-129.

Mori, M., and Shiio, I. (1983). Glutamate transport and production in *Brevibacterium flavum*. *Agric. Biol. Chem. 47*: 983-990.

Mori, M., and Shiio, I. (1984). Production of aspartic acid and enzymatic alteration in pyruvate kinase mutant of *Brevibacterium flavum*. *Agric. Biol. Chem. 48*: 1189-1197.

Mori, M., and Shiio, I. (1985). Synergistic inhibition of phosphoenolpyruvate carboxylase by aspartate and 2-oxoglutarate in *Brevibacterium flavum*. *J. Biochem. 98*: 1621-1630.

Morinaga, Y., Takagi, H., Ishida, M., Miwa, K., Sato, T., Nakamori, S., and Sano, K. (1987). Threonine production by coexistence of cloned genes coding homoserine dehydrogenase and homoserine kinase in *Brevibacterium lactofermentum*. *Agric. Biol. Chem. 51*: 93-100.

Morioka, H., Ishii, K., and Takinami, K. (1977). Microbial production of L-arginine. II. Mechanisms of L-arginine production of L-arginine producing mutants of *Brevibacterium flavum*. *Abst. Ann. Meet. (Agric. Chem. Soc. Jpn.)*, p. 213.

Murata, T., Kato, J., and Chibata, I. (1979). Continuous production of NADP by immobilized *Brevibacterium ammoniagenes* cells. *Biotechnol. Bioeng. 21*: 887-895.

Nakamori, S., and Shiio, I. (1972). Microbial production of L-threonine. Part III. Production by methionine and lysine auxotrophs derived from α-amino-β-hydroxyvaleric acid resistant mutants of *Brevibacterium flavum*. *Agric. Biol. Chem. 36*: 1209-1216.

Nakamori, S., and Shiio, I. (1973). Production of L-threonine by mutants resistant to both α-amino-β-hydroxyvaleric acid and S-(2-aminoethyl)-L-cysteine derived from *Brevibacterium flavum*. *Agric. Biol. Chem. 37*: 653-659.

Nakamori, S., Morioka, H., Yoshinaga, F., and Yamanaka, S. (1982). Fermentative production of L-proline by DL-3,4-dehydroproline-resistant mutants of glutamate producing bacteria. *Agric. Biol. Chem. 46*: 487-491.

Nakamori, S., Ishida, M., Takagi, H., Itoh, K., Miwa, K., and Sano, K. (1987). Improved L-threonine production by the amplification of the gene encoding homoserine dehydrogenase in *Brevibacterium lactofermentum*. *Agric. Biol. Chem. 51*: 87-91.

Nakanishi, T., Yokote, Y., and Taketsugu, Y. (1973). Conversion of L-glutamic acid fermentation to L-proline fermentation by *Corynebacterium glutamicum*. *J. Ferment. Technol. 51*: 742-749.

Nakanishi, T., Taketsugu, Y., and Nakajima, J. (1977). Conversion of L-glutamic acid fermentation to N-acetylglutamine fermentation by *Corynebacterium glutamicum*. *J. Ferment. Technol. 55*: 224-232.

Nakayama, K., and Araki, K. (1982). Fermentative method for lysine production. Japanese Patent Application (Kokoku) No. 57-14,839.

Nakayama, K., and Hagino, H. (1966). Accumulation of L-citrulline by a mutant of glutamate producing bacterium. *J. Agric. Chem. Soc. Jpn. 40*: 377-380.

Nakayama, K., and Yoshida, H. (1972). Fermentative production of L-arginine. *Agric. Biol. Chem. 36*: 1675-1684.

Nakayama, K., Kitada, S., and Kinoshita, S. (1961a). L-Valine production using microbial auxotrophs. *J. Gen. Appl. Microbiol. 7*: 52-69.

Nakayama, K., Kitada, S., and Kinoshita, S. (1961b). Studies on lysine fermentation. I. The control mechanism on lysine accumulation by homoserine and threonine. *J. Gen. Appl. Microbiol. 7*: 145-154.

Nakayama, K., Sato, Z., Tanaka, H., and Kinoshita, S. (1968). Production of nucleic acid-related substances by fermentative processes. XVII. Production of NAD and nicotinic acid mononucleotides with *Brevibacterium ammoniagenes*. *Agric. Biol. Chem. 32*: 1331-1336.

Ogata, K. (1976). Production of nucleosides and nucleotides. In *Microbial Production of Nucleic Acid-Related Substances*, K. Ogata, S. Kinoshita, T. Tsunoda, and K. Aida (Eds.). Halsted Press, New York, pp. 239-247.

Okumura, S., Tsugawa, R. Tsunoda, T., Kono, K., Matsui, Y., and Miyachi, N. (1962). Studies on the L-glutamic acid fermentation. Part I. The new bacteria of the genus *Brevibacterium* isolated from the nature to produce L-glutamic acid. *J. Agric. Chem. Soc. Jpn. 36*: 141-159.

Okumura, S., Shibuya, M., Yoshinaga, F., Maeyashiki, I., and Konishi, S. (1968). Production method of L-ornithine. Japanese Patent 527,567.

Ôtsuka, S., and Shiio, I. (1968). Fatty acid composition of cell wall cell membrane fraction from *Brevibacterium flavum*. *J. Gen. Appl. Microbiol. 14*: 135-146.

Ôtsuka, S., Ishii, R., Shiio, I., and Katsuya, N. (1964). Effect of penicillin on the production of glutamic acid by non-biotin requiring bacteria. *J. Gen. Appl. Microbiol. 10*: 179-180.

Ozaki, H., and Shiio, I. (1983). Production of lysine by pyruvate kinase mutant of *Brevibacterium flavum*. *Agric. Biol. Chem. 47*: 1569-1576.

Ozaki, A., Katsumata, R., Oka, T., and Furuya, A. (1984). Functional expression of genes of *Escherichia coli* in gram positive *Corynebacterium glutamicum*. *Mol. Gen. Genet. 196*: 175-178.

Ozaki, A., Katsumata, R., Oka, T., and Furuya, A. (1985). Cloning of the genes concerned in phenylalanine biosynthesis in *Corynebacterium glutamicum* and its application to breeding of a phenylalanine producing strain. *Agric. Biol. Chem. 49*: 2925-2930.

Sandoval, H., Aguilar, A., Paniagua, C., and Martin, J. F. (1984). Isolation and physical characterization of plasmid pCC1 from *Corynebacterium callunae* and construction of hybrid derivatives. *Appl. Microbiol. Biotechnol. 19*: 409-413.

Sano, K., and Shiio, I. (1967). Microbial production of L-lysine. I. Production by auxotrophs of *Brevibacterium flavum*. *J. Gen. Appl. Microbiol. 13*: 349-358.

Sano, K., and Shiio, I. (1970). Microbial production of L-lysine. III. Production by mutants resistant to S-(2-aminoethyl)-L-cysteine. *J. Gen. Appl. Microbiol. 16*: 373-391.

Sano, K., and Shiio, I. (1971). Microbial production of L-lysine. IV. Selection of lysine-producing mutants from *Brevibacterium flavum* by detecting threonine sensitivity or halo-forming method. *J. Gen. Appl. Microbiol. 17*: 97-113.

Santamaria, R., Gil, J. A., Mesas, J. M., and Martin, J. F. (1984). Characterization of an endogenous plasmid and development of cloning vectors and transformation system in *Brevibacterium lactofermentum*. *J. Gen. Microbiol. 130*: 2237-2246.

Sato, K., and Furuya, A. (1975). Fermentative accumulation of guanosine polyphosphate by *Brevibacterium ammoniagenes*. *Appl. Microbiol. 30*: 480-482.

Satoh, K., Itoh, H., Matsui, K., Miwa, K., Sano, K., Tanaka, T., Nakamori, S., and Enei, H. (1985). Amplification of DAHP synthase of *Brevibacterium lactofermentum*. *Abst. Ann. Meet. (Agric. Chem. Soc. Jpn.)*, p. 123.

Shiio, I. (1978). Microbial production of amino acids and metabolic regulation. *Seikagaku 50*: 1-16.

Shiio, I. (1982). Metabolic regulation and overproduction of amino acids. In *Overproduction of Microbial Products*, V. Krumphanzl, B. Sikyta, and Z. Vanek (Eds.). Academic Press, London, pp. 463-472.

Shiio, I. (1983). Biochemical conditions for amino acid overproduction. *Proc. IVth Int. Sym. Gen. Industrial Microorg., 1982*, Y. Ikeda and T. Beppu (Eds.), Kodansha Ltd., Tokyo, pp. 242-246.

Shiio, I., and Miyajima, R. (1969). Concerted inhibition and its reversal by end products of aspartate kinase in *Brevibacterium flavum*. *J. Biochem. 65*: 849-859.

Shiio, I., and Nakamori, S. (1970). Microbial production of L-threonine. Part II. Production by α-amino-β-hydroxyvaleric acid resistant mutants of glutamate producing bacteria. *Agric. Biol. Chem. 34*: 448-456.

Shiio, I., and Ozaki, H. (1970). Regulation of nicotinamide adenine dinucleotide phosphate-specific glutamate dehydrogenase from *Brevibacterium flavum*, a glutamate-producing bacterium. *J. Biochem. 68*: 633-647.

Shiio, I., and Sano, K. (1969). Microbial production of L-lysine. II. Production by mutants sensitive to threonine and methionine. *J. Gen. Appl. Microbiol. 15*: 267-287.

Shiio, I., and Sugimoto, S. (1976). Altered prephenate dehydratase in phenylalanine-excreting mutants of *Brevibacterium flavum*. *J. Biochem. 79*: 173-183.

Shiio, I., and Sugimoto, S. (1978). Altered regulatory mechanism for tryptophan synthesis in fluorotryptophan-resistant mutants of *Brevibacterium flavum*. *J. Biochem. 83*: 879-886.

Shiio, I., and Sugimoto, S. (1979). Two components of chorismate mutase in *Brevibacterium flavum*. *J. Biochem. 86*: 17-25.

Shiio, I., and Sugimoto, S. (1981a). Effect of enzyme concentration on regulation of dissociable chorismate mutase in *Brevibacterium flavum*. *J. Biochem. 89*: 1483-1492.

Shiio, I., and Sugimoto, S. (1981b). Regulation at metabolic branch points of aromatic amino biosynthesis in *Brevibacterium flavum*. *Agric. Biol. Chem. 45:* 2197-2207.

Shiio, I., and Uchio, R. (1969). Microbial production of amino acids from hydrocarbons. IV. L-Glutamic acid production by *Corynebacterium hydrocarboclastus* R-7. *J. Gen. Appl. Microbiol. 15*: 65-84.

Shiio, I., and Ujigawa, K. (1978). Enzymes of the glutamate and aspartate synthetic pathway in a glutamate-producing bacterium, *Brevibacterium flavum*. *J. Biochem. 84*: 647-657.

Shiio, I., Narui, K., Yahaba, N., and Takahashi, M. (1962a). Free intracellular amino acid of a glutamate-forming bacterium, *Brevibacterium flavum* No. 2247: Analysis and release. *J. Biochem. 51*: 109-111.

Shiio, I., Ôtsuka, S., and Takahashi (1962b). Effect of biotin on the bacterial formation of glutamic acid. I. Glutamate formation and cellular permeability of amino acids. *J. Biochem. 51*: 56-62.

Shiio, I., Ôtsuka, S., and Katsuya, N. (1963). Cellular permeability and extracellular formation of glutamic acid in *Brevibacterium flavum*. *J. Biochem. 53*: 333-340.

Shiio, I., Momose, H., and Ôyama, A. (1969). Genetic and biochemical studies on bacterial formation of L-glutamate. I. Relationship between isocitrate lyase, acetate kinase, and phosphate acetyltransferase levels and glutamate production in *Brevibacterium flavum*. *J. Gen. Appl. Microbiol.* 15: 27-40.

Shiio, I., Miyajima, R., and Nakamori, S. (1970a). Homoserine dehydrogenase genetically desensitized to the feedback inhibition in *Brevibacterium flavum*. *J. Biochem.* 68: 859-866.

Shiio, I., Miyajima, R., and Sano, K. (1970b). Genetically desensitized aspartate kinase to the concerted feedback inhibition in *Brevibacterium flavum*. *J. Biochem.* 68: 701-710.

Shiio, I., Miyajima, R., and Nakagawa, M. (1972a). Regulation of aromatic amino acid biosynthesis in *Brevibacterium flavum*. I. Regulation of anthranilate synthetase. *J. Biochem.* 72: 1447-1455.

Shiio, I., Satô, H., and Nakagawa, M. (1972b). L-Tryptophan production by 5-methyltryptophan-resistant mutants of glutamate producing bacteria. *Agric. Biol. Chem.* 36: 2315-2322.

Shiio, I., Sasaki, A., Nakamori, S., and Sano, K. (1973). Production of L-isoleucine by AHV resistant mutants of *Brevibacterium flavum*. *Agric. Biol. Chem.* 37: 2053-2060.

Shiio, I., Sugimoto, S., and Miyajima, R. (1974). Regulation of 3-deoxy-D-*arabino*-heptulosonate-7-phosphate synthetase in *Brevibacterium flavum*. *J. Biochem.* 75: 987-997.

Shiio, I., Sugimoto, S., and Nakagawa, M. (1975). Production of L-tryptophan by mutants of *Brevibacterium flavum* resistant to both tryptophan and phenylalanine analogues. *Agric. Biol. Chem.* 39: 627-635.

Shiio, I., Mori, M., and Ozaki, H. (1982a). Amino acid aminotransferases in an amino acid-producing bacterium, *Brevibacterium flavum*. *Agric. Biol. Chem.* 46: 2967-2977.

Shiio, I., Ozaki, H., and Mori, M. (1982b). Glutamate metabolism in a glutamate-producing bacterium, *Brevibacterium flavum*. *Agric. Biol. Chem.* 46: 493-500.

Shiio, I., Ozaki, H., and Ujigawa-Takeda, K. (1982c). Production of aspartic acid and lysine by citrate synthase mutants of *Brevibacterium flavum*. *Agric. Biol. Chem.* 46: 101-107.

Shiio, I., Sugimoto, S., and Kawamura, K. (1982d). Production of L-tryptophan by azaserine-resistant mutants of *Brevibacterium flavum*. *Agric. Biol. Chem.* 46: 1849-1854.

Shiio, I., Sugimoto, S., and Kawamura, K. (1984a). Production of L-tryptophan by sulfonamide-resistant mutants. *Agric. Biol. Chem.* 48: 2073-2080.

Shiio, I., Sugimoto, and Toride, Y. (1984b). Studies on mechanisms for lysine production by pyruvate kinase-deficient mutants of *Brevibacterium flavum*. *Agric. Biol. Chem.* 48: 1551-1558.

Shiio, I., Toride, Y., and Sugimoto, S. (1984c). Production of lysine by pyruvate dehydrogenase mutants of *Brevibacterium flavum*. *Agric. Biol. Chem.* 48: 3091-3098.

Shiio, I., Yokota, A., Sugimoto, S. (1987). Effect of pyruvate kinase deficiency on L-lysine productivities of mutants with feedback-resistant aspartokinases. *Agric. Biol. Chem. 51*: 2485-2493.

Shimizu, S., Esumi, A., Komaki, R., and Yamada, H. (1984). Production of coenzyme A by a mutant of *Brevibacterium ammoniagenes* resistant to oxypantetheine. *Appl. Environ. Microbiol. 48*: 1118-1122.

Somerson, N. L., and Phillips, T. (1961). Production of glutamic acid. Belgian Patent No. 593,807.

Sugimoto, S., and Shiio, I. (1974). Regulation of prephenate dehydratase in *Brevibacterium flavum*. *J. Biochem. 76:* 1103-1111.

Sugimoto, S., and Shiio, I. (1977). Enzymes of the tryptophan synthetic pathway in *Brevibacterium flavum*. *J. Biochem. 81*: 823-833.

Sugimoto, S., and Shiio, I. (1983). Regulation of tryptophan biosynthesis by feedback inhibition of the second-step enzyme, anthranilate phosphoribosyltransferase in *Brevibacterium flavum*. *Agric. Biol. Chem. 47*: 2295-2305.

Sugimoto, S., and Shiio, I. (1985). Enzymes of common pathway for aromatic amino acid biosynthesis in *Brevibacterium flavum* and its tryptophan-producing mutants. *Agric. Biol. Chem. 49:* 39-48.

Sugimoto, S., Nakagawa, M., Tsuchida, T., and Shiio, I. (1973). Regulation of aromatic amino acid biosynthesis and production of tyrosine and phenylalanine in *Brevibacterium flavum*. *Agric. Biol. Chem. 37*: 2327-2336.

Takagi, H., Morinaga, Y., Miwa, K., Nakamori, S., and Sano, K. (1986). Versatile cloning vectors constructed with genes indigenous to a glutamic acid-producer, *Brevibacterium lactofermentum*. *Agric. Biol. Chem. 50:* 2597-2603,

Tanaka, H., Sato, Z., Nakayama, K., and Kinoshita, S. (1968). Production of nucleic acid-related substances by fermentative processes. XV. Formation of ATP, GTP and their related substances by *Brevibacterium ammoniagenes*. *Agric. Biol. Chem. 32*: 721-726.

Tanaka, K., Iwasaki, T., and Kinoshita, S. (1960). Studies on L-glutamic acid fermentation. Part V. Biotin and L-glutamic acid accumulation by bacteria. *J. Agric. Chem. Soc. Jpn. 34*: 593-600.

Teshiba, S., and Furuya, A. (1982). Mechanisms of 5'-inosinic acid accumulation by permeability mutants of *Brevibacterium ammoniagenes*. I. Genetical improvement of 5'-IMP productivity of a permeability mutant of *Brevibacterium ammoniagenes*. *Agric. Biol. Chem. 46*: 2257-2263.

Tosaka, O., and Takinami, K. (1978). Pathway and regulation of lysine biosynthesis in *Brevibacterium lactofermentum*. *Agric. Biol. Chem. 42*: 95-100.

Tosaka, O., Hirakawa, H., Takinami, K., and Hirose, Y. (1978a). Regulation of lysine biosynthesis by leucine in *Brevibacterium lactofermentum*. *Agric. Biol. Chem.* *42*: 1501-1506.

Tosaka, O., Takinami, K., and Hirose, Y. (1978b). L-Lysine production by S-(2-aminoethyl)-L-cysteine and α-amino-β-hydroxyvaleric acid-resistant mutants of *Brevibacterium lactofermentum*. *Agric. Biol. Chem.* *42*: 745-752.

Tosaka, O., Takinami, K., Hirose, Y. (1978c). Production of L-lysine by leucine auxotrophs derived from AEC resistant mutant of *Brevibacterium lactofermentum*. *Agric. Biol. Chem.* *42*: 1181-1186.

Tosaka, O., Yoshihara, Y., Ikeda, S., and Takinami, K. (1985). Production of L-lysine by fluoropyruvate-sensitive mutants of *Brevibacterium lactofermentum*. *Agric. Biol. Chem.* *49*: 1305-1312.

Tsuchida, Y., and Momose, H. (1975). Genetic changes of regulatory mechanisms occurred in L-leucine and L-valine producing mutants derived from *Brevibacterium lactofermentum* 2256. *Agric. Biol. Chem.* *39*: 2193-2198.

Tsuchida, T., Yoshinaga, F., Kubota, K., Momose, H., and Okumura, S. (1974). Production of L-leucine by a mutant of *Brevibacterium lactofermentum* 2256. *Agric. Biol. Chem.* *38*: 1907-1911.

Tsuchida, T., Yoshinaga, F., Kubota, K., and Momose, H. (1975). Production of L-valine by 2-thiazolealanine-resistant mutants derived from glutamate producing bacteria. *Agric. Biol. Chem.* *39:* 1319-1322.

Tsuchida, T., Kubota, K., and Yoshinaga, F. (1986). Improvement of L-proline production by sulfaguanidine-resistant mutants derived from L-glutamic acid-producing bacteria. *Agric. Biol. Chem.* *50*: 2201-2207.

Tsuchida, T., Kubota, K., Morinaga, Y., Matsui, H., Enei, H., and Yoshinaga, F. (1987a). Production of L-phenylalanine by a mutant of *Brevibacterium lactofermentum* 2256. *Agric. Biol. Chem.* *51*: 2095-2101.

Tsuchida, T., Kubota, K., Yoshihara, Y., Kikuchi, K., and Yoshinaga, F. (1987b). Fermentative production of L-glutamine by sulfaguanidine-resistant mutants derived from L-glutamate producing bacteria. *Agric. Biol. Chem.* *51*: 2089-2094.

Tsunoda, T., Shiio, I., and Mitsugi, K. (1961a). Bacterial formation of L-glutamic acid from acetic acid in the growing culture medium. I. Culture conditions. *J. Gen. Appl. Microbiol.* *7*: 18-29.

Tsunoda, T., Shiio, I., and Mitsugi, K. (1961b). Bacterial formation of L-glutamic acid from acetic acid in the growing culture medium. II. Growth and L-glutamate accumulation in a chemically defined medium and some other fermentation products. *J. Gen. Appl. Microbiol.* *7*: 30-40.

Tsunoda, T., Tsuchiya, T., Kinoshita, K., Okada, H., and Kawamoto, A. (1961c). Studies on the accumulation of L-glutamine in L-glutamic acid fermentation. *J. Agric. Chem. Soc. Jpn. 35:* 269-275.

Yamaguchi, R., Terabe, M., Miwa, K., Tsuchiya, M., Takagi, H., Morinaga, Y., Nakamori, S., Sano, K., Momose, H., and Yamazaki, A. (1986). Determination of the complete nucleotide sequence of *Brevibacterium lactofermentum* plasmid pAM330 and analysis of its genetic information. *Agric. Biol. Chem. 50:* 2771-2778.

Yamatodani, T., Suzuki, M., and Nakao, Y. (1967). Production of L-proline by auxotrophic mutants of microorganisms. I. Conditions for L-proline production by a histidine requiring mutant of *Brevibacterium* sp. No. 7966. *Amino Acid Nucleic Acid 16*: 126-133.

Yoshida, H., Araki, K., and Nakayama, K. (1979). Mechanisms of L-arginine production by L-arginine producing mutants of *Corynebacterium glutamicum. Agric. Biol. Chem. 43*: 105-111.

Yoshihama, M., Higashiro, K., Rao, E. A., Akedo, M., Sanabruch, W. G., Folletie, M. T., Walker, G. C., and Sinskey, A. J. (1985). Cloning vector system for *Corynebacterium glutamicum. J. Bacteriol. 162*: 591-597.

Yoshinaga, F. (1969). Studies on the fermentative production of L-proline. IV. Mechanisms of L-proline production in *Brevibacterium flavum* 2247 No. 14-5. *J. Gen. Appl. Microbiol. 15*: 387-398.

4

Bioprocess Technology in Industrial Fungi

RICHARD P. ELANDER
Bristol-Myers Company, Industrial Division, Syracuse, New York

I. INTRODUCTION

Human exploitation of fungi has been going on for a long time—starting with the wine and fermented foods of Sumeria and Babylonia in the sixth century B.C.—to the recent cloning of hepatitis B virus outer coat genes in brewer's yeast. Industrial bioprocess technology with fungi is in active ferment and can be characterized using biotechnology jargon as being in a state of high agitation and turbulence and in increasing technology transfer. Recent advances in our understanding of the molecular biology of fungi will further catalyze the growth of fungal technology in both fundamental and applied aspects.

A recent review (Demain and Solomon, 1981) pointed out that what is often lacking in discussions of recombinant DNA and genetic engineering is that biotechnology is not just a new entrepreneurial activity; rather, it is a well-established factor in the present world economy, responsible for annual production in excess of $15 billion (Hacking, 1986).

It is the intent of this chapter to highlight some of the major industrial fungal fermentations with the following major objectives:

- Describe both traditional and current genetic methodologies that have been successful in the commercial exploitation of industrial fungi
- Describe bioprocess development and fermentation scaleup of a variety of industrial processes based on modern genetic and bioengineering technologies

Identify particular areas of biotechnology in which molecular biology methods can be applied to develop and further exploit filamentous fungi

Describe new advances in modern genetic cloning technology that can be used for the synthesis of improved, novel fungal metabolites

II. MAJOR FUNGAL FERMENTATION PRODUCTS

Taxonomically, filamentous fungi comprise a heterogeneous collection of microorganisms (Bennett, 1986). Members of this group occupy a wide range of micro- and macrohabitats, from the heterotrophic saprophytic through symbiotic associations, as in mycorrhyzae and lichens, to parasites.

One factor associated with the biological success of filamentous fungi is their metabolic diversity. This has allowed adaptation to, and exploitation of, unique environments inhospitable to other organisms. This is classically displayed by the extreme specificity of the many fungal pathogens and the uniqueness of the various antibiotics and other secondary metabolites produced by different saprophytic species. Such metabolic diversity has long been identified and well documented (Turner, 1971), but commercial exploitation by the chemical industry has concentrated on a small number of metabolites from an equally small number of species (see Table 1). These limitations are the consequence of several factors. Obviously, the need for a compound must exist at a level that makes profitable production possible. Another, and probably a more limiting factor, has been the nature of the species producing the metabolite. All the species exploited thus far are able to grow vigorously in a defined, inexpensive, artificial culture and are capable of empirical genetic manipulation. Many fungal species do not meet these criteria, and the gene pools of these have been unavailable to researchers. The ability to clone genes from one host and express them in another opens a whole new avenue of exploitation and potential commercialization.

III. FUNGAL GENETICS AND STRAIN IMPROVEMENT

In contrast to the viruses, bacteria, and streptomycetes, the yeasts and filamentous fungi are eukaryotic in that they possess true delimited nuclei with discernible chromosomes. Moreover, a true sexual meiotic mechanism with crossing over and classic Mendelian genetic ratios comparable to those existing in the higher plants and animals occurs in the yeasts and filamentous fungi (Elander and Demain, 1981).

Table 1 Some Examples of Fungi and Fungal Products

Product	Organism
Antibiotics	
Penicillins G and V	*Penicillium chrysogenum*
Cephalosporin C	*Cephalosporium acremonium*
Penicillin N	*Emericellopsis* sp.
Fusidic acid	*Fusidium coccineum*
Variotin	*Paecilomyces varioti*
Griseofulvin	*Penicillium patulum*
Cyclosporin	*Tolypocladium inflatum*
Metabolites other than antibiotics	
Citric acid	*Aspergillus niger*
Itaconic acid	*A. terreus*
Mycophenolic acid	*Penicillium stoloniferum*
Alkaloids	*Claviceps purpurea*
Riboflavin	*Ashbya gossypii*
Gibberellins	*Fusarium moniliforme*
Kojic acid	*Aspergillus oryzae*
β-Carotene	*Phycomyces blakesleeanus*
Fumaric acid	*Rhizopus* sp.
Flavors and fragrances	*Ceratocystis virescens*
Enzymes	
α-Amylase	*Aspergillus orygae*
β-Glucanase	*A. niger*
Cellulase	*Trichoderma reesei*
Dextranase	*Penicillium* sp.
Glucoamylase	*Aspergillus* sp.
Glucose oxidase	*Aspergillus niger*
Lactase	*A. niger*
Lipase	*Aspergillus* sp.
Pectinase	*Aspergillus* sp.
Penicillin acylase	*Fusarium* spp.
Pentosanase	*Aspergillus* sp.
Protease (acid, neutral, alkaline)	*Aspergillus oryzae*
Rennet	*Mucor* sp.
Miscellaneous compounds	
Microbial protein	*Fusarium* sp.
Cheese manufacture	*Penicillium roquefortii*
Microbial insecticides	*Metarrhizium anisopliae*
Steroid conversions	*Rhizopus nigricans*
Nonsteroid conversions	Many species
Double-stranded RNA	*Penicillium chrysogenum*

Table 1 (continued)

Product	Organism
Miscellaneous compounds	
Schizophyllan	*Schizophyllum commune*
Insulin	*Neurospora crassa*
Interferons	*Saccharomyces cerevisiae*
Vaccines	*S. cerevisiae*
Polyols	*Aspergillus* sp.
Ethanol	*Saccharomyces* sp.
Glycerol	*Trigonopsis* sp.
Pullulan	*Aureobasidium pullulans*

A. Life Cycles and Sexuality in Filamentous Fungi

True diploid cells are produced in the life cycles of most fungi. Unlike the partial diploids of bacteria, the homologous chromosome pairs are made up of complete chromosomes. The fungi as a group exhibit a great variety of life cycles that may be categorized into three major types: haploid, diploid, and dikaryon. The dikaryon is found as a transient phase in the sac fungi (Ascomycetes) but may represent the dormant phase in many mushroom and toadstool fungi (Basidiomycetes). The dikaryon results from the delayed fusion of nuclei following anastomosis of two haploid cells.

After cytoplasmic fusion, the binucleate dikaryotic cells divide mitotically to produce two daughter cells, each with the same dikaryotic nucleus.

One of the most extensively studied genetic models is the fungus, *Neurospora crassa*. *N. crassa* is an ascomycetous fungus with a haploid vegetative phase. It is heterothallic and requires two mating types to carry out its complete life cycle. The vegetative mycelium is haploid and can be propagated almost indefinitely by serial transfers of hyphal fragments or asexual spores (conidia) to fresh nutrient media. In *N. crassa*, two types of spores are produced, macroconidia and microconidia. Of the two mating types, A or a, either will produce protoperithecia on appropriate media. After gametangia are produced, the sexual cycle is initiated by fertilization with spermatia or mycelium of the mating type opposite to that of the protoperithecium.

After the nucleus has migrated through a specialized structure known as a trichogyne into the protoperithecium, a dikaryon is formed that is reproduced mitotically to form a large number of dikaryotic cells that become zygotes upon fusion of pairs of the dikaryotic cells. Each fused nucleus undergoes meiosis (reduction division) to form four haploid nuclei, which in turn undergo two successive mitotic divisions to form eight ascospores that are contained linearly in an ascus sac. A perithecium may contain 2-300 asci, each derived from a fused nucleus, and when the perithecia mature, the spores are forcibly discharged from the sac. The ascospores generally require a heat shock that induces germination.

The common baker's yeast, *Saccharomyces cerevisiae*, is a heterothallic ascomycetous budding fungus that has been used for many important genetic studies. Many yeasts are thought to be degenerate sac fungi that exist vegetatively as budding unicells in which the vegetative cell also acts as the sexual cell (ascus). *Saccharomyces* strains reproduce asexually by a process called budding, which is a mitotic division in which equal chromosomal but unequal cytoplasmic division occurs. As in *Neurospora* spp., the diploid zygote of a common yeast undergoes meiosis within a closed sac; hence all products of single meiotic division are kept together. Polyploidy is also common in yeast and arises as the result of a fusion of a haploid and a diploid or of diploid cells. Genetic analyses indicate that polody levels of up to the tetraploid (4N) condition are possible.

Another important ascomycetous fungus that displays a variety of genetic mechanisms is *Aspergillus nidulans*. Like *Neurospora* spp., its vegetative mechanisms confer upon it certain advantages as a tool in genetic investigations. In addition to its regular asexual and sexual cycles, *A. nidulans* possesses unique heterokaryotic and parasexual mechanisms.

The normal asexual cycle consists of the germination of uninucleate conidia that produce typical haploid vegetative mycelia, a portion of which are differentiated into specialized conidial heads. The mature conidia separate and germinate, thereby repeating the cycle.

In contrast to neurosporae, which are usually heterothallic and have two mating types, *A. nidulans* is homothallic and possesses no mating types. Hence, a conidium can initially fertilize an ascocarp. In contrast to *Neurospora*, the ascocarp is a closed, thin-walled globose structure called a cleistothecium. Each cleistothecium contains numerous globose asci containing eight ascospores, which are not arranged in a definite order.

B. Parasexuality

Although *A. nidulans* is generally considered an organism with a haploid vegetative phase, it can diploidize. The ability to form rare diploid spores conveys upon it certain advantages not found in species of *Neurospora*. The unusual diploid phase in the asexual cycle of the fungus is a result of the fusion of somatic vegetative

Table 2 A Comparison of the Sexual and Parasexual Cycles in Fungi

Step	Sexual cycle	Parasexual cycle
1	Hyphal anastomosis (plasmogamy)	Hyphal anastomosis (plasmogamy)
2	Nuclear exchange to form heterokaryon	Nuclear exchange to form heterokaryon
3	Nuclear fusion (karyogamy) in specialized sexual structures to yield "selfed" or "hybrid" ascocarps	Rare nuclear fusion in vegetative cells; heterozygotes recognized by color or nutrition; homozygotes, if formed, are not detected
4	*Diploidization*: Zygote persists for one nuclear generation	*Diploidization*: Zygote may persist through many mitotic divisions
5	Recombination at *meiosis*; crossing over at four-strand stage, in all chromosome pairs, random assortment of each chromosome pair followed by reduction to *haploid* state	Recombination by rare accident of *mitosis*: (1) *mitotic crossing over* at four-strand stage and usually only one exchange in a single chromosome arm; (2) *haploidization*: independent of crossing over and random assortment of each chromosome pair
6	Products of meiosis readily recognized and isolated	Recombinants occur among vegetative cells and recognized only by use of suitable markers and selection techniques

Source: From Elander and Espenshade (1976).

nuclei in the heterozygous condition. Unusual mitotic crossing over occurs at the four strand stage, segregation is random and mitotic, and the expected reciprocal products are found. The somatic recombination cycle has been terms *parasexuality* by Pontecorvo and his associates.

Parasexuality has been discovered in a number of industrially important fungi, including the penicillin fungus, *Penicillium chrysogenum*. The unusual cycle makes it possible to test the effects of genes in the heterozygous condition and has been used to map genes in *A. nidulans*. With respect to recombination efficiency, however, somatic recombination is 0,002 as effective as sexual recombination in *A. nidulans* (Elander and Espenshade, 1976). A comparison of these two important recombinational mechanisms is presented in Table 2.

Filamentous microorganisms also display another unusual method for genetic variation. Many fungi show hyphal anastomosis when they are actively growing in culture. Fusion results in a mixing of both the cytoplasm and the nuclei of the fused hyphae. If the hyphae represent genetically different strains (biochemically deficient, drug-resistant, or conidial color mutants), the fusion may result in the production of a strain that will have nuclei of both strains. Such a strain is called a *heterokaryon*. If each parental strain has a different nutritional requirement due to gene mutation, the heterokaryon will have the wild-type allele of each and will grow on a minimal medium (prototrophic). Frequently, heterokaryotic growth is better than that obtained from homokaryotic strains (Elander, 1979).

Heterokaryosis is of great significance in organisms in which it is found. Many fungi that are capable of forming heterokaryons are probably heterokaryotic in their natural habitats. Heterokaryosis is also an important consideration in the cultivation and maintenance of important higher yielding strains of industrially important microorganisms. Mutation to lower levels of productivity often occurs, which results in "rundown" and lower production yields of valuable production strains. Such strains must be repurified by single-spore selection.

Heterokaryosis is most significant genetically since it allows the establishment of balanced genetic systems that have a selective advantage over homokaryotic strains. Heterokaryosis may be considered a process midway between syntrophism and heterozygosity in a diploid organism. Synthrophism is a relationship between two different strains of an organism resembling symbiosism, which enables both strains to survive in a particular environment whereas neither would survive alone.

The discovery of diploid nuclei in aspergilli and penicillia has resulted in a more profound understanding of recombination potential among fungal organisms. In order to detect rare heterozygous diploid nuclei, an environment that selects for a unique phenotype from an extremely large population of parental strains is necessary. A method commonly employed to demonstrate parasexual recombination is to propagate complementary biochemically deficient strains on a nonselective medium.

Hyphal segments of copropagated mycelium are then plated on a selective medium, a condition that selects for heterokaryotic mycelium. Following anastomosis and fusion of the haploid mutant strains, the conidia of the heterokaryotic mycelium are then examined for the parental types employed for the initial mating. If parental types are found, the mycelium is heterokaryotic. Rare heterozygous

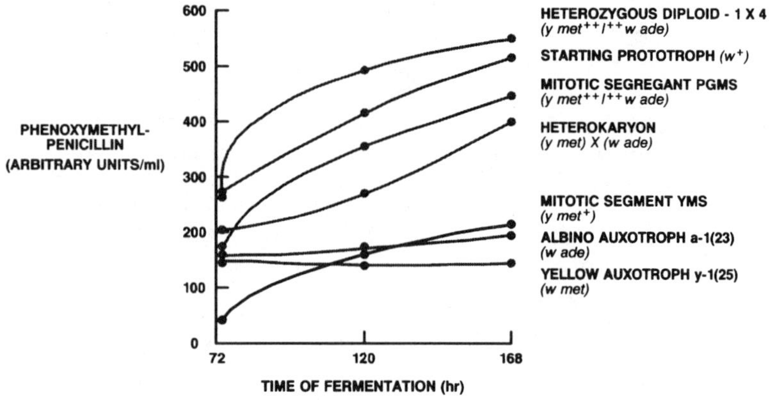

Figure 1 Penicillin production by mutant and recombinant strains of *P. chrysogenum*. (From Elander and Chang, 1979.)

conidia may also be found. Their frequency of occurrence is 1 in 1 million. The spores are characteristically larger than their haploid parents, they may be more resistant to ultraviolet (UV) radiation than haploid cultures, their deoxyribonucleic content is doubled. their conidial color is usually of the wild phenotype, and they are prototrophic with respect to nutritional requirements. The heterozygous diploid nuclei may give rise to recombinants, either by vegetative haploidization (via random assortment of chromosomes) or by mitotic crossing over (Elander and Chang, 1979).

The penicillin productivity of a highly stable diploid strain of *P. chrysogenum* derived from a haploid production strain E-15 is shown in Fig. 1. Spontaneous variation in the diploid and haploid strain was 9.2 and 31.9%, respectively, as assessed by the color types that segregated. The difference was even more striking following exposure to UV irradiation, in which the ratios of the diploid to haploid-derived variants were 11.6 and 41.1%, respectively. In this case, the diploid strain was far more stable than its haploid progenitor.

In contrast, Ball (1982) reported that diploids show instability greater than that of their haploid progenitors but within the same range. The diploid showed twice as many poorly sporulating types as did the haploid, whereas densely sporulating types were increased 10-fold.

Macdonald and Holt (1976) have suggested the use of parental haploids that grow poorly in complex fermentation media to select against the parental segregants arising during fermentation. Ball and Azevedo (1976) suggested the induction of recessive lethals in the diploid so that haploid segregants would be eliminated. It has been suggested that the Lilly diploid may have been a balanced lethal diploid in that it was extremely stable, but even after treatment with UV irradiation, the proportion of viable segregants it produced was only slightly increased. Provided diploid stability and unrestricted parasexual recombination can be achieved, it should be possible to make further selection progress. Elander (1967) described a spontaneous segregant from the Lilly diploid that produced nearly 25% more antibiotic than its parent, which itself yielded better than the production haploid E-15 strain. Ball (1982) has also described recombinant segregants with improved yields.

C. Mutation and Selection

Historically, the strategies used for the manipulation and screening of microorganisms to generate strains that produce large amounts of antibiotics of commercial importance have included, primarily, random mutation and selection. These methods, although successful in most cases, are based on empirical experience and have, in general, a moderate probability of success. They are "hit or miss" methods that require brute force effort, persistence, and skill in the art of microbiology. These older technologies were developed at a time when knowledge of the genetics and biochemistry of antibiotic producer organisms was very limited.

A number of the classic methods have, in recent years, been replaced by more "rational" (that is, directed) selection techniques, such as (1) direct colony selection following overlay biossay, (2) selection for mutants resistant to toxic precursors or to the toxicity of end product, (3) use of analog-resistant mutants that overproduce rate-limiting biosynthetic intermediates, (4) selection for mutants resistant to metallic ions that complex antibiotic molecules, (5) selection of auxotrophic strains deficient for biosynthetic intermediates followed by prototrophic reversion, (6) selection for improved strains after protoplast fusion, and (7) use of antibiotic selection for the isolation of multiple plasmid copy strains.

Since aminoadipic acid, cysteine, valine, and methionine have been shown to be either direct biosynthetic precursors or stimulants of the important β-lactam antibiotics of the penicillin and cephalosporin class, mutants of *Penicillium* and *Cephalosporium* that were resistant to analogs of these and other related amino acids have been isolated and their β-lactam antibiotic productivity examined. In comparison with random selection, mutants resistant to certain methionine or lysine analogs yielded a much higher frequency of superior strains. Although methionine is not a precursor for cephalosporin C biosynthesis, it was hoped that mutants resistant to methionine analogs, such as trifluoromethionine or selenomethionine, would overproduce methionine, thus alleviating the need for added methionine.

Ions of heavy metals, such as mercuric ions (Hg^{2+}), cupric ions (Cu^{2+}), and related organometallic ions, are known to complex with β-lactam-producing organisms at high concentrations. One can theorize that many of the mutants that become resistant to these metallic ions do so by overproducing β-lactam antibiotics to detoxify the metallic ions. Mutants have been isolated for resistance to phenylmercuric acetate (PMA) for improved cephamycin production in *Streptomyces lipmanii* (Godfrey, 1973). At Bristol-Myers Company, we have isolated mutants resistant to inhibitory concentrations of Cu^{2+}, PMA, chromate, Mn^{2+}, and Hg^{2+} and tested them for cephalosporin C production in shake flasks. The results showed that the frequency of superior isolates among mutants resistant to mercuric chloride was higher than that for isolates selected randomly among survivors of ultraviolet treatment. The isolation of mutants resistant to organometallic compounds has been used for the selection of specific types of nutritional mutants.

In addition to the selecting mutants resistant to amino acid analogs, mutants resistant to various metabolic inhibitors have been isolated, which may have altered permeability characteristics, thereby affecting the resulation of secondary metabolites. The results of Bristol-Myers' penicillin V and cephalosporin C yield improvement programs based on rational selection procedures are summarized in Table 3.

D. Protoplast Fusion

Cell and protoplast fusion represent recent significant new approaches to gene transfer in microbial and fungal organisms. Microbial protoplasts have an important value in genetic manipulations because of the relative ease with which they can be prepared and their capability for undergoing reversion to the normal cellular form (Pebergy, 1980). Protoplast fusion also offers an approach to bypass the natural barriers to genetic exchange in nonfertile microorganisms or distantly related microorganisms. Poly(ethylene glycol) (PEG)-induced fusion of protoplasts has been used effectively to force cells of differing genotypes to form transient hybrids or diploids, thus allowing chromosomes to rearrange and DNA sequences to recombine. A further advantage is that high-frequency genetic exchange can occur between more than two parents (Hopwood, 1981). This advantage is applicable to industrial strain improvement programs.

Species of *Acremonium chrysogenum* that overproduce cephalosporin C were obtained by protoplast fusion (Hamlyn and Ball, 1979). The value of protoplast fusion in developing hybrid drugs is enhanced

Table 3 Comparison of Random Versus Directed Selection Procedures in Strains of *Penicillium chrysogenum* and *Cephalosporium acremonium*

Selection procedure	Mutagenesis	Organism	Selection method	% Retained[a]
Random	Ultraviolet, NTG	*P. chrysogenum*	None	0.81
		and		
		C. acremonium		
Directed	Ultraviolet, NTG	*P. chrysogenum*	1. Colony plate	1.36
		and	2. Auxotrophs	5.71
		C. acremonium	3. Haploidization agents	1.59
			4. Mitotic inhibitors	1.33
			5. Amino acid analogs	1.41
			6. Sulfur analogs	3.98
		C. acremonium	1. Methionine alalogs	3.64
			2. Increased methionine sensitivity	
			a. Growth	1.62
			b. β-Lactam synthesis	3.85

[a]Superior on both primary and secondary screening tests.

Source: Adapted from Elander and Chang (1979).

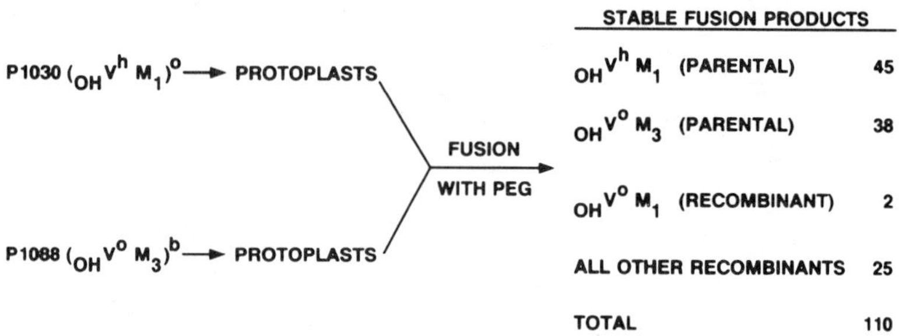

Figure 2 Use of protoplast fusion for the breeding of low ρ-hydroxypenicillin V strains of *Penicillium chrysogenum* (18). M_1 = fast growth; M_3 = slow growth; OHV^h = 8-10% ρ-OHV; OHV^o = 0.05% ρ-OHV; a = strain P1030; b = strain P1088.

by the possibility of interspecific genetic recombination. Interspecific hybrid formation via protoplast fusion was achieved between species of *Aspergillus* and *Penicillium* (Ferenczy, 1981). Fusions between distantly related species have been achieved, indicating that genetic recombination is possible since the morphological and metabolic properties of the hybrids are derived from both parent strains (Peberdy, 1980). Protoplast fusion hybrids of *P. chrysogenum* and *A. chrysogenum* were obtained; they showed, in some cases, *Acremonium* morphology with *Penicillium* outgrowth, suggesting nutritional complementation. Stable isolates were obtained that had *Acremonium* morphology but differed in their response to phenoxyacetic acid compared with the *Acremonium* parent in shake flask fermentations. The prototrophy and altered morphology of these stable fusion products suggest that gene recombination involving *Acremonium* and *Penicillium* has occurred in the hybrids (Peberdy, 1980).

Protoplast fusion technology has also been applied at Bristol-Myers Company for the development of fast-growing, low ρ-hydroxypenicillin V producing strains of *P. chrysogenum* (Chang et al., 1982). Figure 2 shows an application of protoplast fusion methodology for the development of more desirable strains of *P. crysogenum*.

E. Recombinational Breeding

Mutation alters a microorganism's genes. Recombination, the other basic approach to genetic programming, rearranges genes or parts of genes and brings together in an individual organism genetic information from two or more organisms. Recombination results from any of a wide variety of natural processes and laboratory techniques.

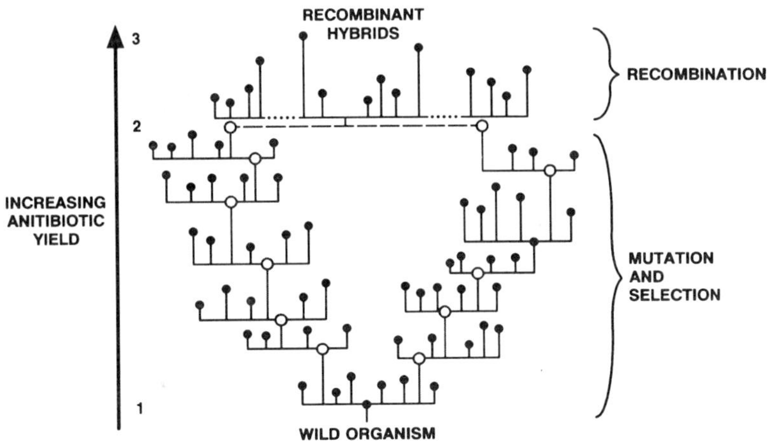

Figure 3 Mutation selection, and recombination for improved antibiotic yield. (From Hopwood, 1981.)

Homologous recombination takes place when bacterial or eukaryotic chromosomes that have similar DNA base sequences, brought together by some mating process, exchange corresponding parts through the breaking and rejoining of DNA. In the case of eukaryotes, sexual reproduction provides another shuffling process, rearrangement, in which the sets of chromosomes derived from two individuals are scrambled.

Homologous recombination is amazingly effective in producing new genotypes or sets of genetic information. If two individuals differ in n genes or parts of genes, recombination between their sets of genes will generate 2^n genotypes. Matings between microorganisms of two strains whose DNA differs in only a dozen base pairs scattered among billions of base pairs can generate 2^{12}, or almost 5000, new genotypes (see Fig. 3). In most cases the parent cells differ more than that, and so astronomical numbers of new combinations arise when they are crossed. Although nearly all microorganisms are probably capable of exchanging genes with related strains, there are rather few instances in which the potential power of natural genetic recombination has been exploited to develop industrial cultures with desirable features derived from more than one strain.

Industrial examples of successes with recombinational breeding following protoplast fusion are the development of low *p*-hydroxypenicillin strains (Chang et al., 1982) and improved cephalosporin C producers (Hamlyn and Ball, 1979).

Table 4 Filamentous Fungal Species for Which DNA-Mediated Transformation has been Reported

Aspergillus nidulans
A. niger
Cephalosporium acremonium
Neurospora crassa
Mucor circinelloides
Penicillium chrysogenum
Podospora anserina

F. Molecular Genetics

Over the past several years, the molecular biology of fungi has been successfully elucidated and molecular techniques have now been refined to the degree that commercial exploitation is now a reality (Elander and Vournakis, 1986; Upshall, 1986).

1. Genetic Transformation and Genetic Engineering Strategies

An essential step in applying recombinant DNA technology to a species is to develop a DNA-mediated transformation system (Table 4). Of these, the most extensively studied are *A. nidulans* and *N. crassa*. Although these are not in themselves commercial species, the methodology developed can and has been transferred across species boundaries and applied to industrial and pathogenic fungi (Kelly and Hynes, 1983). Since most available data are from *Aspergillus* and *Neurospora*, however, this outline primarily concentrates on these two species.

Molecular manipulations are now routine. Efficient DNA-mediated transformation of protoplasts is afforded with both negative and positive selection systems. Genes have been cloned from plasmid (Johnstone et al., 1985) or cosmid libraries (Yelton et al., 1983) following DNA hybridization with equivalent genes from other species by the complementation of yeast mutants with genomic or cDNA libraries and by isolating cDNA clones from induced cultures. Gene disruption, gene replacement, and cotransformation methods have been developed (Kelly and Hynes, 1983). Following transformation and plating on selective medium, transformants are phenotypically of two types: large, well-conidiating and small, producing few conidia. This latter group occasionally grows into the larger type. The large colonies have been interpreted as representing stable transformants in which the transforming DNA has become integrated into the chromosome, usually, but not always, at a homologous location and often in multiple tandem array (Yelton et al., 1984).

There is published evidence for the transformation of two industrial species, *Mucor circinelloides* and *Aspergillus niger* (Heeswisk and Roncero, 1984). In addition, *Cephalosporium acremonium* was recently shown to be transformable (Samson et al., 1985; Penalva et al., 1986). Both *A. niger* and *Aspergillus terreus* have been transformed with the acetamide utilization gene *amdS* (Hynes et al., 1983), a positive selection system. This selection seems to have the potential to initiate the development of DNA transformation systems for a wide range of fungi.

2. Plasmids and Cloning Vectors

One of the major problems in applying recombinant DNA technology to model fungal systems, that is, *Neurospora, Aspergillus*, and others, has been the lack of vectors suitable for the development of efficient cloning systems, that is, plasmids that can replicate autonomously in *Neurospora, Aspergillus*, and *Escherichia coli*. Of particular interest is the development of shuttle vectors in order to carry out gene manipulations, such as the substitution of modified cloned sequences for resident chromosomal sequences.

Stohl and Lambowitz (1984) have constructed a potentially useful shuttle vector, *pALS-1*, which consists of a mitochondrial plasmid, pBR325, and the *ga-2*$^{\pm}$ gene of *Neurospora*. Another shuttle vector, pDV1001, has been reported, but it has been difficult to recover from *Neurospora* transformants (Paietta and Marzluf, 1986). A stable vector, pJP102, containing an autonomously replicating sequence (*ars*) has been described (Stohl et al., 1984). The vector has made possible specific gene disruption using linear fragments carrying the *am*$^{+}$ sequence disrupted by the selectable marker *ga-*$^{2+}$. Weaknesses of the current *Neurospora* system are inherent instability of the plasmids and a low transformation frequency.

Transformation systems and cloning vectors for *Aspergillus* are also available (Yelton et al., 1983; Turner et al., 1985). In contrast to the *Neurospora*, the *Aspergillus* system is characterized by an increased transformation efficiency using an oligomycin resistance gene, *oliC31*, as the selectable marker (Turner et al., 1985). A vector carrying a chromosomal sequence was used to isolate the isocitrate lyase (*acu D*) and fawn (*fwA*) genes by complementation of mutants.

Direct cloning by transformation can be achieved even with moderate transformation frequencies (1000 transformants per μg DNA) now routinely obtained in most laboratories with *A. nidulans*. Multimarked strains are transformed and then screened for complementation of individual nonelected markers. It appears that with the current transformation frequencies reported for *A. nidulans*, any structural or regulatory, metabolic or developmental gene can be cloned in this organism. The technique available for *Aspergillus* cloning should now be applicable to most industrial fungi (Turner et al., 1985).

An efficient transformation system and cosmid cloning vector has been recently reported for the cephalosporin C fungus, *C. acremonium* (Samson et al., 1985). The isopenicillin N synthetase (cyclase) gene was isolated with a 20 mer probe using hygromycin phosphate transferase gene (*HPT*) as the selectable marker. The final cyclase gene cloning vector consisted of an *HPT* gene cloned from *E. coli* placed under control of the phosphoglycerokinase promoter from *S. cerevisiae*, a fragment of *Cephalosporium* mitochondrial DNA that functioned as an *ARS* sequence in yeast and pBR322 DNA, including an origin of replication for easy isolation in *E. coli*. Expression of the cyclase gene in *E. coli* was magnified to the point that nearly 20% of the *E. coli* protein was the cyclase enzyme. When recloned back into *Cephalosporium*, there was an interesting gene dosage effect with a concomitant increase in cephalosporin synthesis. The fungal gene was also expressed in cephamycin-producing strains of *S. clavuligerus* (Samson et al., 1985).

IV. BIOPROCESS TECHNOLOGY

The development of a typical biotechnology process (Fig. 4) includes three steps: elaboration on a laboratory scale, scaleup to a pilot plant level, and final adjustment and run of the process on an industrial scale. A standard production recommendation consists of six trials, usually in one of the smaller production fermentors (10,000-20,000 gal).

A. Laboratory Fermentations

The goals of laboratory research are to develop

1. Culture preservation procedures
2. Culture and propagation on suitable media
3. Nutrient medium of suitable composition
4. Techniques of laboratory-scale culture

The medium for propagation of the strain has to meet several criteria. It has to ensure good growth and/or sporulation and, at the same time, assure satisfactory stability of the strain. Both the strain and its culture conditions, that is, the composition of the nutrient medium, incubation temperature, time of incubation, and so on, should be conceived as forming an integral system that can be severely disturbed by changing only some of the conditions.

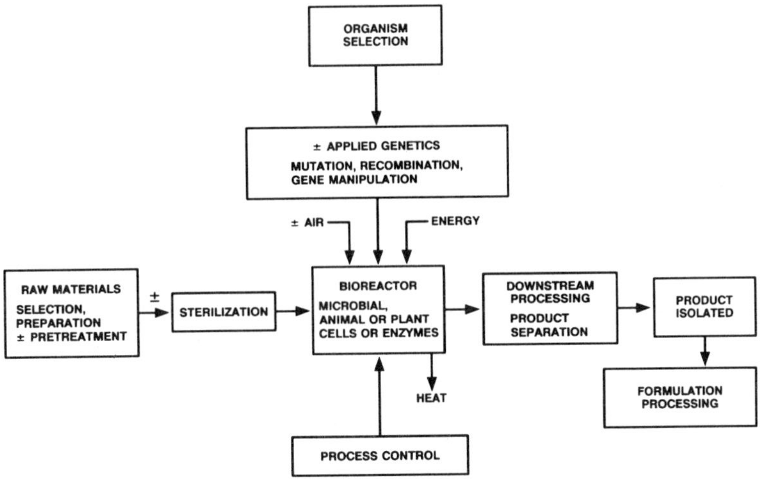

Figure 4 Schematic overview of a biotechnological process.

The appropriate preservation techniques are to be determined by studying the stability of the production strain preserved by different techniques in long-term experiments. Each type of stock culture should be prepared in a sufficient amount even when the preserved strain is not yet optimal for production. This procedure ensures the supply of a uniform inoculum material for sufficient periods of time; the stability of the strain and the efficiency of preservation must be examined in long-term studies over several years.

The composition of the nutrient medium is determined at the initial stages of research; the medium should, above all, ensure reproducible yields. The quality and quantity of raw materials used and the technique of media preparation and sterilization should be adjusted accordingly.

The next step is a detailed study of the composition of the medium that provides the optimum quantitative ratios of individual components and ensures the maintenance of their optimum proportions throughout the process. Apart from productivity, a number of additional factors influence the suitability of a medium for industrial purposes. These include the accessibility and price of individual raw materials, their sterilizability, requirements for an economic isolation procedure, and final purification of the product.

Another step is the elaboration of the process on a laboratory scale when the culture is usually grown on rotary shakers to provide higher aeration of the culture. The optimum aeration regimen for laboratory cultures can be attained by using shakers with different types of motions and changing the shape and closure of the culture flasks and the volume of the nutrient medium. The aeration intensity is to some extent inversely proportional to the amount of the medium in the flask. With optimum aeration intensity defined, other factors are determined that could affect the product yield. Among these are the technique of inoculation with spores or vegetative cells, the size and age of inoculum, and the length of time necessary to attain maximum yield. Another important facet is the determination of strain stability during vegetative transfers.

This scheme forms a basis for the development of a tentative laboratory process. When this is set up successfully, the factors exhibiting a crucial effect on production yields are studied in detail. This cyclic sequence of research operations is widely used for gradually stepping up the production of a given substance to an economically desirable level and, consequently, for determining optimum conditions for industrial production. The immediate goal of the procedure is to provide a basis for the scaleup of the process to the pilot plant level.

B. Pilot Plant and Production-Scale Fermentations

The first task in laboratory and pilot plant development is to generate procedures affording reproducible results. Absolute production yields are of secondary importance, the main emphasis being put on the expected standard course of the microbial process with the attendant changes in growth, pH, consumption of nutrients, production of metabolites, and the physiological state of the microorganism. On attaining reproducible results, the search for individual optimum conditions can begin.

The first factors to be explored are usually aeration and agitation. Next to be examined is the effect of changes in the ratio of individual components of the nutrient medium, with special reference to the source of carbon and nitrogen and their mutual proportions. The effect of different types and amounts of antifoam agents is also studied since the physical action of these agents can affect the dispersion of dissolved gases to the culture and the overall respiration rate of the microbial mass; some antifoam agents can be utilized by the microorganisms, thus changing the metabolism of the culture, the pH of the medium, and the course of the process. The culture fluid from laboratory and pilot plant fermentors can be used for a tentative isolation of metabolites and preliminary tests of their quality.

Further studies include the appraisal of factors crucial for pilot plant and industrial processes, that is, the increase in the mass of seed culture. This entails a study of the amount of the spore or vegetative inoculum and its propagation in several stages. The propagation of a sufficient amount of inoculum is usually achieved by two- or three-stage culture in which the volume of the medium is increased in each subsequent stage. The inoculum produced in the final stage should have sufficient quality to ensure the highest yield in the shortest possible time. During these experiments the composition of the medium should be rechecked.

With a sufficient amount of inoculum in the seed fermentor and with sterile medium kept at the appropriate temperature in the main fermentor, the production fermentor can be seeded with the inoculum. To prevent contamination of the whole system during this operation, the liquids are transferred from one vessel to another using pressurized sterile air. Throughout the inoculation and subsequent growth the contents of the fermentor are stirred and intensively aerated; the volume of air passing through the culture should correspond to 0.3-1.0 volume min^{-1} per volume of the culture. The aeration intensity depends on a number of factors, such as the size and shape of the impeller and agitator speed, the number of baffles in the fermentor, and the composition of the medium.

Under suitable conditions after inoculation, the culture proliferates and the pH may shift spontaneously from values optimal for rapid growth to levels sustaining the biosynthesis of the product. Any irregularity in the process attests to a disturbance of the desired equilibrium between aeration, agitation, and the composition of the medium. In batch culture, the lag phase usually takes up 15-30% of the time needed to attain maximum product formation. When product formation is unconnected with growth, the maximum yield is reached 10-20 hr earlier; thus during the biosynthesis of penicillin, production begins after 15-20 hr and the maximum rate of penicillin accretion attained after 70-75 hr. Although seemingly advantageous, prolonged culture leading to maximum yields is often uneconomical; the economic aspects of both the yields and prolonged culture have to be taken into account when devising a technologically suitable procedure.

The optimum temperature for the culture, which depends on the type of microorganism, should usually be maintained throughout the process. The optimum temperature for many fungi is 24-28°C. The composition of the nutrient medium should be controlled and its individual components always brought to the optimum level. Suitable medium composition ensures desirable pH trends, although if the pH deviates from the optimum level, it can be adjusted by acids or alkalies.

The testing and measuring techniques and procedures used to control the course of the process include

1. Monitoring the fermentation parameters, such as medium temperature, aeration, impeller function, pH, defoaming, and dissolved oxygen.

2. Checking culture development by microscopic techniques. The course of the process may also be appraised from changes in the composition of the medium and from the rate of synthesis of a given product.

C. Problems Associated with Scaleup

The scaleup of culture procedures from the laboratory flask level to pilot plant vessel and eventual production fermentors involves many problems, one of which is the decrease in yields observed in fermentors compared with the laboratory process. The main reason is the difference in conditions: aeration and agitation techniques, sterilation of the medium, inoculation, foam breaking, and type of construction materials. Whereas hundreds of experiments can be carried out in a flask culture, the repetition of experiments in fermentors is usually somewhat limited.

To be able to reproduce laboratory results in fermentors in the shortest possible time, the reason for the differences in yields must be established. Parallel trials are therefore carried out in flasks on shakers and in fermentors, a close check being kept on the differences between the two runs. A prerequisite of a successful parallel experiment is satisfactory reproducibility of results in the two systems and the use of uniform raw materials, salts, water, and so on. It is therefore expedient to perform these experiments in all cases in the same standard apparatus.

Possible differences may form a basis for setting up the following schedule for an optimum scaleup:

1. Standard laboratory procedures.
2. Checking of preparation of the nutrient medium; medium is prepared in fermentors, sterilized in the laboratory, and inoculated with shake flask inoculum.
3. Checking of medium sterilization; medium is prepared and sterilized in fermentors and inoculated with laboratory inoculum prepared in flasks on shakers.
4. Checking of inoculum prepared in fermentors; medium is prepared and sterilized in the laboratory and inoculated with seed prepared in fermentors.
5. Checking of the culture process; medium is prepared and sterilized in fermentors and inoculated with inoculum prepared in fermentors.

Samples of the culture are taken at regular intervals and further cultured in flasks. Such procedures are often referred to as satellite experiments and can also be carried out in pilot-scale vessels. Scaleup problems are encountered in the following data set:

Step	1	2	3	4	5
Production (%)	100	100	60	100	60

These results indicate that the lower yields in fermentors may be due to differences in the mode of medium sterilization; medium sterilization in fermentors should therefore be checked. The conventional procedure in this case is to measure the time necessary for sterilizing the medium and to check the sterilization temperature. If sterilization is carried out batchwise in a central vessel, samples of the medium are taken during the run, inoculated, and cultured in the laboratory. An analogous procedure is adopted when the decrease in yields is brought about by other factors, such as differences in the preparation of the seed culture; samples are taken during the growth of the culture in the seed fermentor and used as inoculum for a medium in flasks for a subsequent test on a shaker.

Reduced yields are sometimes due to several factors; likewise, the yields in all experimental variants may be lower than in the standard laboratory procedure. In this case, additional factors should be checked, such as foam formation and breaking. Careful parallel fermentations can aid in the scaleup of results from the laboratory to the pilot plant and industrial levels with minimum time and cost.

V. REPRESENTATIVE FUNGAL FERMENTATIONS

A. Enzymes

1. Cellulase

This enzyme is produced by a number of fungi. *Trichoderma reesei* has been the most extensively studied from the point of view of mutant isolation (Montencourt and Eveleigh, 1977). Agar plate screening techniques were developed to allow the isolation of a range of mutants of *T. reesei* capable of synthesizing cellulase under conditions of high catabolite repression (Montencourt and Eveleigh, 1977). One of these mutants (NG-14) produced five times the cellulase degrading activity per milliliter of culture medium and twice the specific activity per milligram of excreted protein in submerged culture when compared with its ancestor. This mutant was isolated in the presence of 5% glycerol on agar plates, but in similar liquid medium, NG-14 exhibited only partial derepression of the cellulase complex. Since the proportions of filter paper activity and various enzymes of the cellulase complex were not the same in the mutant as in the ancestral strain, and the yields of each enzyme under conditions repressive for cellulase synthesis were different, differential control of each enzyme of the cellulase complex was implied.

In later studies, Montencourt and Eveleigh isolated a catabolite repression-resistant mutant (C-30) from strain NG-14 (Montencourt and Eveleigh, 1985). C-30 synthesized cellulose degrading enzymes, cellobiohydrolase, endoglucanase, and β-glucosidase when grown in combination with 1% cellulose and either 5% glucose or 5% glycerol as catabolite repressors. C-30 did show enhanced cellulase production relative to its immediate parent NG-14 but still required cellulose as an inducer and was thus not a constitutive cellulase producer.

Recently there have been exciting biochemical advances related to fungal cellulases. The complete amino acid sequence of one of the cellobiohydrolases (CBHI) has now been determined (Fägerstam et al., 1984). The CBHI gene of *T. reesei* has been cloned, and its complete nucleotide sequence has been determined (Shoemaker et al., 1983). The gene was found to contain two introns within the coding region.

2. Protease

A particularly notable example of the use of genetic breeding to increase protease yield is the study of Ikeda and coworkers (1957) in which protease yields in *Aspergillus oryzae* were increased at least 30% in certain strains of higher ploidy, namely, in diploid, in one triploid, and in one tetraploid strain. Again, this work, carried out at the same time as the kojic acid study, was done with sister strains so the improvements observed were probably due to higher ploidy per se or to some unusual interaction of the markers used in the cross.

3. Glucoamylase and α-Amylase

In a study with *Aspergillus foetidus* (Chang and Terry, 1973), diploidization did not five rise to improved yields of enzyme. In other studies with the same fungus, an instability in the strain was described with analogy with instabilities described in *A. nidulans* (Chang and Tuveson, 1975).

Ball and coworkers (1978) studied glucoamylase production by *A. niger*. After two cycles of recombination, they produced a strain with improved industrial potential. A high-yielding strain was crossed, parasexually, with a low-yielding strain from the same genealogy. Recombinants were produced having the efficient broth filtration characteristics of the low-yielding strain. One such segregant was superior for the industrial production of amyloglucosidase to the best parent strain previously used for that purpose.

Innis and coworkers (1985) have recently cloned and expressed the glucoamylase gene from *Aspergillus awamari* in *S. cerevisiae*.

A. niger has been shown to have a single glucoamylase gene that results in two products, G1 and G2 (Kurushima et al., 1972). The G1 and G2 producing genes were poorly expressed in yeast and required the presence of a special promoter and terminator.

Kurushima and coworkers (1972) obtained a mutant of *Aspergillus cinnamoneus* with a sixfold increase in production of acid-stable α-amylase; Thorbek and Eplov were unsuccessful in increasing yields in *A. oryzae* by diploidization (Thorbek and Eplov, 1974).

4. Enzymes Associated with β-Lactam Antibiotic Manufacture

Immobilized fungal enzymes from *Bovista* and *Fusarium* (Lowe et al., 1986) have now been established as important manufacturing tools in the large-scale production of β-lactam antibiotics (Lowe, 1985). Immobilized penicillin acylase enzymes currently account for more than half the 6-aminopenicillanic acid (6-APA) produced and can yield up to 250 kg 6-APA per kg enzyme activity used. Similar enzyme systems can be used to yield 7-aminodesacetoxycephalosporanic acid (7-ADCA) from ring-expanded penicillins. Penicillin acylases and associated enzymes can also function in the reverse direction to synthesize semisynthetic penicillins and cephalosporins (Savidge, 1984). 7-Aminocephalosporanic acid (7-ACA) appears to be more difficult to obtain by enzymic means, but industrially useful enzymes capable of hydrolyzing the fermented cephalosporin C to 7-ACA have been reported in Japan. Enzyme activities have been patented that yield 7-ACA from N-acylated cephalosporin C and from a deaminated product, glutaryl-7-ACA (Shibuya et al., 1981). The ester group on cephalosporin C can also be hydrolyzed by enzymic means to yield the desacetyl derivatives, useful intermediates in the synthesis of certain injectable cephalosporins.

B. Organic Acids and Other Miscellaneous Metabolites

1. Citric Acid

Citric acid is the most important organic acid produced by industrial fermentation and is widely used in the food and pharmaceutical industries. It serves as an acidulant in soft drinks, a buffer in jellies, jams, and desserts, an antioxidant of fats, an emulsifying and stabilizing agent, a detergent, a blood anticoagulant, and a component of effervescent tablets and powders. Other industrial uses of citric acid involve electroplating processes, plasticizers, water conditioners, and secondary oil recovery.

Since Wehmer's (1903) first unsuccessful attempts to commercially exploit the fermentation of sugar to citric acid by penicillia, industrial citric acid production has evolved from a process dependent on the precipitation of 10,000 tons/year of the calcium salt from pressed citrus fruit to large-scale fungal fermentations that today generate more than 283,000 metric tons annually in the United States, Western Europe, and Japan (Strauss, 1981). The ability to form citric acid is not uncommon among fungi, especially the penicillia and aspergilli. A patent by Zaborsky first revealed *A. niger*, a conidial fungus originally described in connection with gallic acid production, to be a citric acid producer and let to refinements that allowed yields of over 60% after several weeks of fermentation in surface culture. Present-day fermentations with selected strains of *A. niger*, however, employ submerged cultures that substantially reduce fermentation times, utilize cheap raw carbohydrate-based media with meticulously controlled phosphorus, metal ion, and nitrogen concentrations, and yield up to 90% (w/w) citric acid from sucrose (Kubicek and Röhr, 1982).

Although many *Aspergillus* species produce some citric acid, *A. niger*, which represents a morphologically and biochemically diverse group, is generally accepted as the most useful producer. Ikeda (1961) reported a 60% improvement in a tetraploid strain, and Shcherbakova and Rexvaya (1977) showed a 50% increase using diploid strains of *A. niger* induced by colchicine. In no cases were the constructed strains superior to the most productive haploid strains. Based on crosses with low-yielding auxotrophic haploid strains of *A. niger*, Chang and Terry (1973) concluded that nuclear genes are primarily responsible for citric acid production and that gene dosage is not responsible for the observed increases in citric acid yield in somatic diploids. They proposed that the increased yields in diploids or heterokaryons were due to intergenic complementation that corrected for the effects of pleiotropic mutations causing not only the nutritional deficiencies but also impaired citric acid metabolism in the parental haploids. Significant atrain improvement has been achieved after mutagenesis of *A. niger* and then screening for elevated production of citric acid.

Selective breeding after chemical mutagenesis and five stages of selection has generated *A. niger* strains that produce 70% more citric acid but only one-half the amount of oxalic acid, an undesirable by-product (Golubtsova et al., 1979). It is surprising, however, that after many decades of research no selection procedures for *A. niger* or genetic loci have been described that involve specific steps of citric acid metabolism in this organism. More recent approaches with morphological, auxotrophic, and benomyl-resistant mutants and their revertants may eventually allow association of genetic loci with the citric acid excretion phenotype (Kubicek and Röhr, 1986).

The advent of new genetic technology will undoubtedly have an impact on the exploitation of *A. niger* for industrial chemicals and enzymes. Genetic engineering approaches should eventually make citric acid production more economical. The transfer to *A. niger* of genes involved in the breakdown of cellulose, starch, or other inexpensive carbon sources would substantially lower the cost of organic acid production. Equally important would be the identification and cloning of fungal genes that play a role in citric acid overproduction and the construction of new production strains by plasmid transfer or cell fusion (Kubicek and Röhr, 1986).

2. Itaconic Acid

Itaconic acid, a substituted methyl acrylic acid formed from *cis*-aconitic acid by various fungi, is marketed as a refined product or as esters and has been used in the manufacture of plastics such as resin coatings, and synthetic detergents. It can also be obtained by pyrolysis of citric acid. The microbial synthesis of itaconic acid was first described by Kinoshita in *Aspergillus itaconicus* and later in *A. terreus, Ustilago zeae*, and *Helicobasidium mompa* (Bigelis, 1985). Succinic acid and tartaric acid are common by-products of itaconic acid production. Fermentation processes for itaconic acid have been reviewed elsewhere (Kubicek and Röhr, 1982, 1986). High-yielding strains of *A. terreus*, the organism used in industrial fermentations, have been isolated from nature and also improved after irradiation of conidia and screening. More recent studies report the formation of itaconic acid by wild-type and mutant strains of *Candida* and raise the possibility of industrial production by submerged cultures of yeast (Tabuchi et al., 1981).

3. Gluconic Acid

About 45,000 metric tons of gluconic acid, an oxidation product of glucose, are manufactured annually worldwide, mainly as sodium gluconate, a versatile sequestering agent, or as the calcium salt, a pharmaceutical used in treating calcium deficiencies. It is also made as the free acid, a noncorrosive cleaning agent, or δ-gluconolactone, a slow-acting acidulant. Fermentation processes are competitive with electrochemical synthesis.

Gluconic acid is produced by aspergilli and penicillia, in addition to other fungi. The first large-scale production of gluconic acid used *Penicillium purpurogenum* var. *rubri-sclerotium* and then *P. chrysogenum* in surface culture. *A. niger*, the organism first found to produce this compound, is the one utilized today in submerged fermentations. A recent review on strain improvement reports that present strains function at near-maximum theoretical efficiency (Kundu and Das, 1982).

4. Mycophenolic Acid

This compound is produced by *Penicillium stoloniferum*. Polyene antibiotic-resistant mutants in this fungus with altered yields of mycophenolic acid and ergosterol have been described (Queener et al., 1982). The precursor of both compounds is farnesyl pyrophosphate, which also serves as a precursor for ergosterol, the major sterol in fungi and yeasts. Increased ergosterol levels were correlated with increased mycophenolic acid production (Queener et al., 1982).

Mycophenolic acid is an interesting metabolite in that it has antifungal, antibacterial, antiviral, and antitumor activity. It has also been reported to have potential for the treatment of psoriasis (Queener et al., 1982).

5. Ergot Alkaloids

Mutation work to increase the yields of alkaloids produced by different strains of *Claviceps* sp. has been described (Rehaak, 1985). Spalla and Marnati (1978), working with different strains, highlight a problem of heterokaryosis necessary for alkaloid production. They describe protoplast fusion experiments between strains of *Claviceps purpurea* that produce high levels of ergocristine with those that produce a high level of ergocornine and ergocryptine. The cell fusion culture following protoplast fusion produced intermediate levels of these alkaloids. The fusion product was considered a heterokaryon.

6. Gibberellic Acid

A description of mutants with enhanced yields of gibberellin phytohormones in *Fusarium moniliforme* has been given (Jefferys, 1970), and work with this fungus, attempting to exploit the parasexual cycle to improve the yield of gibberellic acid, has been reported with the claim that recombinants with improved yields can be obtained (Calam et al.,1973). The gibberellins are derived from gibberellan, and over 45 analogs have been discovered. The most important compound, GA_3, is better known as gibberellic acid. *F. moniliforme*, the imperfect stage of the fungus, *Gibberella fujikuroi*, also produces the antibiotic bikaverin along with the gibberellins during active fermentation. Six physiologically distinct phases have been characterized for the production process:

1. Lag phase
2. Growth phase (24-36 hr) without nitrogen limitation, low production of gibberellin
3. Glycine limitation, little cell reproduction, slow gibberellin production
4. Glycine absent, remaining glucose catabolized, strong gibberellin production
5. Glucose content zero, slow accumulation of gibberellins
6. Lysis of cells, increase in pH

A low nitrogen content (<0.2%) and a mixture of different carbon sources are two important conditions for gibberellin production. Yields in production-scale fermentors after 120-150 hr are 1-2 g/liter of total gibberellin at 25°C, with the aeration rate of 0.5 vvm. Continuous processes have been described, but have not become widely used.

C. Vitamins

An abundant journal and patent literature documents the use of microorganisms for the production of vitamins, especially those of the B group, including biotin, folic acid, pantothenic acid, pyridoxine, riboflavin, thiamine, and vitamin B_{12}. Of all microbial processes that have been described, however, only two have been successfully commercialized: the production of vitamin B_{12} by certain bacteria and the production of riboflavin by the ascomycete fungi *Eremothecium ashbyii* and *Ashbya gossypii* (Yameda et al., 1971; Perlman, 1979).

1. Riboflavin

Present-day microbiological production of riboflavin, or vitamin B_2, with *A. gossypii* competes commercially with chemical synthesis. The annual production of more than 1.25 million k is used to supplement the human diet and feedstuffs. A number of bacteria and fungi can produce riboflavin, but *E. ashbyii* and *A. gossypii* are the most important since these fungi can make up to 2.5 and 7 g/liter, respectively, in industrial fermentations (Perlman, 1979). Details regarding the processes and pathways for microbial production of this vitamin have been discussed elsewhere (Perlman, 1979).

Hyperproducing mutants of *A. gossypii* used in industrial fermentations have internal concentrations of the flavin nucleotides similar to those in wild-type strains but have much more intracellular riboflavin. Early researchers generated such flavinogenic strains with various mutagens and reported a correlation between pigmentation and overproduction. Analysis of mutants has also indicated a correlation between sporulation and flavinogenesis. The best strains sporulated well, and in one study, all mutants incapable of overproducing riboflavin did not sporulate at all, suggesting that common factors may play a role in regulating both processes. Early studies employed sodium dithionate to eliminate low-yielding mutants in selection experiments, recommended frequent reisolation of flavinogenic strains, and advised immediate preservation of useful strains to cope with instability inherent in riboflavin-producing mutants.

Extensive studies on fermentation processes have complemented genetic manipulation of *A. gossypii*. Production by flavinogenic mutants is enhanced by a number of compounds, among them purines and surface-active agents (Lago and Kaplan, 1981).

2. Vitamin A

Carotenoids, a class of pigments considered secondary metabolites, are often precursors of vitamin A when ingested by animals and are added to a variety of foods and feed as a coloring agent. Although such additives are synthetic, much interest has focused on *Blakeslea trispora*, a heterothallic fungus of the Murcorales group that is extremely rich in mycelial β-carotene, as a potential source of carotenoids. The yield of β-carotene is as much as 1 g/liter of medium or 20 mg/g of dry mycelia when both sexual forms are grown together. *Phycomyces blakesleeanus*, a fungus amenable to genetic manipulation, also produces substantial amounts of β-carotene. Heterokaryotic strains produce 0.5 mg of β-carotene per gram dry weight of cells but are unstable, segregating the components homokaryotically. Recently, superproducing strains of *P. blakesleeanus* have been isolated that synthesize 25 mg of β-carotene per gram dry weight of cells (Murillo et al., 1983). Future markets and prices will dictate the feasibility of microbiological production of carotenoids as a source of coloring agents and vitamin A.

3. Vitamin D

Sterols, which are secondary metabolites but essential for cellular functions, are widely distributed among the fungi. Ergosterol, a commercially important lipid, can be extracted from dehydrated yeast, irradiated with ultraviolet light, and thereby converted to vitamin D. Efficient ergosterol-producing yeast strains have been isolated by screening natural isolates, hybridization and polyploidization, and mutagenesis followed by enrichment with echinocandin and nystatin (Parks et al., 1982).

VI. FUNGAL β-LACTAM ANTIBIOTIC FERMENTATION: PENICILLINS AND CEPHALOSPORINS

A. Biology of the Producer Organisms

The initial discovery of Fleming's contaminating *Penicillium* fungus and its remarkable antibacterial properties now date back more than 50 years. Although the Oxford group proposed in 1943 that penicillin contained a fused β-lactum ring, its structure was not generally accepted until it was confirmed by the x-ray crystallographic studies of Hodgkin and Low in 1945. The reactivity of the β-lactam ring is undoubtedly one of the major factors responsible for its intrinsic antibacterial activity.

Exploitation of the second major class of β-lactam substances from fungi began with the discovery of the cephalosporin-producing organism, *A. chrysogenum*, by Brotzu in Sardinia some 30 years ago and the chemical isolation of cephalosporin C as a contaminant in a crude preparation of penicillin N by Newton and Abraham. Cephalosporin C was found to be resistant to acid and to enzymatic hydrolysis and was shown to be more active than the natural penicillins against gram-negative bacteria. These characteristics attracted the attention of clinicians because of the increasing incidence of bacterial resistance to penicillins. A major impetus for the commercial exploitation of cephalosporin C was the discovery of a practical method for chemically splitting cephalosporin C to 7-aminocephalosporanic acid (7-ACA). This stimulated the development of the semisynthetic cephalosporins, which was analogous historically to the rapid progress in chemotherapy induced by the discovery of 6-aminopenicillanic acid nucleus (6-APA) and its semisynthetic derivatives. Similarly, the rapid start in the development of superior semisynthetic cephalosporins over the last decade parallels the history of the semisynthetic penicillins over the past two decades.

In the past 40 years, penicillins have been developed from unstable, low-purity laboratory extracts into multiproduct, multinational, multibillion dollar industries using the services of over 20,000 people in their direct manufacture. Today there is sufficient fermentation and production capacity worldwide to provide each individual on this planet with sufficient penicillin for one therapeutic treatment per year.

The market for penicillins and cephalosporins has reached maturity in the developed countries, with worldwide sales in excess of $5 billion (U.S.) in 1983 (Hacking, 1986). Future expansion will depend upon penetration into developing countries, which at the moment is restricted by currency exchange and low per capita income. With patent expiration and the increase in the availability of high-producing mutant strains and relevant manufacturing technology, developing countries should be able to establish their own processes.

The occurrence of microbial metabolites containing the β-lactam ring is widespread. The distribution of microorganisms known to elaborate β-lactam antibiotics is large and now includes a variety of bacterial genera (Elander, 1983).

B. Strain Improvement

Pharmaceutical companies have carried out extensive mutation and selection programs to screen for mutant cultures of *P. chrysogenum* with higher productivities. These independent programs have created a wide variety of subspecies, each unique in its ability to perform optimally in its specific manufacturing environment. Strain improvement programs are considered proprietary information. However, some details have been made available in the reviews (Elander and Vournakis, 1986; Lein, 1986). Panlabs, Inc., has had a nonexclusive penicillin strain development program in operation for many years. Nearly all of today's industrial strains were derived ultimately from the Wisconsin Q-176 and Wisconsin 51-20 improved strains.

Mutations in *P. chrysogenum* are generally induced by ultraviolet irradiation or *N*-methyl-*N'*-nitro-*N*-nitrosoguanidine (NTG). The former has the advantage of being cleaner to use and avoids the handling of potent carcinogenic material. However, NTG has an extremely high mutagenic effect with relatively low kill and its mechanism of mutagenesis complements that of UV radiation. Combinations of these treatments can also be effective.

Emphasis in strain improvement has been directed toward titer increase, with secondary emphasis on mycelial vigor and sporulation and germination efficiency. The latter properties are important for strain preservation and inoculum propagation. Initially, selection techniques were empirical, relying on the use of large numbers to identify positive mutations. However, more rational screening procedures are now considered more effective. Such techniques are based on known or probable biochemical mechanisms, and many can be adapted to agar prescreen technology (Elander and Chang, 1979). Typical selection techniques are summarized in Table 3. Today's cultures are capable of producing in excess of 30 mg/ml penicillin G (Lein, 1986).

A strain improvement program was initiated in the 1950s to improve the low levels of cephalosporin C in Brotzu's strain of *A. chrysogenum*. Mutagenesis of this isolate resulted in the selection of a mutant, M-8650, which was the progenitor strain for many industrial programs. An improved mutant (CW-19) developed at Eli Lilly and Co. produced threefold more antibiotic than the Brotzu culture. When CW-19 was fermented under more favorable conditions, the culture synthesized 15 times more antibiotic than the progenitor strain. The CW-19 variant also had a significantly improved cephalosporin C-penicillin N ratio. The CW-19 mutant has been used for a number of biosynthetic studies. Biometric considerations of the data using normal populations of UV survivor strains showed an 11:1 advantage for "normal" clones versus "abnormal" clones in searching for mutants producing 20% higher antibiotic titers. The probability statements were calculated for many survivor populations on a statistical basis (Elander, 1983). The improved UV variants differed markedly from the progenitor strains in cultural and biochemical properties. Untreated populations of the improved variants showed a progressive reduction in colony diameter, decreased vegetative development, and decreased sporulation vigor, features that were also characteristic of the improved penicillin variants (Elander, 1979). The genealogy of improved industrial strains of *C. acremonium* is presented in Fig. 5. Strain CW-19 was developed

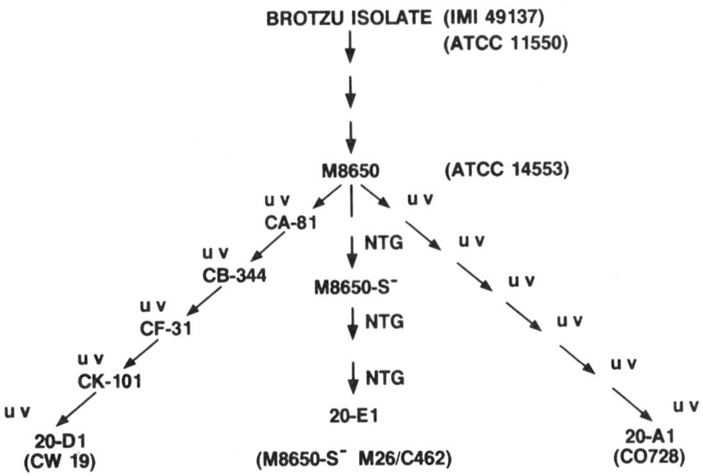

Figure 5 Genealogy of three raised titer strains of *Cephalosporium acremonium*. (Codes in brackets are strain numbers allocated at the place of origin of the strains; from Perez-Martinez and Peberdy, 1985.)

at Eli Lilly (Elander, 1983); strains 20 E-1 and 20 A-1 were developed at Ciba-Geigy and Glaxochem, Ltd., respectively (Perez-Martinez and Peberdy, 1985).

C. Strain Maintenance

Highly mutated industrial strains of *P. chrysogenum* and *C. acremonium* have inherent genetic instabilities, and it is of prime importance to maintain both primary stock cultures and working cultures under conditions in which phenotypic changes are minimized. This normally entails storage at the temperature of the gaseous (-167°C) or liquid (-196°C) phase of nitrogen. Long-term preservation procedures used at Bristol-Myers have been reported (Elander, 1983). Squibb reported the storage of *P. chrysogenum* for 3 years in liquid nitrogen with no loss in conidial viability and minimal loss in productivity (Fortney and Thoma, 1977).

D. Biosynthesis and Regulation

1. Penicillin

The β-lactam thiazolidine ring of penicillin is constructed from L-cysteine and L-valine. Biosynthesis occurs in a nonribosomal process by means of a tripeptide composed of these two amino acids and L-α-aminoadipic acid (L-α-AAA). The first product of the

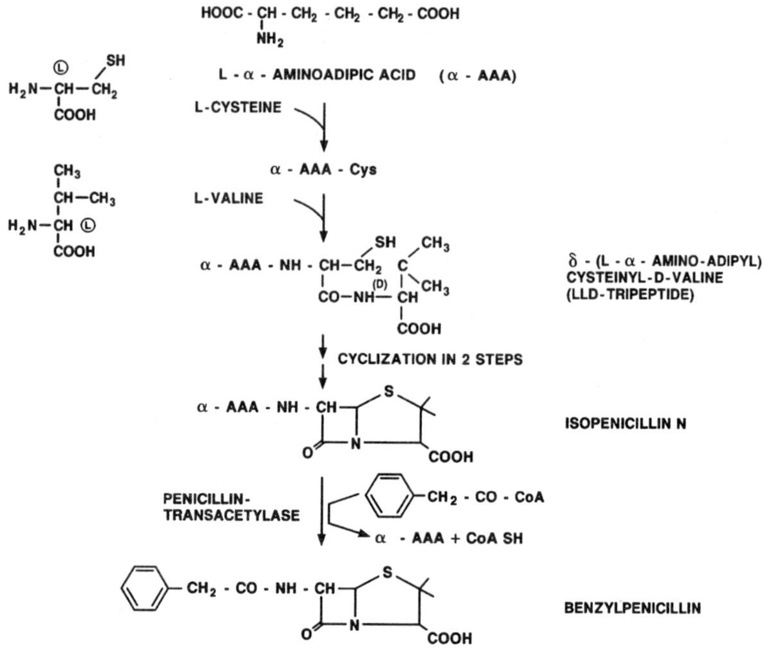

Figure 6 Biosynthesis of penicillin by *Penicillium chrysogenum*.

cyclization of the tripeptide that can be isolated is isopenicillin N, but the biochemical reactions leading to this intermediate are not completely understood at this time.

Benzylpenicillin is produced in the exchange of L-α-AAA with activated phenylacetic acid (Fig. 6). 6-APA, which is not an intermediary product of biosynthesis, is excreted in the absence of a side-chain precursor.

Several regulatory mechanisms are known in penicillin biosynthesis. The amino acid lysine is synthesized from a pathway that involves L-α-aminoadipic acid, so that penicillin and lysine share a common branched biosynthetic pathway. Lysine inhibits penicillin synthesis because it is a feedback inhibitor of homocitrate synthase, an enzyme involved in L-α-AAA synthesis. If L-α-AAA is deficient, penicillin cannot be synthesized, but the addition of small amounts of L-α-AAA stimulates penicillin synthesis. The inhibition by lysine of homocitrate synthase is less than the inhibition of penicillin formation, however, so that additional regulatory processes are postulated.

Penicillin biosynthesis is affected by phosphate concentration and also shows a distinct catabolite repression, particularly by

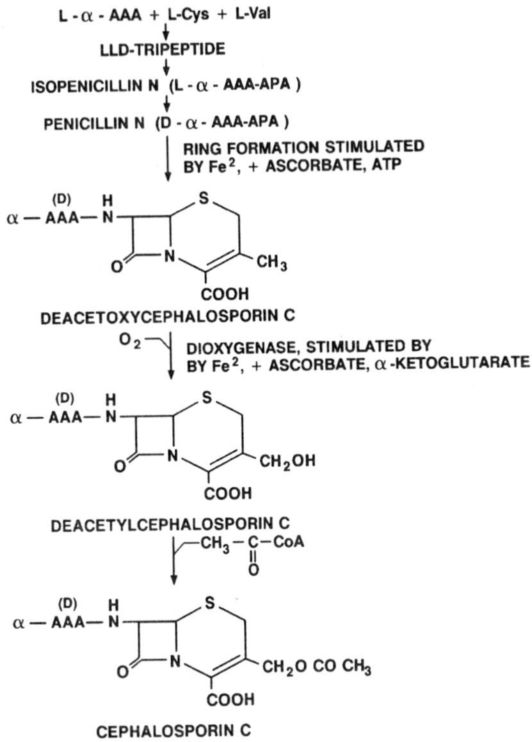

Figure 7 Biosynthesis of cephalosporin C by *Cephalosporium acremonium*.

glucose. Therefore, penicillin fermentations were originally done only with the slowly metabolizable sugar lactose. Moreover, penicillin seems to regulate its own production, since the concentration of penicillin which inhibits further synthesis in a given strain, is usually similar to the production potential of the respective strain.

2. Cephalosporin C

As in the case of benzylpenicillin, the first stages of cephalosporin biosynthesis (Fig. 7) proceed from δ-(α-aminoadipyl)-L-cysteinyl-D-valine to isopenicillin N.

The cyclization activity was first discovered in a cell-free system of *C. acremonium*. In the next stage, penicillin N is produced by transformation of the L-α-AAA side chain into the D form via the action of a very labile racemase. By ring expansion of penicillin N, deacetoxycephalosporin C is produced, and after oxidation

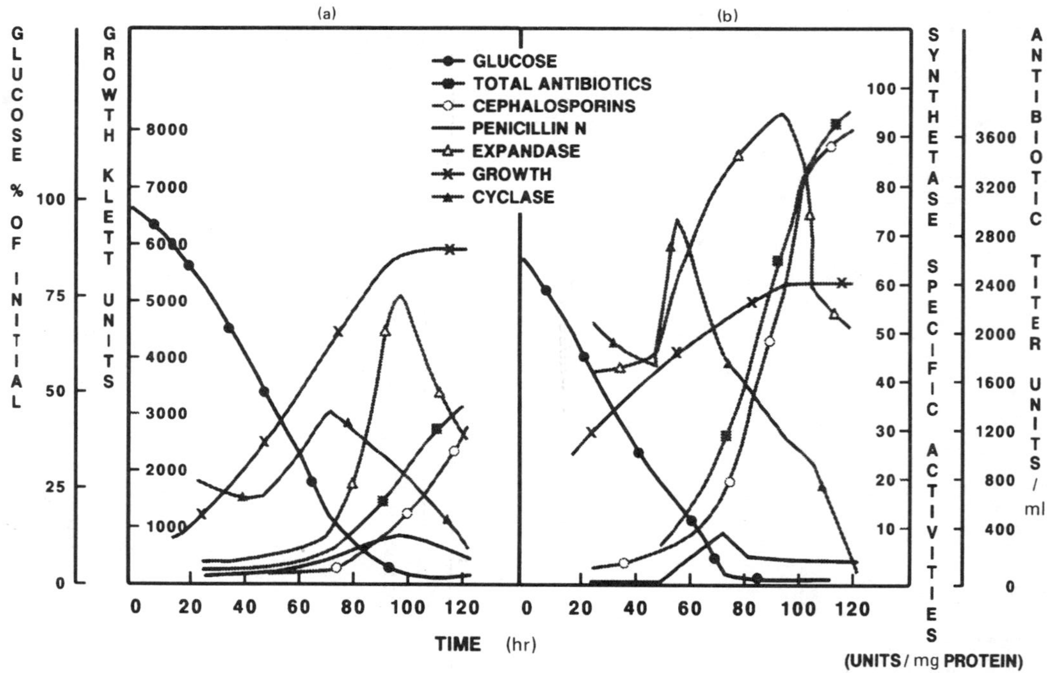

Figure 8 Formation of cyclase and expandase enzymes during fermentation in chemically defined medium. (a) *Cephalosporium acremonium* strain CW-19; (b) *Cephalosporium acremonium* strain C-10. (From Shen and coworkers, 1986.)

and acetylation it is converted into cephalosporin C. In contrast to penicillin, the D-α-AAA moiety cannot be changed with precursor feeding.

Cephalosporin production is affected by carbohydrate catabolite repression and phosphate regulation. The addition of lysine in low concentrations promotes cephalosporin production, whereas the inhibition caused by higher amounts of lysine can be overcome by addition of DL-α-AAA. Methionine stimulates cephalosporin synthesis in *C. acremonium*.

Cyclase and expandase enzymes have been shown to be most important in the regulation and control of cephalosporin overproduction (Shen et al., 1986). A comparison of the enzymes in low- and high-yielding mutants is shown in Fig. 8. The cyclase enzyme is capable of generating new penicillins and cephalosporins when tripeptide analogs are added to cell-free systems (Baldwin et al., 1984). The entire cyclase gene has been recently isolated, sequenced, and cloned in bacteria and streptomycetes (Samson et al., 1985).

E. Fermentation Development

Companies involved in penicillin and cephalosporin production generally have development groups whose responsibilities are to generate new strains with improved productivities, develop new processes to take advantage of new raw materials or modern control and fermentation concepts, and troubleshoot production-specific problems. At the laboratory stage such operations, of necessity, rely on the use of shake flask fermentations to evaluate new strains and culture conditions. Analytical services must be available to assay for antibiotic production and for other required testing. Successful shake flask culture conditions are transferred to a pilot plant to scale up to a stirred tank process for thorough evaluation before recommendation for production trials. Scaleup studies are the most exacting and demanding of development skills and the most critical in the development of successful new strains and culture conditions.

F. Assay Development

Microbiological bioassay can be used, particularly as a preliminary assay, but such techniques are labor intensive and usually lack the precision of automated chemical assays. Automated assays, such as Technicon, have been used extensively over the past 20 years using hydroxylamine (Boxer and Everett, 1949) or penicillanic acid detection (Bundgaard and Ilver, 1972). Over the past 10 years the use of high-pressure liquid chromatography (HPLC) has become common, using reversed-phase ^{18}C columns and UV detection (Miller and Neuss, 1983). High-pressure liquid chromatography has the advantage of high accuracy and good precision and can give excellent quantitative results for penicillin, precursor, and side products. Analytical departments are now using automated or semiautomated systems to improve the productivity and accuracy of their measurements. Laboratory robots, such as the Zymate (Zymark Corp., Hopkinson, Massachusetts) are now capable of sample preparation, for example, automated weighing, diluting, mixing, and filtering. Such machines even have their own automated injection station for HPLC analysis. The majority of HPLC manufacturers supply automated sample injection systems together with automated data acquisition, which can greatly facilitate sample throughput by running unattended overnight.

G. Production Fermentation

Vessels of 30,000-200,000 liters are used for penicillin and cephalosporin C fermentations. Agitation is provided by turbine impellers at a power input of 1-4 W/liter. Volumetric oxygen uptake rates of a rapidly growing culture can attain 0.4-0.8 mmol/liter per min and it is necessary to supply air at rates of 0.5-1.0 v/v per min. Compressed air is normally sterilized by using depth filters or membrane filters. The majority of the carbohydrate is supplied by feeding glucose, sucrose, corn syrup, or some other similar carbon source throughout the fermentation at predetermined rates to provide a limited but sustained growth during the early stages of the fermentation and restricted growth during the latter stage to allow maximum penicillin production. Typical raw material requirements are cornsteep solids, 22 g/liter; $(NH_4)_2SO_4$, 10 g/liter; KH_2PO_4, 3 g/liter; sodium phenylacetate, 16 g/liter; and glucose (or equivalent), 280 g/liter (Swartz, 1979). Foaming is controlled by the addition of vegetable oil or an antifoam agent, such as polypropylene glycol. Because of the feeding of sugar solutions, volume expansion can be very high and mini-harvest schedules can be used to obtain high productivity rates.

The fermentation environment inside the tanks is controlled either by using in situ sensors or by sampling the broth and manually measuring the necessary parameters. In situ, on-line analysis can easily be applied to pH, dissolved oxygen, and exit gas CO_2 by the use of the appropriate probes and analytical instruments. On-line analysis is necessary to operate feedback control of the fermentation broth. Monitoring of pH enables the fermentations to be run at predetermined pH profiles, the actual pH being adjusted by the addition of acid, as sulfuric acid, and alkali, as ammonia gas bled into the air input line. Dissolved oxygen can be used to regulate the airflow in order to use the minimum amount of air necessary to maintain predetermined dissolved oxygen levels. Exit gas analysis, by infrared for CO_2 or mass spectrometry for

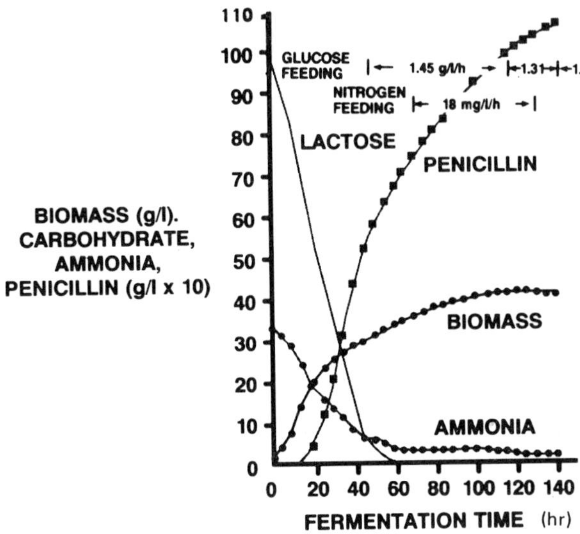

Figure 9 Penicillin fermentation with *Penicillium chrysogenum*. (From Swartz, 1979.)

CO_2, O_2, H_2O, and N_2, can provide metabolic information to regulate the feed rates. Periodic broth sampling is always necessary to determine concentrations of soluble penicillin, precursor, sulfate, amino nitrogen, carbohydrate, and dry cell weight. Such values can be used to regulate the addition of precursor, ammonia, or sulfate and to provide detailed fermentation patterns that can be recorded for retrospective analysis. A typical fermentation profile for penicillin G is shown in Fig. 9.

The fermentation of cephalosporin is similar to that of penicillin. Complex media with cornsteep liquor, meat meal, sucrose, glucose, and lard oil are used. Fermentations are carried out at pH 7.0 at temperatures between 25 and 28°C. In the main growth phase (up to 90 hr), a high aeration rate is necessary, but during the production phase (90-160 hr), O_2 consumption decreases sharply.

Considerable effort, both in industry and in university research, has been directed toward the various relationships between mycelial growth, metabolic pathways, nutrient assimilation, and penicillin production. Recent topics are summarized in Table 5.

Today's industrial fermentations are concerned with obtaining the highest penicillin productivities (Fig. 10) related to minimum capital, raw material, and labor and service outlays. The economics of penicillin fermentation have been reviewed by Swartz (1979).

Table 5 Bioprocess Control in Penicillin Fermentation

Topic	Reported by
Conversion yields for sugar to penicillin	Hersback et al., 1984; Cooney, 1979
Carbon and nitrogen limitations on growth and penicillin formation	Hersbach et al., 1984; Sanchez et al., 1981
Quantitative physiology, specific uptake rates for sugar, NH_3, PO_4, SO_4	Hersback et al., 1984; Lee et al., 1978
Enzyme electrodes and thermistor analyses	Hersbach et al., 1984; Enfors and Nilsson, 1979
Growth modeling and on-line monitoring for process optimization	Hersbach et al., 1984; Mou and Cooney, 1983
Use of pelleted growth in specific reactors	Hersbach et al., 1984; Van suijdam et al., 1984
Automation and computer control	Hersbach et al., 1984; Nelligan and Calam, 1983
Cycling dissolved oxygen tension	Hersbach et al., 1984; Mou and Cooney, 1983

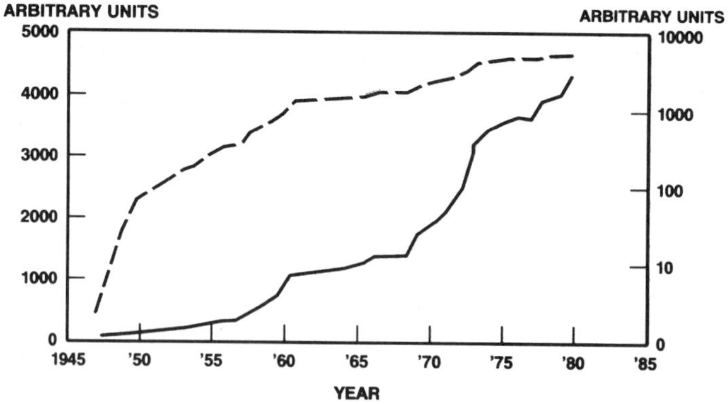

Figure 10 Mean penicillin G productivity of Gist-Brocades (fermentation production per unit volume and per unit time). (From Hersback and coworkers, 1984.)

Considerable cost savings can be obtained by using the least expensive raw materials, provided that titers are not reduced. Today, agricultural crop prices are extremely variable owing to both political and climatic conditions far beyond the control of the fermentation industry, and considerable effort is directed toward developing fermentations that can effectively use a variety of raw materials.

VII. HARVEST BROTH CHARACTERISTICS VERSUS DOWNSTREAM PROCESSING

The extraction and purification of fermentation products may be difficult and costly. Ideally, one attempts to obtain a high-quality product as quickly as possible at an efficient recovery rate using minimum plant investment operated at minimal costs. Unfortunately, recovery costs of microbial products may vary from as low as 20% to as high as 60% of the total manufacturing costs (Swartz, 1979). Obviously, the chosen process will depend on the specific products and the available extraction equipment.

If a fermentation broth is analyzed at the time of harvesting, it will be discovered that the specific product may be present at a low concentration in an aqueous suspension that contains intact microorganisms, cell fragments, soluble and insoluble medium components, and other metabolic products. The product may also be intracellular, heat labile, and easily broken down by contaminating microorganisms. All these factors tend to increase the difficulties of product recovery. To ensure good recovery or purification, speed of operation may be the overriding factor because of the labile nature of a product. The processing equipment must therefore be of the correct type and also the correct size to ensure that the harvested broth can be processed within a satisfactory time limit.

The choice of recovery process is based on the following criteria:

1. Intracellular or extracellular location of the product
2. Concentration of the product in the fermentation broth
3. Physical and chemical properties of the desired product as an aid to selecting separation procedures
4. Intended use of the product
5. Minimal acceptable standard of purity
6. Impurities in the fermentor broth
7. Marketable price for the product
8. Available facilities

The main objective of the first stage for the recovery of an extracellular product is the removal of large solid particles and microbial cells by centrifugation or filtration (Fig. 11). In the next

BROTH FROM FERMENTOR

↓

REMOVAL OF SOLIDS

↓

PRIMARY ISOLATION OF PRODUCT

↓

PURIFICATION AND CONCENTRATION

↓

FINAL PRODUCT ISOLATION

Figure 11 Isolation stages in the recovery of a soluble extracellular product from a harvested fermentation broth.

stage, the broth is fractionated or extracted into major fractions using adsorption or ion-exchange chromatography, liquid-liquid solvent extraction, or precipitation. Afterward, the product-containing fraction is purified by fractional precipitation, chromatography, crystallization, and other procedures.

Attempts to simplify this outline extraction procedure for antibiotic recovery using "whole-broth" processing have met with limited success. The technique of whole-broth processing involves initial removal of large particles, which is then followed by passage of the broth (including cells) through well-mixed ion-exchange columns or countercurrent solvent extraction units to extract the antibiotic directly.

It may be possible to modify the handling characteristics of the broth so that it can be handled faster with simpler equipment making use of a number of techniques:

1. Selection of a microorganism that does not produce pigments or undesirable metabolites
2. Modification of the fermentation conditions to reduce the production of undesirable metabolites
3. Precise timing of harvesting
4. pH control after harvesting
5. Temperature treatment after harvesting
6. Addition of flocculating agents
7. Use of enzymes to attack cell walls

It must be remembered that the fermentation and product recovery are integral parts of an overall process. Because of the interactions between the two, neither stage should be developed independently, which might result in problems and unnecessary expense.

Large volumes of supernatants containing extracellular enzymes need immediate processing, and harvesting times and enzyme yields may not be predictable (Lee et al., 1980). This can make recovery programs difficult to plan. Pigment production also renders some recovery procedures difficult, that is, when the pgiment binds to the same resin as the enzyme. Changes in fermentation conditions may also reduce pigment formation. Certain antifoams remain in the supernatant and affect ultrafiltration or ion-exchange resins used in recovery stages. Major reviews on penicillin and cephalosporin recovery and extraction processes are numerous (Podbielniak and coworkers, 1979; Zhukovskaya, 1972; Todd, 1972; Flynn, 1972; Savidge and Cole, 1975; Vandamme, 1980; Atkinson and Sainter, 1982; Kroner et al., 1978; Lacey and Loeb, 1972; Dombrowski and Munday, 1968; Karr et al., 1980).

Future bioproducts derived from fungal fermentations will be proteins and enzymes, and many of the new developments in fermentation technology and genetic engineering are already having considerable impact on downstream processing. Most of the important industrial fungal enzymes (including amylases, proteases, and penicillin acylases) are actively secreted by the producer fungus, thus making the downstream processing much easier. It may well be possible to genetically engineer the excretion of normally intracellular proteins, making their purification less complicated (Priest, 1984). Similarly, it may be possible to selectively degrade contaminating proteins and nucleic acids by controlled genetic activity. Finally, it may be possible to transfer the genes of product formation to other organisms with more suitable downstream processing characteristics (Fish and Lilly, 1984; Flemming and Steinschneider, 1983).

VIII. FUTURE PROSPECTS FOR INDUSTRIAL FUNGI

This chapter has examined the application of modern genetics and biotechnology to industrial fungal fermentations. The volume and variety of metabolites produced by such fermentations clearly indicate that in specific cases, traditional genetic approaches have been quite successful in generating fungal strains that can efficiently produce marketable substances. The disadvantages of industrial organisms have been overcome by persistence and ingenuity, and some industrial fermentations may operate at near maximal capability. Now, industrial strain improvement programs are embracing newer genetic technology and seeking even better strains and processes and also new products.

Tremendous progress has been made over the past 5 years in applying genetic engineering strategies to filamentous bacteria and fungi producing antibiotics, the present largest market for biotechnology products. Gene amplification and cascade hybridization strategies offer exciting new strain improvement techniques for increasing the levels or modifying the regulation of the important -lactam synthase enzymes (AC synthetase, cyclase, expandase, acyltransferase, and others) in fungi producing penicillins and cephalosporins. Such programs can lead to the isolation and cloning of entire antibiotic biosynthetic pathways (Stanzak et al., 1986) in different microbial strains and to the synthesis of novel, more useful, "hybrid" antibiotics (Omura et al., 1986; Fayerman, 1986).

Can strain improvement programs for important metabolites be expected to match the recent dramatic successes in producing valuable proteins in genetically engineered yeast? A number of such proteins, many with medical applications, have been synthesized in *S. cerevisiae* using cloned mammalian or viral genes. These scarce substances will eventually be marketed as high-value specialty chemicals or will be used to manufacture important new therapeutic products. Examples of some of these substances are human leukocyte interferon-D, human interferons-α^1, β^1, and γ, which can be secreted; human epidermal growth factor, urogastrone; human proinsulin; human preproparathyroid hormone; human interleukin-2; human blood factor VIII; human α^1-antitrypsin; human β-endorphin; calf chymosin; bovine growth hormone; and hepatitis B virus surface antigen. This list will certainly grow as more genes important in medicine, agriculture, and industry are cloned and expressed in suitable hosts.

The advantages of using fungi as expression systems and the problems that need to be alleviated are common to all types of proteins. In outline, filamentous fungi, especially commercial strains, secrete large amounts of proteins. Exact levels are hard to assess, but they are obviously high and commercially rewarding. Cleavage of signal peptides to generate mature active protein is dependent on amino acid signals common to many species. Proteins are modified by glycosylation with mannose sugars attached to asparagine-linked *N*-acetylglucosamine core molecules in a fashion similar to mammalian systems.

The requirements for mRNA processing are also akin to those present in higher eukaryotes. Intervening sequences present in the primary transcript carry consensus splice donor, splice acceptor, and internal splice recognition signals. Fungal systems thus offer all features necessary for mRNA transcription and translation and for the processing of both mRNA and protein. But these are not enough in themselves. The product must be produced at high economic levels. In other systems, the approach to achieve this goal has been based on gene amplification using high copy number plasmid vectors. In filamentous fungi, the stable, and hence manipulable, transformants are the result of the integration of the DNA into the chromosome. Although several copies can be integrated, the very high copy number characteristics of plasmid-based systems are rarely achieved. A key question is whether high copy number vectors are absolutely necessary.

A consideration must be given to the problem seen by many as the major stumbling block to the exploitation of filamentous fungi—proteases. *A. oryzae* is a major organism commercially used for the production of all types of protease. Copious amounts of these protein-degrading enzymes are produced, especially under derepressed conditions. There is much literature on the presence of proteases and some on their activity, but they are genetically undefined. It is possible that the industrial strains of fungi selected for high levels of enzyme secretion are deficient in specific proteases or in a number of them, except, of course, *A. oryzae*. This is another strong reason for using the ready prepared genotypes in any experiments testing heterologous protein expression. Undoubtedly, the protease problem will have to be addressed, but its magnitude has yet to be determined. An empirical, initial solution may be to ensure that the culture medium contains sufficient quantities of the agents known to repress protease production.

In conclusion, there is mounting evidence that filamentous fungi are emerging as candidates for strong hosts capable of expressing primary and secondary metabolite foreign proteins. In addition, rDNA methods are also being applied to improve the long-standing existing processes and to widen their use in order to expand the range of products of fungal origin currently being produced.

The historical development of the fungal fermentation industry and the emergence of fungal biotechnology signal a bright future for industrial mycology. One cannot easily predict directions in a field that is changing so rapidly. Undoubtedly, however, fungal biotechnology will improve industrial fermentations, allow the economical production of new substances, and afford opportunities unimagined today.

REFERENCES

Atkinson, B., and Sainter, P. (1982). Development of downstream processing. *J. Chem. Tech. Biotechnol. 32*: 100-108.

Baldwin, J. W., Aalington, R. M., Derome, A. E., Ting, H-H., and Turner, N. J. (1984). Penicillin biosynthesis: Multiple pathways from a modified substrate. *J. Chem. Soc. Commun.*: 1211-1214.

Ball, C. (1982). Genetic approaches to overproduction of β-lactam antibiotics in eukaryotes. In *Overproduction of Microbial Products*, V. Krumphanzl, B. Sikta, and Z. Vanek (Eds.). Academic Press, London, pp. 515-534.

Ball, C., and Azevedo, J. L. (1976). Genetic instability in parasexual fungi. In *Second International Symposium on the Genetics of Industrial Microorganisms*, K. D. Macdonald (Ed.). Academic Press, London, pp. 243-251.

Ball, C., Lawrence, A. J., Butler, J. M., and Morrison, K. B. (1978). Improvement in amyloglucosidase production following genetic recombination of *Aspergillus niger* strains. *Eur. J. Appl. Microbiol.* *5*: 95-102.

Bennett, J. W. (1986). Molds, manufacturing and molecular genetics. In *Molecular Genetics of Filamentous Fungi*, W. E. Timberlake (Ed.). Alan R. Liss, New York, pp. 345-366.

Bigelis, R. (1985). Primary metabolism and industrial fermentation. In *Genetic Manipulations in Filamentous Fungi*, J. W. Bennett and L. L. Lasure (Eds.). Academic Press, New York, pp. 357-401.

Boxer, G. E., and Everett, G. E. (1949). Colorimetric determination of benzyl penicillin. *Anal. Chem.* *21*: 670-675.

Bundgaard, H., and Ilver, K. (1972). A new spectrophotometric method for the determination of penicillins. *J. Pharm. Pharmacol.* *24*: 790-794.

Calam, C. T., Daglish, L. B., Gaitskell, W. S. (1973). Hybridization experiments with *Penicillium patulum* and *Fusarium moniliforme*. In *Genetics of Industrial Microorganisms*, vol. II, Z. Vanek, Z. Hostalek, and J. Cudlin (Eds.). Elsevier, Amsterdam, pp. 265-277.

Chang, L. T., and Terry, C. A. (1973). Intergenic complementation of glucoamylase and citric acid production in two species of *Aspergillus*. *Appl. Microbiol.* *25*: 890-895.

Chang, L. T., and Tuveson, R. W. (1975). An unstable state of *Aspergillus foetidus* segregating proline auxotrophs. *J. Appl. Microbiol.* *30*: 140-146.

Chang, L. T., Terasaka, D. T., and Elander, R. P. (1982). Protoplast fusion in industrial fungi. *Dev. Ind. Microbiol.* *23*: 21-19.

Cooney, C. L. (1979). Conversion yields in penicillin production: Theory vs practice. *Proc. Biochem.* *14*: 31-33.

Demain, A. L., and Solomon, N. A. (1981). Industrial microbiology. In *Industrial Microbiology and the Advent of Genetic Engineering*, A Scientific American Book. W. H. Freeman, San Francisco, pp. 3-11.

Dombrowski, N., and Munday, G. (1968). Spray drying. In *Tropics in Enzyme and Fermentation Biotechnology*, vol. 5, N. Blakebrough (Ed.). Academic Press, London, pp. 209-320.

Elander, R. P. (1967). Enhanced penicillin synthesis in mutant and recombinant strains of *Penicillium chrysogenum*. In *Induced*

Mutations and Their Utilization, H. Stübbe (Ed.). Akademie-Verlag, Berlin, pp. 403-423.

Elander, R. P. (1979). Mutations affecting antibiotic synthesis in fungi producing β-lactam antibiotics. In *Genetics of Industrial Microorganisms*, O. K. Sebek and A. I. Laskin (Eds.). American Society for Microbiology, Washington, D.C., pp. 21-35.

Elander, R. P. (1983). Strain improvement and preservation of β-lactam producing microorganisms. In *Antibiotics Containing the β-Lactam Structure*, A. L. Demain and N. A. Solomon (Eds.). Springer-Verlag, New York, pp. 97-146.

Elander, R. P., and Chang, L. T. (1979). Microbial culture selection. In *Microbial Technology: Fermentation Technology*, vol. II, H. J. Peppler and D. Perlman (Eds.). Academic Press, New York, pp. 244-302.

Elander, R. P., and Demain, A. L. (1981). Genetics of microorganisms in relation to industrial requirements. In *Biotechnology: Microbial Fundamentals*, vol. 1, H. J. Rehm and G. Reed (Eds.). Verlag-Chemie, Weinheim, pp. 237-277.

Elander, R. P., and Espenshade, M. A. (1976). The role of microbial genetics in industrial microbiology. In *Industrial Microbiology*, B. M. Miller and W. Litsky (Eds.). McGraw-Hill, New York, pp. 192-256.

Elander, R. P., and Vournakis, J. N. (1986). Genetic aspects of overproduction of antibiotics and other secondary metabolites. In *Over-production of Microbial Metabolites*, Z. Vanek and Z. Hostalek (Eds.). Butterworths, Boston, pp. 63-79.

Enfors, S. O., and Nilsson, H. (1979). Design and response characteristic of an enzyme electrode for measurement of penicillin in fermentation broths. *Enzyme Microb. Technol. 1*: 260-265.

Fägerstam, L. G., Pettersson, G. M., and Engstrom, J. A. (1984). The primary structure of 1,4-β-glucan cellobiohydrolase from the fungus, *Trichoderma reesei*. *FEBS Lett. 167*: 309-314.

Fayerman, J. T. (1986). New developments in gene cloning in antibiotic producing microorganisms. *Bio/Technology 4*: 786-789.

Ferenczy, L. (1981). Microbial protoplast fusion. In *Genetics as a Tool in Microbiology*, S. W. Glover and D. A. Hopwood (Eds.). Cambridge University Press, London, pp. 1-34.

Fish, N. M., and Lilly, M. D. (1984). The interactions between fermentation and protein recovery. *Bio/Technology*: 623-627.

Flemming, R. J., and Steinschneider, A. (1983). Fermentation economics in relation to genetic engineering. *Bio/Technology*: 773-776.

Flynn, E. H. (1972). *Cephalosporin and Penicillins: Chemistry and Biology*. Academic Press, New York.

Fortney, K. L., and Thoma, R. W. (1977). Stabilization of culture productivity. *Dev. Ind. Microbiol. 18*: 319-325.

Godfrey, O. W. (1973). Isolation of regulator mutants of the aspartic and pyruvate acid families and their effect on antibiotic production in *Streptomyces lipmannii*. *Antimicrob. Agents Chemother. 4*: 73-79.

Golubtsova, V. M., Shcherbakova, E. Y., Runkovshaya, L. Y., and Ermakova, V. P. (1979). Change in the ratio of citric and oxalic acids in *Aspergillus niger* under the influence of mutagenic factors. *Mikrobiologiya 48*: 1061-1065.

Hacking, A. J. (1986). *Economic Aspects of Biotechnology*. Cambridge University Press, London.

Hamlyn, P. F., and Ball, C. (1979). Recombination studies with *Cephalosporium acremonium*. In *Genetics of Industrial Microorganisms*, O. K. Sebek and A. I. Laskin (Eds.). American Society for Microbiology, Washington, D.C., pp. 285-191.

Heeswisk, R., and Roncero, M. I. G. (1984). High frequency transformation of *Mucor* with recombinant plasmid DNA. *Carlsberg Res. Commun. 49*: 691-701.

Hersbach, G. I. M., van der Beek, C. P., and van Dijk, P. W. M. (1984). The penicillins: Properties, biosynthesis and fermentation. In *Biotechnology of Industrial Antibiotics*, E. J. Vandamme (Ed.). Marcel Dekker, New York, pp. 46-140.

Hopwood, D. A. (1981). The genetic programming of industrial microorganisms. In *Industrial Microbiology and the Advent of Genetic Engineering*, A Scientific-American Book, W. H. Freeman, San Francisco, pp. 31-42.

Hynes, M. J., Corrick, C. M., and King, J. A. (1983). Isolation of genomic clones containing the *amd S* gene of *Aspergillus nidulans* and their use in the analysis of structural and regulatory mutations. *Mol. Cell. Biol. 3*: 1430-1439.

Ikeda, Y. (1961). Potential application of parasexuality in breeding fungi. *Recent Adv. Bot.*: 383-386.

Ikeda, Y., Nakamura, K., Uchida, K., and Ishitani, C. (1957). Two attempts upon improving an industrial strain of *Aspergillus oryzae* through somatic recombination and polyploidization. *J. Gen. Appl. Microbiol. 3*: 93-102.

Innis, M. A., Holland, M. J., McCabe, P. C., Cole, G. E., Wittman, V. P., Tal, R., Watt, K. W., Gelfand, D. H., Holland, J. P., and Meade, J. H. (1985). Expression, glycosylation and secretion of an *Aspergillus* glucoamlase from *Saccharomyces cerevisiae*. *Science 228*: 21-26.

Jefferys, E. G. (1970). The gibberellin fermentation. *Adv. Appl. Microbiol. 13*: 283-296.

Johnstone, Y. L., Hughes, S. G., and Clutterbuck, A. J. (1985). Cloning of an *Aspergillus* developmental gene by transformation. *EMBO J. 5*: 1307-1311.

Karr, A. E., Gebert, W., and Wang, M. (1980). Extraction of whole fermentation broth with a Karr reciprocating plate extraction column. *Can. J. Chem. Eng. 58*: 249-252.

Kelly, J. M., and Hynes, M. J. (1983). Transformation of *Aspergillus niger* by the *amdS* gene of *Aspergillus nidulans*. *EMBO J. 4*: 475-479.

Kroner, K. H., Hustedt, S., Granda, S., and Kula, M. R. (1978). Technical aspects of separation using aqueous two-phase systems. *Biotech. Bioeng. 20*: 1967-1988.

Kubicek, C. P., and Röhr, M. (1982). Novel trends in physiology and technology of citric acid production. In *Overproduction of Microbial Products*, V. Krumphanzl, B. Sikyta, and Z. Vanek (Eds.). Academic Press, New York, pp. 253-262.

Kubicek, C. P., and Röhr, M. (1986). Citric acid fermentation. *Crit. Rev. Biotechnol. 3*: 331-374.

Kundu, P. N., and Das, A. (1982). Calcium gluconate production by a nonconventional fermentation method. *Biotechnol. Lett. 4*: 365-368.

Kurushima, M., Fujii, T., and Kitachara, K. (1972). Enzymatical and morphological variations by ultra-violet irradiation. II. Studies on fungal amylases. *Agr. Biol. Chem. 36*: A26.

Lacey, R. E., and Loeb, S. (1972). *Industrial Processing with Membranes*. Wiley-Interscience, New York.

Lago, B. D., and Kaplan, L. (1981). Vitamin fermentations: B_2 and B_{12}. In *Advances in Biotechnology: Fermentation Products*, vol. 3, C. Vezina and K. Singh (Eds.). Pergamon, New York, pp. 241-257.

Lee, J. S., Shin, K. C., Yang, H. S., and Ryu, D. D. Y. (1978). Effects of carbon sources and other process variables in fed-batch fermentations of penicillin. *Misaengmul Hakkoe Chi. 16*: 21-29.

Lee, S. W., Hur, K. C., Lee, K. S., and Kim, W. S. (1980). Extra cellular production of β-lactamase by *Penicillium chrysogenum*. *Hanzu Kyum. Chi. 8*: 63-68.

Lein, J. (1986). The Panlabs penicillin strain improvement program. In *Overproduction of Microbial Metabolites*, Z. Vanek and Z. Hostalek (Eds.). Butterworths, Boston, pp. 105-139.

Lowe, D. A. (1985). Industrial importance of biotransformations of β-lactam antibiotics. *Dev. Ind. Microbiol. 26*: 143-155.

Lowe, D. A., Romancik, G., and Elander, R. P. (1986). Enzymatic hydrolysis of penicillin V to 6-aminopenicillanic acid by *Fusarium oxysporum*. *Biotechnol. Lett. 8*: 151-156.

Macdonald, K. D., and Holt, G. (1976). Genetics of biosynthesis and overproduction of penicillin. *Sci. Prog. (Oxford) 63*: 547-573.

Miller, R. D., and Neuss, N. (1983). High performance liquid chromotography of β-lactam antibiotics. In *Antibiotics Containing the Beta-Lactam Structure*, A. L. Demain and N. A. Solomon (Eds.). Springer-Verlag, New York, pp. 301-308.

Montencourt, B. S., and Eveleigh, D. E. (1977). Preparation of mutants of *Tricholderma reesei* with enhanced cellulase production. *Appl. Environ. Microbiol. 34*: 777-786.

Montencourt, B. S., and Eveleigh, D. E. (1985). Fungal carbohydrases, amylases and cellulases. In *Gene Manipulations in Fungi*, W. E. Timberlake (Ed.). Academic Press, New York, pp. 491-512.

Mou, D. G., and Cooney, C. L. (1983). Growth monitoring and control through computer aided on-line mass balancing in a fed-batch penicillin fermentation. *Biotechnol. Bioeng. 25*: 225-255.

Murillo, F. J., Calderon, I. L., Lopez-Diaz, I., and Cerda-Olmedo, E. (1983). Carotene super-producing strains of *Phycomyces*. *Appl. Environ. Microbiol. 36*: 639-642.

Nelligan, I., and Calam, C. T. (1983). Optimal control of penicillin production using a mini-computer. *Biotechnol. Lett. 5*: 561-565.

Omura, S., Ikeda, H., Malpartida, F., Keiser, H. M., and Hopwood, D. A. (1986). Production of new hybrid antibiotics, mederrhodins B, by a genetically engineered strain. *Antimicrob. Agents Chemother. 29*: 13-19.

Paietta, J., and Marzluf, G. A. (1986). Development of shuttle vectors and gene manipulation techniques for *Neurospora crassa*. In *Molecular Genetics of Filamentous Fungi*, W. E. Timberlake (Ed.). Alan R. Liss, New York, pp. 1-14.

Parks, L. W., Rodriguez, R. J., and McCammon, M. T. (1982). Sterols of yeast: A model for biotechnology in the production of fats and oils. *J. Am. Chem. Soc. 59*: 294A.

Peberdy, J. F. (1980). Protoplast fusion—a tool for genetic manipulation and breeding of industrial microorganisms. *Enzyme Microb. Technol. 2*: 23-39.

Penalva, M. A., Tourino, A., Patino, C., Sanchez, F., Fernandez Sousa, J. M., and Rubio, V. (1986). Studies on transformation of *Cephalosporium acremonium*. *Molecular Genetics of Filamentous Fungi*, W. Timberlake (Ed.). Alan R. Liss, New York, pp. 59-68.

Perez-Martinez, G., and Peberdy, J. F. (1985). Production of cephalosporin C, and its intermediates, by raised titre strains of *Acremonium chrysogenum*. *Enzyme Microb. Technol. 7*: 389-394.

Perlman, D. (1978). Vitamins. In *Primary Pruducts of Metabolism, Economic Microbiology*, vol. 2, A. H. Rose (Ed.). Academic Press, New York, pp. 303-326.

Perlman, D. (1979). Microbial process for riboflavin production. In *Microbial Technology*, vol. I., 2nd ed., *Microbial Processes*, H. J. Peppler and D. Perlman (Eds.). Academic Press, New York, pp. 521-527.

Podbielniak, W. J., Kaiser, H. R., and Ziegenhorm, G. J. (1979). The history of penicillin production. *Chem. Eng. Symp. Ses. 66*: 44-49.

Priest, F. G. (1984). *Extracellular Enzymes*. Aspects of Microbiology-9, Van Nostrand-Reinhold, Wokingham.

Queener, S. W., Wilkerson, S. G., and Nash, C. H. (1982). Sterol content and titer of mycophenolic acid in polyene antibiotic-resistant mutants of *Penicillium stoloniferum*. In *Overproduction of Microbial Products*, V. Krumphanzl, B. Sikyta, and Z. Vanek (Eds.). Academic Press, London, pp. 535-548.

Rehaak, Z. (1985). Biosynthesis of ergot alkaloids. In *Biotechnology: Applications and Research*, P. M. Cheremisinoff and R. P. Ouellette (Eds.). Technomic Publishing Co., Lancaster, Pennsylvania, pp. 565-585.

Samson, S. M., Belagaje, R., Blankenship, ., Chapman, J. L., Perry, D., Skatrud, P. L., VanFrank, R. M., Abraham, E. P., Baldwin, J. E., Queener, S. W., and Ingolia, T. D. (1985). Isolation, sequence determination and expression in *Escherichia coli* of the isopenicillin N synthetase gene of *Cephalosporium acremonium*. *Nature 318*: 191-194.

Sanchez, S., Paneagua, L., Matess, R. C., Lara, F., and Mora, J. (1981). Nitrogen regulation of penicillin G biosynthesis in *Penicillium chrysogenum*. *Adv. Biotechnol. 3*: 147-156.

Savidge, T. A. (1984). Enzymatic conversions used in the production of penicillins and cephalosporins. In *Biotechnology of Industrial Antibiotics*, E. J. Vandamme (Ed.). Marcel Dekker, New York, pp. 171-224.

Savidge, T. A., and Cole, M. (1975). Penicillin acylase. *Methods Enzymol. 43*: 705-711.

Shcherbakova, E. Y., and Rezvaya, M. N. (1977). Formation of diploids in *Aspergillus niger* and their biosynthesis of citric acid. *Mikrobiologiya 46*: 1064-1069.

Shen, Y. Q., Wolfe, S., and Demain, A. L. (1986). Levels of isopenicillin N symthetase and deacetoxycephalosporin C synthetase in *Cephalosporium acremonium* producing high and low levels of cephalosporin C. *Bio/Technology 4*: 61-63.

Shibuya, Y., Matsumoto, K., and Fujii, T. (1981). Isolation and properties of 7-(4-carboxybutanamido) cephalosporanic acid acylase-producing bacteria. *Agr. Biol. Chem. 45*: 1561-1567.

Shoemaker, S., Schweikant, V., Ladner, M., Gelfand, D., Kwok, S., Myanko, K., and Innis, A. M. (1983). Molecular cloning of exocellobiohydrolase I derived from *Trichoderma reesei* strain L27. *Bio/Technology 1*: 691-696.

Spalla, C., and Marnati, M. P. (1978). Genetic aspects of the formation of ergot alkaloids. In *Antibiotics and Other Secondary Metabolites*, R. Hutter, T. Leisinger, J. Nüesch, and W. Wehrli (Eds.). Academic Press, London, pp. 219-228.

Stanzak, K., Matsushima, R. H., Baltz, R. H., and Rao, R. N. (1986). Cloning and expression in *Streptomyces lividans* of clustered erythromycin biosynthesis genes from *Streptomyces erythreus*. *Bio/Technology 4*: 229-232.

Stohl, L. L., and Lambowitz, A. M. (1984). Construction of a shuttle vector for the filamentous fungus, *Neurospora crassa*. *Proc. Natl. Acad. Sci. USA 80*: 1058-1062.

Stohl, L. L., Akins, R. A., and Lambowitz, A. M. (1984). Characterization of an autonomously replicating *Neurospora* plasmid. *Nucleic Acids Res. 12*: 6169-6178.

Strauss, E. S. (1981). Citric acid. In *Chemical Economics Handbook*. SRI International, Menlo Park, California, pp. 5021-5023P.

Swartz, R. W. (1979). The use of economic analysis of penicillin G manufacturing costs in establishing priorities for fermentation process improvement. *Annu. Rep. Ferment. Proc. 3*: 75-110.

Tabuchi, T., Sugisawa, T., Ishidori, T., Nakahara, T., and Sugiyama, J. (1981). Itaconic acid fermentation by a yeast belonging to the genus *Candida*. *Agr. Biol. Chem. 45*: 475-479.

Thorbek, L., and Eplov, P. (1974). The formation, general characteristics and production of α-amylase in heterokaryons and diploids of *Aspergillus oryzae*. *J. Appl. Bacteriol. 37*: 549-556.

Todd, D. B. (1972). Improving performance of centrifugal extractors. *Chem. Eng. 79*: 153-156.

Turner, G., Ballance, D. J., Ward, M., and Beri, R. K. (1985). Development of cloning vectors and a marker for gene replacement techniques in *Aspergillus nidulans*. In *Molecular Genetics of Filamentous Fungi*, W. E. Timberlake (Ed.). Anal R. Liss, New York, pp. 15-28.

Turner, W. B. (1971). *Fungal Metabolites*. Academic Press, London.

Upshall, A. (1986). Filamentous fungi in biotechnology. *Biotechniques 4*: 158-166.

Vandamme, E. J. (1980). Penicillin acylases and β-lactamase. In *Economic Microbiology*, vol. 5, A. H. Rose (Ed.). Academic Press, New York, pp. 467-522.

Van Suijdam, J. C., Hols, H., and Kossen, N. W. F. (1984). Unstructured model for growth of mycelial pellets in submerged cultures. *Biotechnol. Bioeng. 24*: 177-184.

Yameda, K., Nakahara, T., and Fukui, S. (1971). Petroleum microbiology and vitamin production. In *Biochemical and Industrial Aspects of Fermentation*, K. Sakaguchi, T. Umura, and S. Kinoshita (Eds.). Kodansha, Tokyo, pp. 62-90.

Yelton, M. M., Timberlake, W. E., and van den Hondel, C. A. M. J. J. (1983). A cosmid selecting genes by complementation in *Aspergillus nidulans*: Selection of the developmentally regulated by A locus. *Proc. Natl. Acad. Sci. USA 82*: 834-838.

Yelton, M. M., Hamer, J. E., and Timberlake, W. E. (1984). Transformation of *Aspergillus nidulans* by using a *trpC* plasmid. *Proc. Natl. Acad. Sci. USA 81*: 1470-1474.

Zhukovskaya, S. A. (1972). Selection of effective conditions for penicillin extraction. *Khim. Farm. Zh. 66*: 42-47.

5

Using Mammalian Cells to Produce Products

C. L. PROPST

Affiliated Scientific, Inc., Ingleside, Illinois

RANDALL J. VON WEDEL

CytoCulture International, Inc., San Francisco, California

ANTHONY S. LUBINIECKI*

Genentech, Inc., South San Francisco, California

I. INTRODUCTION

The use of cells to produce products is not a new technology. Until recently, however, the cells most often employed were microorganisms, usually fungi or bacteria. Because of their complexity mammalian cells were rarely used except for the production of viral vaccines. In the last decade, however, mammalian cells have begun to be used for the production of a wide variety of complex products. This is especially true when the products are high value-added compounds that are difficult or impossible to make synthetically or with other cell-based technologies (Ratafia, 1987; McCormick, 1987).

The ability to use mammalian cells to produce a variety of products has resulted from many breakthroughs in cell culture technology. These have allowed mammalian cell culture to become much easier and consequently more cost effective and have occurred at each major step in the production process. This chapter discusses how mammalian cell culture is used to produce large quantities of products important to humans.

**Present affiliation*: Smith, Kline & French Laboratories, King of Prussia, Pennsylvania.

Table 1 Milestone Product Approvals of Cell Culture Biologicals for Parenteral Use in the Last Decade

Product	Cell substrate	Country	Significance	Year
Urokinase (Abbokinase)	HEK	USA, UK, others	First therapeutic licensed from human cell culture	1978
Poliovirus vaccine (inactivated)	Vero	France, Belgium	First biological licensed from continuous cell line	1982
Lymphoblastoid interferon (Welferon)	Namalva	UK, Germany, Austria	First therapeutic licensed from human tumor cell line	1986
OKT-3 monoclonal antibody (Orthoclone OKT3)	Hybridoma	USA	First licensed therapeutic monoclonal antibody	1987
Tissue plasminogen activator (Activase)	CHO	Philippines, New Zealand, France, others	First therapeutic licensed from continuous cell line modified by rDNA technology	1986

A. Characteristics of Mammalian Cells

Although mammalian cell growth technology is based to a great extent on the experience gained from growing microbes, the use of mammalian cells to produce most products has lagged behind the use of less complicated organisms. This is primarily because mammalian cells are fastidious and more difficult to grow than bacteria, yeasts, or fungi. Microbes and mammalian cells differ substantially in their biology, and this translates into difficulties in using these cells for production purposes. Mammalian cells lack the cell walls of microbes, which provide protection from physical damage caused by shear and extremes of osmolarity. Bacteria and yeasts have short doubling times and can grow to high cell densities on a variety of inexpensive media. In contrast, mammalian cells grow slowly, with doubling times from 12 to 48 hr. They grow to low cell densities, rarely exceeding 10^7 cells per ml. In addition, they require complex nutrients for growth. Genetically, microbes are simpler in the organization of their cellular components. In mammalian cells

Table 2 Partial List of Current Mammalian Cell Culture Products in Human Clinical Trial[a]

Molecule of interest	Type	Comment
HIV proteins (various)	Vaccine	Recombinant and conventional
Hepatitis B surface antigen	Vaccine	Recombinant
Erythropoeitin	Hormone	Recombinant
Human growth hormone	Hormone	Recombinant
Factor VIII	Antihemophilic	Recombinant
GM-CSF	Wound healing	Recombinant
Monoclonal antibodies	Antitumor	Hybridoma
Tumor necrosis factor	Antitumor	Natural

[a]In addition to clinical trials for products listed in Table 1 and conventional licensed products designed to obtain approval for additional countries or indications.

the nuclear membrane, large chromosome number, multiple copy genes, and the existence of introns all contribute additional problems in working with these cells.

Despite these problems, however, there are compelling reasons for using mammalian cells to produce certain kinds of products. The most persuasive occurs when products cannot be made efficiently by any other means. This is the case for most virus vaccines and many complex biologically active molecules of mammalian origin when little genetic information is available. The use of mammalian cells is also often required when the product needs correct post-translational modifications, such as cleavage, glycosylation, carboxylation, hydroxylation, disulfide bond formation, or special folding and subunit assembly to preserve biological activity, normal antigenicity, or stability. The failure of most microbes to secrete mammalian products, coupled with the toxicity of certain of these products to microorganisms and the inherent pyrogenicity of some microbial cell walls, makes mammalian cells still more desirable. Table 1 provides a historic perspective of "firsts" in the use of mammalian cells to produce products for human use. Table 2 provides a partial list, growing quickly, of mammalian cell products currently used in clinical trials.

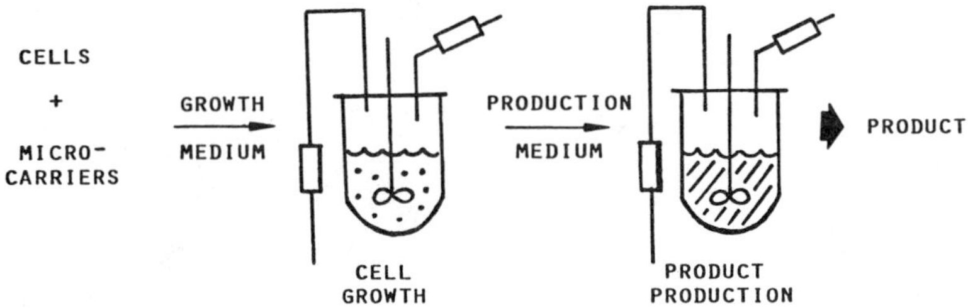

Figure 1 General scheme for the production of products by anchorage-dependent mammalian cells.

B. Products from Mammalian Cells

Growing mammalian cells in culture to produce a desired final product involves a series of steps. The specifics of this process, however, depend on the cell type and the type of product. Nevertheless, the general process is essentially the same. The basic steps are schematically diagrammed in Fig. 1 using a microcarrier-based process as an example. Briefly, after acquiring the appropriate mammalian cells, these cells are inoculated into a temperature-controlled, aerated growth vessel containing sterile growth medium. An attachment surface for the cells is also provided if the cells are anchorage dependent. The cells are allowed to grow until they reach the population size and state at which they can produce product. The growth medium may then be replaced by a cell maintenance (production) medium more appropriate to the accumulation of product. Finally, the cells produce the desired product, which is harvested and purified. After formulation and testing, products may be released for use. Each step in the production of products from mammalian cells is discussed in detail in subsequent sections of this review.

II. CELLS

The first steps in using mammalian cells to produce products are to acquire and propagate the appropriate cells in culture.

A. Types of Cells

The initial source of mammalian cells is the living animal. Cells may be removed from a variety of tissues and organs, which may be normal or abnormal. Such cells have been classified (Hayflick and Moorhead, 1961; Hayflick, 1965) into three major groups based primarily on life expectancy. These groups include (1) primary cells,

Table 3 Characteristics of Different Types of Mammalian Cells

Type	Karyotype	Life span (doublings)	Growth	Example
Primary	Normal, diploid	<10	Anchorage dependent	Primary monkey kidney
Strains	Normal, diploid	Approximately 50	Anchorage dependent	HEK, WI-38, FS-4
Lines	Abnormal, heterodiploid	Infinite	Suspension (or anchorage dependent)	HeLa Bowes, hybridomas

which are derived directly from the living mammal and have a finite life span of 1-5 doublings; (2) diploid cell strains, which have a longer finite life span of about 50 doublings; and (3) established cell lines, which have unlimited life spans. The life expectancy in these different groups of cells is also correlated with a variety of other growth characteristics, which are important for product production. These characteristics are summarized in Table 3.

For cell culture purposes, an important distinction is the requirement for a growth support surface. Cells are characterized as capable of growing free in suspension (anchorage independent) or requiring a growth support surface (anchorage dependent). Certain cells under certain conditions have the capacity to grow in either mode, although this characteristic is not common.

Cells are further distinguished by their tissue or origin, such as fibroblast, epithelium, endothelium, lymphoblast, or hybridoma, which frequently imparts characteristic cell morphology and behavior under culture conditions.

B. Cell Screening

Initial attempts to use mammalian cells to produce products required substantial effort to locate the most appropriate cells. In the past this was normally accomplished by screening existing and new cells in order to identify those with the natural ability to produce the desired product. In addition, since different isolates of even the same cell type may produce product differently, the screening process was often quite lengthy.

Since the cells by necessity had to produce the right product naturally, other characteristics, such as how well they grew in culture and how much product was produced were secondary—albeit

Table 4 Frequently Used Cells and Their Characteristics

Name	Species	Source	Type	Culture mode	Morphology[a]	Major Use[b]
Vero	African green monkey	Kidney	Line	Attachment	fib	Human vaccine
HeLa	Human	Cervix	Line	Suspension	epi	Hormones
CHO	Chinese hamster	Ovary	Line	Both	epi	t-PA, others
Mouse L	Mouse	Skin	Line	Attachment	fib	Hormones
MRC-5	Human	Lung	Strain	Attachment	fib	Human vaccines
FS-4	Human	Foreskin	Strain	Attachment	fib	β-IFN
Bowes	Human	Melanoma	Line	Both	epi	t-PA
C127	Mouse	Breast	Line	Attachment	fib/epi	Human growth hormone
MDCK	Dog	Kidney	Line	Attachment	epi	Veterinary vaccines
BHK-21	Baby hamster	Kidney	Line	Attachment	fib	Veterinary vaccines
Namalva	Human	Lymphoma	Line	Suspension	lym	α-IFN
WI-38	Human	Lung	Strain	Attachment	fib	Human vaccines
3T3	Mouse	Skin	Strain	Attachment	fib	Research
HEK	Human embryonic	Kidney	Strain	Attachment	endo	Urokinase
Primary	Various	Various	Primary	Attachment	various	Vaccines
SP2/O AG14	Mouse	Myeloma	Line	Suspension	lym	Hybridomas/MABs
P3x63Ag8	Mouse	Myeloma	Line	Suspension	lym	Hybridomas (OKT3 MAB)

[a]Morphology: fibroblastic (fib), epithelial (epi), lymphoblastic (lym), endothelial (endo).

[b]Abbreviations: t-PA (tissue plasminogen activator), IFN (interferon), MABs (monoclonal antibodies).

important—considerations. Thus the next steps taken to improve product yields after cell selection were growth and media optimization for the specific cells being used. This two-step procedure exhausted the basic techniques for yield improvement.

Despite the trial-and-error approach of the screening process, many useful types of cells have been discovered over time, and their growth and productivity characterized and optimized. Some examples of the major types of cells used in production are listed in Table 4.

C. Genetic Engineering

Recent advances in the genetic engineering of mammalian cells have resulted in major improvements in the process of selecting cells for producing products and, likewise, in the productivity of these cells. The most dramatic impact that the genetic engineering of mammalian cells has had is that it has allowed greater flexibility in choosing cells as substrates. Human protein production is no longer restricted to production in cells that produce the human product naturally, or in fact to production in human cells at all. The ability to genetically engineer proteins across tissue and species barriers has allowed host cell lines to be chosen based on practical considerations and then genetically engineered to produce the protein of choice.

Cells selected as preferred substrates are those whose history is well known. Thus, they are often those same cell lines originally isolated by screening methods and listed in Table 4. Host cells are chosen, among other reasons, for their ease of use, ability to be genetically engineered, ability to process the chosen product correctly, and relative safety, preferably with the absence of endogenous viruses (e.g., retroviruses) and tumorigenicity. The practicality of employing genetic engineering in mammalian cells to produce products normally requires that the host cell substrate be an immortal cell line. Since cell lines in general are derived from transformed cells, the two safety criteria listed above require considerable effort to meet. (See Sec. VII.) Recently it has been shown that the life span of normal cells can be extended by the incorporation of SV40 oncogenes. This technology has allowed primary human embryonic kidney cells to be subcultivated for more than 100 generations (Goochee et al., 1987; van Brunt, (1987a). The ability to choose a host cell for its biological characteristics has resulted in improved production for a number of mammalian cell products. For example, human tissue plasminogen activator, once produced in human melanoma cells (Bowes) (Kluft et al., 1983), is now also produced in Chinese hamster ovary (CHO) and other nonhuman cells (United Kingdom Patent Application, 1983; Orr, 1988).

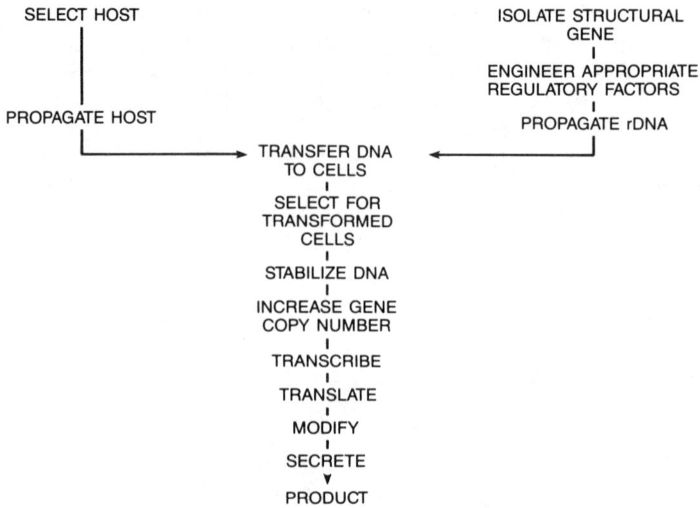

Figure 2 General scheme for genetically engineering a cell line to produce a desired product.

Once the host cell is selected, the gene(s) of choice are inserted along with the appropriate genetic control systems to ensure high levels of protein production. In general, the processes used to genetically engineer mammalian cells are similar in concept, and often in practice, to those used with microbes. The specifics of these processes have been reviewed recently by Weymouth and Barsoum (1986). Figure 2 illustrates schematically the general sequence of events needed to establish a genetically engineered cell line producing a desired product.

A key problem, once the genes of choice have been stably transferred to their new host cell, is to encourage the cell to produce large quantities of the product. Two principal strategies have been employed to increase cell productivity. These are gene regulation and gene amplification.

Gene regulation aims to increase the number of times a single gene is transcribed and then translated into product. Regulatory stimulation may be external, such as the use of poly(rI)-poly(rC) to increase human β-interferon production in mouse cells (Canaani and Berg, 1982), or internal, such as the engineering in of specific nucleotide enhancer sequences to increase the number of RNA transcripts per gene in lymphoid cells (Gillies et al., 1983).

The second major strategy involves gene amplification. Since it is generally true that the amount of product produced is dependent on the number of gene copies for that product, gene amplification aims to increase the number of genes that are available for transcription to produce the product. There are two major types of systems in use for increasing product by increasing gene number. Amplification may be accomplished through the use of genetically engineered "latent" viruses, which exist as multicopy plasmids, or alternatively through the use of certain DNA amplifying drugs.

Bovine papilloma virus (BPV) is the virus most often used for virus-based gene amplification. BPV is a small DNA virus that can be used to introduce genetically engineered genes into rodent cells, particularly those of mice and rats (Law et al., 1981). Cell transformation serves as a marker for cells receiving the foreign DNA (Hsiung, 1985). The DNA itself is maintained as an episome (Law et al., 1981) and is naturally amplified to a level of 30-60 copies (Law et al., 1981), which are often maintained stably even without selective pressure (Zinn et al., 1982). This technology has been used for the production of a number of proteins in mouse cells, including human β-interferon (Zinn et al., 1982) and human growth hormone (Pavlakis and Hamer, 1983).

Alternatively, genes may be amplified by exposing cells to increasing concentrations of specific drugs. The archetype of this is the use of methotrexate (MTX). Methotrexate is an analog of folic acid and an inhibitor of the enzyme dihydrofolate reductase (DHFR). Cells that are resistant to MTX normally produce elevated levels of DHFR due to an increase in gene copy number (Alt et al., 1978). Under appropriate conditions engineered genes will be coamplified with the DHFR gene (Kaufman and Sharp, 1982). Stable cell lines, if amplified in a stepwise selection process, can produce as much as 1000-fold amplification (Kaufman and Sharp, 1982; Shimke, 1984). This technique is frequently used for production of products in CHO cells.

Although genetic engineering of mammalian cells has allowed major advances in the use of these cells to produce products, much room exists for improvements in this new technology. The process of matching cell line, gene engineering and regulatory technology, and product type remains a challenging endeavor. Nevertheless, it is clear that in the future the most cost effective human cell products produced in mammalian cells will likely come from genetically engineered cells of infinite life span and nonhuman origin.

III. MEDIA

Mammalian cells in the body receive nutrients from the blood and lymphatic systems that allow them to grow and reproduce. In culture, cells receive these nutrients from the culture medium.

Table 5 Some Media in General Use

Medium	Cells	Reference
BME: basal medium eagle	BHK-21	Eagle, 1955
F-12: Ham's F-12 medium	CHO	Ham, 1965
MEM: minimum essential medium Eagle	VERO, MRC-5 HeLa, MDCK	Eagle, 1959
DMEM: Dulbecco's modification of MEM	3T3	Dulbecco and Freeman, 1959
IMDM: Iscove and Melchers' modification of DMEM	Hybridomas	Iscove and Melchers, 1978
RPMI-1640: Roswell Park Memorial Institute 1640	Namalva	Morton, 1970
L-15: Leibowitz's L-15	Vero	Leibowitz, 1963
Medium 199	LLC-RK	Morgan et al., 1950

A. Growth Media

Historically, there have been two major categories of media: growth media and production (maintenance) media. Growth media, as the name implies, are designed to support the growth and multiplication of cells. The majority of these media were formulated decades ago. Since most were initially developed for use with a specific cell type, they have undergone many modifications during their optimization for use with different cells. Because of this, and because most growth media are purchased from media supply houses, the Tissue Culture Association has established standard formulas for many of them. Table 5 lists some of the frequently used growth media.

In general, growth media are composed of an aqueous solution of salts, vitamins, amino acids, and carbohydrates. Required exotic nutrients and growth factors are usually provided by supplementation with 5-15% serum but, when known, may be added directly (see Sec. III.D). Because of the presence of serum, the cost of growth media is very high, with serum accounting for 75-98% of the cost. This is a major reason that growing cells is the single most expensive step in mammalian cell product production. In addition, these sera, which are normally of bovine or equine origin and frequently from fetal animals, are periodically unavailable. The requirement for serum also makes growth media subject to lot-to-lot variation, with high protein burdens that lead to purification difficulties and foaming. Additional problems associated with sera include the presence of inhibitors, adventitious agents, product degrading enzymes, bacterial endotoxins, veterinary drugs, nucleases, and other unwanted substances.

For growth media the carbohydrate most frequently used is glucose (Eagle, 1955). Under conditions in which acid production is a problem, however, fructose (or galactose) is sometimes substituted (Imamura et al., 1982; Reitzer et al., 1979), since glucose is believed to be extensively metabolized through glycolysis to lactic acid. In HeLa cells only about 5% of the available glucose enters the citric acid cycle and produces energy (Reitzer et al., 1979). Fructose, in contrast, is almost totally processed through the pentose phosphate cycle with little or no resulting energy (Reitzer et al., 1979). In such cultures the primary function of the sugar appears to be in providing precursor compounds.

In addition, most cells require at least 13 amino acids, including arginine, cyst(e)ine, glutamine, histidine, isoleucine, leucine, lysine, methionine, phenylalanine, threonine, tryptophan, tyrosine, and valine (Eagle, 1955, 1959). Of these, glutamine is the most heavily consumed, for reasons that are not fully understood. When glucose is not present in large quantities, however, it is believed that glutamine may also serve as the cell's major energy source (Reitzer et al., 1979).

Growth media also normally contain most of the B complex vitamins, but usually not vitamins A, C, D, E, or K (Bettger and Ham, 1982). Standard vitamins in media include pyridoxine, riboflavine, thiamine, vitamin B_{12}, nicotinamide, inositol, folic acid, choline, pantothenate, and biotin.

B. Production Media

Cells are normally grown in growth medium until they reach the appropriate concentration for product accumulation. Because of the problems and expense associated with growth medium and because cells often produce product as well or better under conditions in which cell growth and product synthesis are uncoupled, the actual production phase of cell culture is often carried out in a different type of medium than that used to grow the cells.

Production media are selected depending on whether the desired product is produced better by growing or stationary-phase cells. The major goal of these media is to maintain the cells in their most productive phase. In general, production media are less rich than growth media, usually having less protein and complex nutrients. Decreases in the concentration of protein, especially in the form of serum or serum albumin, and complex nutrients vastly simplify product recovery while decreasing the cost of the medium. Media costs are especially important during the actual production phase, when cells may be maintained for months actively producing secreted products by constantly perfusing them with fresh medium (Kluft et al., 1983). The perfusion process, in which fresh medium is pulsed or continuously fed into the system and older medium is removed, serves to feed the cells, remove waste materials, and allow product to be continually harvested (see Sec. VI).

Media used for production in general vary more than media used for growth because of the specialized nutritional needs for individual product production by the cells. Often these media are based on the "powdered" portion of the growth media listed in Table 5, supplemented as needed by constituents that optimization studies have shown to preserve or increase productivity.

C. Serum-Free and Reduced-Serum Medium

Recently, new media have begun to be developed. Collectively these new media are known as serum-free and reduced-serum media. As their names imply, they have little or no added serum. Many are a mixture of the powdered portion of older, conventional media, supplemented with newly discovered growth and attachment factors (see Sato et al., 1982, Sec. III.D). In particular, a 1:1 mixture of DMEM and F12 has been used extensively as the starting point in the development of these new media (see Table 5; Flam, 1982; Barnes and Sato, 1980).

Serum-free media (SFM) have a great appeal for a variety of reasons. A defined medium would allow better control of the growth and production processes. In addition, the elimination of serum helps reduce a number of serum-associated problems, including lot-to-lot variation, the presence of inhibitors, potential contamination by adventitious agents, and product purification difficulties due to high protein burdens. An important goal of these new media is to provide a single medium for both growth and production phases, which would decrease the labor associated with changing from growth to production medium. Finally, these media have the potential to be less expensive than conventional media. Serum-free media are also of use as selective media, such as for the isolation and preferential growth of transitional epithelial cells from the bladder (Detrisac and Moon, 1986).

In general, serum-free media have performed very well for both growth and production for a wide variety of cells (Propst, 1987; Wolfe et al., 1988). The successful use of these media, however, requires at least the same level of care in their selection and modification as for their more complex predecessors. As with the older media, there are many types of serum-free media and the best medium for a particular cell line must be selected. Supplementing the basic medium with additional additives may also be desirable. In some cases cells need several passages for adaptation to the new medium, in order to allow for cell adjustment and the expression of new enzymes. When these practices are applied to using serum-free media, as they routinely are to conventional media, these new media usually permit growth and production at or near normal levels.

Serum-free medium is only now beginning to live up to its economic potential. The reasons are more related to business and sales than science and technology, however. Serum-free media achieve cell growth and product production by providing the cells with the defined additives the cells need. Since many of these supplements are not available in large quantities, their cost is high, making the price of the medium high. With more commercial demand, prices will fall, since the majority of supplements can be produced cost effectively through chemical techniques or genetically engineered microorganisms. Alternatively the gene(s) for the required supplement can be engineered into the host cell along with the gene(s) for the desired product. An example of this is engineering the von Willebrand factor gene into CHO cells to allow production of active Factor VIII in SFM (Kaufman et al., 1987). In the short term, where costs are a factor, substantial savings over conventional media are possible, with most of the advantages of serum-free media, by using reduced-serum media containing 0.25-1.5% serum for growing cells (Propst, 1987).

In the future it seems likely that successful, cost-effective mammalian cell production processes will utilize serum-free medium with its additives for cell growth, followed by product production in the same base medium with or without the addition of supplements. Such media will be designed for use in conjunction with the large-scale bioreactors and feeding techniques described in Section VI.

D. Growth and Attachment Factors

The use of defined media has focused attention on the special growth and attachment factors required by cells. Once provided incognito by serum, their absence is clearly signaled by the failure of cells to thrive in unsupplemented SFM.

In addition to those factors routinely included in conventional growth media (i.e., water, salts, a carbohydrate, vitamins, and amino acids), growth factors of a variety of types are needed by cells (see Sato et al., 1982). Foremost among these are insulin, a polypeptide involved in glucose metabolism, and transferrin, a protein from serum involved in iron transport. Both are required almost universally by cells growing in SFM (Bettger and Ham, 1982; Butler, 1986). Other growth factors that are often required for cells include epidermal growth factor (Barnes and Sato, 1980) and somatostatin (Ambesi-Impiombato et al., 1980).

In addition, cells sometimes need one or more of a variety of steroid-type hormones. These include such compounds as dexamethasone, testosterone, progesterone, and hydrocortisone (Mather and Sato, 1979; Hoshi et al., 1982).

Cells also may require lipid-based factors for growth, including phospholipids, cholesterol, and sphingomyelin (Bettger and Ham, 1982), as well as vitaminlike substances, such as ubiquinone (Bettger and Ham, 1982), and vitamins, such as vitamin E (Bettger and Ham, 1981).

Many early and some current SFM contain large quantities of bovine serum albumin (BSA). Although undesirable for many of the same reasons that serum is, BSA—the cruder the better—was often a necessary early requirement for growing cells under low-serum or serum-free conditions. It is now known that BSA functions predominantly as a lipid carrier (Yamane, 1978). It has been suggested that the fatty acids bound to albumin are important in maintaining the stability of cell membranes under culture conditions (Iscove and Melchers, 1978). In addition, BSA may also carry small amounts of other factors required for growth, as well as remove inhibitors in cell culture (Mendelsohn et al., 1982).

Finally, anchorage-dependent cells often have specific requirements relating to cell attachment (Barnes, 1984, 1987; Barnes et al., 1984). A major factor needed by many cells for attachment to surfaces is fibronectin (Yamada and Olden, 1978). Fibronectin is a glycoprotein of high molecular weight naturally found on the surface of some cells and in blood plasma. In the absence of serum, many anchorage-dependent cells require its addition as a supplement in order to attach to culture vessels for subsequent growth (Orly and Sato, 1979). Fibronectin is not usually needed for attachment to microcarriers. It has been suggested that the failure of transformed cells to attach to surfaces is due to their loss of the ability to produce fibronectin (Vaheri and Mosher, 1978). Another important factor in cell attachment and cell characteristic determination is collagen. This substance is often used to coat microcarriers and is discussed further in Section IV and by Barnes (1984).

E. Other Media Considerations

Although oxygen is not usually considered a nutrient, it is a critical factor in the successful growth of cells. Because the solubility of O_2 in media at 37°C is low (0.2 mmol/liter) and cell usage may be high, oxygen can quickly become a limiting factor. When cells are growing at low concentrations in small-scale culture, such as roller bottles or flasks, it is usually sufficient to provide oxygen by air diffusion, sometimes aided by agitation, as with roller bottles. In large-scale culture oxygen becomes a much more serious problem because of the greater cell numbers and limited air/liquid interface. Potential solutions for maintaining proper oxygen concentrations under larger scale culture conditions include enriching air with pure O_2, sparging, using surface aerators, and using gas-permeable membranes. This area has recently been reviewed by Hu and Wang (1986) and is discussed further in Section V.

In contrast, carbon dioxide is normally considered a by-product of mammalian cell metabolism. It also serves as an intermediate in a number of important cellular synthetic pathways, however, including those that produce fatty acids and nucleic acids. Small-scale cell cultures are routinely grown in CO_2 incubators with 5% CO_2. Many hybridomas, however, produce higher yields of monoclonal antibodies at 7% CO_2. Carbon dioxide dissolves in water to form carbonic acid and can affect the pH of the culture medium.

Mammalian cells are also sensitive to hydrogen ion concentration. Although different cells vary substantially in their pH tolerance, many grow best at a pH between 7.2 and 7.8 (Eagle, 1973). Because of this, buffers are normally included in the media. These are usually bicarbonate or HEPES (Thomas, 1986). To monitor pH, an indicator, such as phenol red, is also included. Along with cell morphology, phenol red color is one of the quickest techniques cell handlers have for gauging the health of a cell culture and its freedom from contamination.

Contamination is a major problem for cell culture. Because of their composition, most media, sera, and other supplements cannot be autoclaved. Sterilization normally occurs by filtration. Since growth media are very rich in nutrients and since mammalian cells grow slowly relative to bacteria and other potential contaminants, any break in aseptic technique may result in contamination of the culture. Therefore, antibiotics are sometimes used in media to prevent overgrowth by microorganisms. Antibiotics used include penicillin or streptomycin, gentamycin, kanamycin, polymixin, erythromycin, neomycin, mycostatin, tylosin, and amphotericin B (Armstrong, 1973). Although antibiotics are useful under certain circumstances, their continuous use is not advantageous for a number of reasons (Coriell, 1973). They can be toxic to cells and decrease productivity. Moreover, their presence in media, directly or as derivatives, may carry over as a contaminant into the final product. This is one reason that penicillin is not used for production purposes (see Sec. VII.E). Finally, antibiotics serve to mask problems in aseptic technique and the presence of cryptic microbial contaminants, which ultimately must be resolved.

Empirically it has been found that cells often grow and produce products better in liquid medium that has been freshly made. In part this may be due to the deterioration with time of important ingredients once the medium is hydrated. Additionally, contaminating microorganisms can grow in liquid media prior to sterilization, using up trace ingredients and releasing undesirable by-products. The presence in media of high levels of endotoxins, pyrogenic cell wall constituents of gram-negative bacteria, will often decrease cell productivity, as well as make product purification more difficult.

Finally, because mammalian cells are derived from warm-blooded animals, it is not unexpected that they prefer physiological temperatures. Most cells grow well at 37°C and are relatively tolerant of temperatures in that range. In contrast to growth, however, certain cells show dramatic decreases in product yields with minor temperature changes of 1-2°C, which are well within the acceptable growth range for these cells. Consequently, for certain products, maintenance of culture temperature within close tolerances can become a critical factor.

IV. GROWTH SUPPORT AND CELL IMMOBILIZATION SYSTEMS

Anchorage-dependent cells must have a suitable substrate upon which to attach for long-term survival and growth. For the vast majority of adherent cell types grown in conventional laboratory flasks or roller bottles, the growth support substrate consists of acid-cleaned glass or one of the many commercially available hydrophilic tissue culture plastics (Nardelli and Panina, 1976). Primary cells and some cell strains prefer, or may even require, the substrate surface to be coated with collagen or other proteins (Barnes, 1984) to more closely simulate the in vivo environment (see Sec. III.D). Fortunately, most adherent cells of commercial interest will readily adapt to a variety of synthetic materials and medium components, selected to accommodate high densities of immobilized cells in bioreactors (Barnes, 1987; Barnes et al., 1984).

The growth support system is a crucial component in the selection of a system for expanding a particular cell in long-term culture (Harper and Juliano, 1981; Juliano and Gagalano, 1977; Kauvala et al., 1981; Kleiman et al., 1987; Liang et al., 1987). Conventional stirred tank or airlift bioreactors, like laboratory spinner flasks, were originally developed as vessels for suspension cultures of microorganisms and, later, were modified to accommodate mammalian cells (Cherry and Hull, 1960; de Bruyne et al., 1974; von Wedel, 1986a; Zwerner et al., 1981). The need to grow large quantities of attachment-dependent cells for the commercial production of virus vaccines and, later, interferons resulted in the development of concomitant technology for growing adherent cells on suspended beads or "microcarriers." The intent was to take full advantage of existing large-scale suspension cell culture equipment to produce products from "suspended" adherent cells. The specific gravity and size of the microcarrier bead were important for proper suspension and convective flow within the bioreactor chamber, whereas the surface chemistry proved critical for promoting the attachment and growth of different kinds of cells. The technology of growing cells on microcarriers has now evolved since its introduction by van Wezel over a decade ago into a competitive field of its own (van Wezel, 1967; Kluft, et al., 1983), offering a wide range of materials and configurations to accommodate different bioreactors and cell types (Feder and Tolbert, 1985a,b; Lydersen, 1987; Spier and Griffiths, 1985, 1988; Swarz, 1985; Thilly, 1986; Fleischaker, 1987). In addition, new perfusion culture bioreactor designs (see following section) are providing the opportunity to immobilize adherent cells on a variety of new materials, to enhance the maintenance of high-density cultures over long periods of time.

A. Dextran Microcarriers

To be effective growth support systems for growing adherent cells in a suspension culture mode, microcarriers had to be nontoxic, biocompatible beads of appropriate density (specific gravity to stay suspended) with surface structure and chemistry suitable for cell attachment and growth. DEAE-dextran beads (Cytodex, Pharmacia Fine Chemicals, Piscataway, NJ) were among the first materials developed to promote the attachment and growth of adherent cells in suspension culture vessels, such as spinner flasks, stirred tank, and airlift bioreactors. Cytodex microcarriers are now available in three forms: positively charged (DEAE) throughout the matrix, surface coated with DEAE, and collagen-coated beads. All forms of the dextran beads are spongelike porous matrices that provide considerable surface area for attachment of cells. The charged surfaces of the beads allow cells in suspension to be attracted electrostatically for rapid initial contact to facilitate the more gradual process of attachment. The density of the derivatized beads has proven to be near optimal for most stirred tank and airlift suspension culture bioreactors. For the isolation of nonsecreted intracellular products, confluent cells may be harvested from the microcarriers using digestive enzymes, although the harvest conditions can prove very harsh to the cells and recoveries are limited. Extensive literature has been published on the application of these microcarriers to the growth and scaleup of a wide variety of adherent cells (Pharmacia/LKB Biotechnology, Inc., 1984). Similar microcarriers have been developed with collagen complexed to a DEAE-dextran bead (Agrobead, Galil, Israel).

Although the dextran microcarrier bead technology provided the first widely accepted means of growing adherent mammalian cells in conventional suspension (tank) culture systems, there have been several drawbacks in using this approach. One is cost, particularly for scaleup considerations. Another is that these beads could not be recycled (cleaned and autoclaved) for repeated use in routine culture applications. In addition, dextran beads consist of three-dimensional matrices that can be penetrated by growing cells. The cells tend to secrete proteases in an attempt to become dislodged from the interstices of the microcarrier. These proteases threaten the integrity of the secreted protein product accumulating in the medium, as well as generate digestion products (e.g., degraded collagen) that lessen the initial purity of the secreted biological. Some cell detachment may also result.

B. Other Microcarrier Technologies

Polystyrene core microcarrier beads were introduced (Biosilon by Nunc; Cytospheres by Lux) as an alternative to the original dextran beads in an attempt to employ the time-proven surface chemistry of plastic culture flasks for immobilizing adherent cells in bioreactors. The beads are not available in glass- or collagen-coated forms. Unfortunately, these plastic beads, which have no density modifications, have proven too heavy (i.e., specific gravity greater than 1.05 g/ml) for most suspension applications.

Solid-core collagen beads were then introduced as alternative microcarriers that would provide attachment surfaces believed to resemble those in vivo. These porous, jellylike beads are made from porcine dermal collagen with appropriate size and density specifications for suspension bioreactors, but unfortunately they appear to dissolve slowly during long-term cultures. As a result, the secreted protein product in the medium becomes contaminated with increasing amounts of degraded collagen.

Other materials, including glass (Brown et al., 1988; Varani et al., 1983; Whiteside and Spier, 1985), polyacrylamide (IBF, France), porous silica, and different cross-linked proteins, were employed to generate microcarrier products with their own limitations in terms of maintaining appropriate density ranges, porosity, surface chemistry, leaching of contaminants, adsorption of cell toxins or desired secreted products, toxicity of residual chemicals in the beads, breakdown during sterilization, problems in the reuse of beads, and difficulty in harvesting viable cells. Some of these problems have been overcome for specific applications by in-house researchers at major research institutions (e.g., the Swedish Biological Laboratory, Stockholm) and biotechnology companies.

The recent commercial introduction of solid plastic microcarrier beads of carefully controlled densities, however, appears to address most of these problems (Solo Hill Engineering, Ann Arbor, MI). The beads have solid cores, so cell attachment is restricted to filopodia adhesion to the outer surface of each sphere. Any limitations this may cause in terms of decreased surface area are more than balanced by the potential for multilayers of cells and the reduction in proteolytic enzyme activity in the cell culture medium. Cells apparently can move around and reattach to the solid surface without having to secrete large amounts of proteases as they would inside the interstices of a porous matrix bead. The plastic beads are available with untreated plastic surfaces (uncharged, so seeding relies on random contact for attachment) or with glass- or collagen-coated surfaces. The bead densities (range 1.02-1.04) can be selected for the intended bioreactor application. Plastic- and glass-coated beads can be cleaned and autoclaved for reuse. Collagen-coated beads are made with porcine dermal collagen that is electrostatically adsorbed to the plastic. Although not covalently bound to the bead, there appears to be little contamination of the medium by collagen, in part because the coating process leaves only a monolayer of the protein on the plastic bead. Harvesting of cells proves to be much more efficient with all three forms of the new plastic beads. Furthermore, the lower cost of these plastic-core beads, combined with their potential for reuse, may accelerate their acceptance for industrial scaleup applications (Varani et al., 1985, 1986, 1988).

C. Applications of Microcarrier Technologies

Originally, microcarriers were intended as a means of scaling up the growth of adherent cells for conventional stirred tank or airlift suspension culture bioreactors for the production of virus vaccines or the recovery of secreted protein products. They have been enormously successful in this regard. Cell densities of over 10^6 cells per ml can be maintained in long-term culture using modified tank bioreactors. Toxicity problems have been eliminated to the extent that densities of microcarriers of up to 40 g/liter can be maintained without adverse effects on cell metabolism or productivity. Scaleup efforts have led to batch microcarrier cultures of over 2000 liters.

Nonetheless, in the interest of generating higher densities of viable cells that in turn lead to higher initial purities of desired product, alternative bioreactor configurations based on perfusion culture have been tried, including hollow fiber cartridges and tangential flow membrane units. Although these perfusion systems offer several advantages in terms of the circulation of nutrients, the removal of metabolites, and the harvest of desired product during long-term cultures, the practical aspects of maintaining the microcarriers in suspension have proven difficult. In contrast, the adaptation of perfusion culture technology to conventional tank bioreactor designs has been very well demonstrated as a means of enhancing the production of labile secreted proteins from long-term cultures of adherent cells. Radial flow bioreactors would also appear promising as a culture technology that could augment the net concentration and biological activity of secreted biologicals from high densities of cells immobilized on microcarriers. All the microcarrier products discussed are commercially available in research- to industrial-scale amounts and are often featured in cell culture contract production services.

D. Microspheres and Fluidized Bed Technology

Although seemingly a variant of microcarrier products, the "microsphere" concept was developed independently by Verax Corp. (Lebanon, NH) for specific application as a matrix to protect cells in specialized large-scale fluidized bed bioreactors (Young and Dean, 1987; van Brunt, 1986; Rundstadler and Young, 1987). The microsphere itself is a highly permeable, spherical collagen lattice within which cells are trapped by a proprietary process. Thus, the cells do not simply adsorb to the collagen, and in fact, the process appears equally suited for suspension as well as attachment-dependent cell types. The 500-600 μm microsphere is weighted with microscopic inert metal pellets to permit proper circulation in the fluidized bed. High densities of cells can be maintained within the microsphere because of the mass transfer properties inherent in the solid-liquid fluidized bed configuration, discussed in more detail later. High cell density, uniformity of the culture environment, and scaleup are among the advantages this technology offers for industrial production of biologicals in spite of its high cost. The technology is available commercially from one company only for very large scale in-house facilities and is not offered as a contract production service.

E. Cell Encapsulation Technologies

An earlier innovation for trapping cells within matrices was the development of sodium alginate beads that could be used to immobilize cells in a porous, inert matrix in the form of a bead, typically around 800 μm in diameter (Lim, 1984; Jarvis and Gardina, 1983; Posillico, 1986; Posillico et al., 1987; Nilsson, 1987; Schreiver et al., 1983). Alginate is a gluronic and mannuronic acid complex extracted from kelp. The alginate bead could then be coated with polylysine so that, upon digesting away the alginate, viable cells would be left growing in suspension within a membrane of polylysine. By controlling the polymerization conditions, polylysine membranes could be formed with different porosities, for example, to exclude high-molecular-weight medium proteins while retaining secreted protein products within the beads. An entire technology was developed around this concept by Damon Biotech (Needham Heights, MA) using stirred tank bioreactors to grow up large batches of cells trapped within the capsules (Jarvis and Gardina, 1983; Rupp, 1985). The secreted product accumulates inside the capsule as the cell density increases (and viabilities presumably decrease). Hence, this culture process is limited to batch mode operations and the production of biologicals that are relatively stabile to proteolytic degradation. Nonetheless, this cell encapsulation technology is being used successfully to produce therapeutic quantities of monoclonal antibodies for cancer treatment. The Encapcell technology is available commerically as a contract production service by Damon Biotech.

A spin-off technology of the cell encapsulation approach was employed by Karyon Technology (Norwood, MA) to overcome the limitations of the original methods. Their approach was to leave the cells trapped in the sodium alginate matrix (no polylysine membrane) and simply perfuse the immobilized cells with medium. The continuous perfusion allowed separation of product from cells and the establishment of long-term cultures, a major advantage for low-productivity cells or for cells secreting labile products. Unfortunately, the company that pioneered this approach no longer offers a contract production service, but a similar modified sodium alginate and gelatin procedure for suspension cells is now available through New Brunswick Scientific (Familletti, 1987; Familletti and Fredericks, 1988).

F. Glass Bead and Packed Bed Technology

Secreted biological products have also been produced successfully from high-density cultures of attachment-dependent cells grown on glass beads packed into vertical columns, somewhat analogous to a chromatography column (Bioresponse Inc., Haywood, CA). Fast flow rates and extensive aeration of incoming medium have been shown to provide sufficient mass transfer of nutrients and oxygen to sustain confluent cultures for over a year. However, there are definite limitations on the available bead surface area and medium flow rates needed to support high densities of adherent cells, so scaleup is restricted to increasing the number of packed bed column modules rather than to increasing the size of each module.

G. Ceramic Matrix and Perfusion Technology

An entirely novel approach to providing an optimal growth support system for attachment-dependent cells, and suspension-type cells as well, is a recently developed perfusion culture system based on a ceramic matrix substrate (Bognar et al., 1983; Lydersen et al., 1985; Putnam, 1987), known as the Opticell Culture System (Charles River Biotechnical Services, Inc., Wilmington, MA). The substrate is actually an extruded ceramic cylinder with tiny square channels running the length of the core. The smooth walls of the channels provide the attachment surface for adherent cells. Suspension cells are grown on a similar cartridge in which the ceramic is much rougher and can physically trap the cells as they settle into the interstices of the matrix. In both cases, the vast majority of viable cells in the perfusion system divide and remain immobilized on the ceramic matrix. Constant circulation of medium with automatic feed-harvest cycles provides a very uniform culture environment for growing cells to very high densities in the matrix. The ceramic matrices have been used to produce secreted products from a wide variety of adherent and suspension types of mammalian cells. The complete perfusion system, which is available for placement in facilities or as a contract production service, is described in greater detail in Section V.

H. Future Trends

The immobilization of mammalian cells on growth support systems facilitates the scaleup production of biologicals from high-density cultures of mammalian cells, particularly attachment-dependent cells. High densities of cells may improve the per cell productivity of a culture, but more importantly, they improve the initial purity of the product, thereby lowering costs and efforts downstream.

Microcarriers and encapsulation beads have enabled conventional stirred tank and airlift bioreactors, still widely used in the industry, to be used for growing adherent cells in a suspension culture environment. The integration of perfusion culture technologies can further improve the productivity of cells grown on microcarriers in these bioreactors.

Collagen microspheres in fluidized beds and ceramic matrix perfusion systems are two examples of new cell immobilization technologies that have improved productivity for biologicals harvested from both adherent and suspension cells. The industry trend in producing biologicals from immobilized cells favors development of growth support systems that are versatile (suitable for adherent or suspension cells), inexpensive (disposable) or reusable, capable of promoting high densities of viable cells for long-term cultures, and suitable for research-, pilot-, and production-scale bioreactor applications. Almost certainly, growth support systems of the future will be designed for use in perfusion culture systems or in tank bioreactors modified with perfusion culture technologies. At the research scale, at least, there will be increased demand for growth support matrices from which cells may be readily detached without proteases or harsh conditions. Future production cell lines, recombinant protein expression lines, or continuous cultures of primary cells may require specialized matrices for maximal expression in culture.

V. LABORATORY-SCALE CULTURE SYSTEMS

A. Conventional Laboratory Tissue Culture

Surprisingly, most research laboratories, and even some pharmaceutical industry production laboratories, still employ conventional roller bottles and spinner flasks to produce larger amounts of a needed biological from mammalian cells. Roller bottle technology has changed little since the 1930s except for the advent of disposable plastic and larger bottles. Spinner flasks have changed little as well. The most notable improvement has been in the development of flasks, impellers, and magnetic stirrers designed specifically to keep microcarriers in suspension without damaging the attached mammalian cells. In their normal mode of operation, both roller bottles and spinner flasks require considerable labor for maintaining cultures. Both approaches were intended for batch cultures, but they can each be taken through a series of feed-harvest cycles to extend cultures over longer periods of time. Cell densities are limited to around 10^6 cells per ml, yet research facilities routinely run 200-300 roller bottles per day to produce only a few hundred milligrams of a biological. Some vaccines are still produced commercially from thousands of roller bottles, in part because the technology was originally specified in the FDA-approved procedures. Similarly, spinner flasks are scaled up to 10 or 20 liters for batch cultures of cells that could have been grown much longer in laboratory-scale bioreactor systems.

B. Flexible Cell Culture Flasks

A marked development in laboratory tissue culture vessels has been the introduction of flexible cell culture flasks made of gas-permeable plastic that are similar in appearance to blood bags (Fenwall Division of Baxter Laboratories, Deerfield, IL). For routine mammalian cell culture work and laboratory scaleup, these bags offer the advantages of closed system transfers (no pipetting), improved oxygen-carbon dioxide gas exchange for high-density cultures, and convenience of easy scaleup within a conventional CO_2 incubator

(Kolanko and Symbol, 1987). Furthermore, they greatly facilitate the seeding of bioreactors through direct septum tubing connections with much lower risks of contamination. Flasks come in 1-3 liter sizes and are available in transparent plastic designed for the attachment of adherent cells.

C. Stirred Tank Bioreactors

The original stirred tank bioreactors used for mammalian cells were, of course, simply microbial fermentors adjusted or slightly modified to minimize damage to fragile cells (Cherry and Hull, 1960; Martin et al., 1987; Lebherz, 1987). These bioreactors consist of tanks with rotating impellers of various configurations that maintain constant agitation for adequate gas, nutrient, and heat exchange. The impeller stir speed and sparging mechanisms were the only factors that were changed initially. Gradually, more sophisticated approaches were taken to redesign the vessel shape, impeller configuration, gas transfer system, constant temperature control, head plate and gasket system, probes, pH controls, level control, and other features of the fermentor to better meet the needs of suspension or microcarrier cultures of mammalian cells. Manufacturers of microbial bioreactors around the world started selling modified equipment for mammalian cell culture. In spite of these changes, however, stirred tank bioreactors can perform little better than a conventional spinner flask (in terms of final cell density or product concentration), even with elegant process controls. There are simple limitations in mass transfer of nutrients, metabolites, and gasses that cannot be overcome without some type of perfusion system. Hence, most laboratory stirred tank bioreactors in batch operation serve as semiautomated spinner flasks or roller bottle assemblies to produce research quantities of biologicals.

Laboratory-scale stirred tank bioreactors are generally considered to have volumes in the range of 1-20 liters and are available as glass, stainless steel, or glass-stainless steel combination vessels. Tanks of under 15 liters are usually made of glass and must be sterilized by autoclaving. Bench top or stand-alone steel units from 7 liters and up are normally designed to be sterilized in situ.

Oxygen is almost universally the rate-limiting factor for cell growth and productivity in high cell density bioreactors. Most stirred tank bioreactors use some form of perforated tubing to bubble (sparge) air or specific gas mixtures directly into the cell suspension. Careful control of bubble size is essential to minimize damage to suspension cells or to cells trying to adhere to microcarriers. Alternative aeration techniques include the use of silicone tubing to diffuse oxygen directly into the medium and injection of oxygen or air into the headspace above the culture (see Sec. III.E).

A recently modified stirred tank system has incorporated a unique gassing-impeller system and perfusion technology to enhance the productivity of suspended (or microcarrier) cultures. The Celligen (New Brunswick Scientific, New Brunswick, NJ) laboratory-scale cell culture system has tried to combine conventional fermentation technology with elements of mammalian cell culture processes to improve gassing and increase medium turnover while minimizing the shearing of fragile cells. The gas exchanger-impeller system consists of a revolving hollow shaft with three tapered discharge ports that pull cells or microcarriers up into the core of the shaft for gassing and then out through the ports for recirculation. Perfusion operation for suspension cultures is achieved by connecting a hollow fiber cartridge into a medium recirculation loop to remove metabolites and harvest secreted product. Microcarrier cultures of adherent cells are maintained by exchanging medium via a simple decanting column installed into the stirred tank headplate. An 80 μm steric barrier screen retains the microcarriers as medium (and product) are pumped vertically out of the vessel.

Another recent modification in stirred tank technology by Pegasus Industrial Specialties (Scarborough, Ontario) that appears promising has been the replacement of the usual rotating impeller with a vertically vibrating agitator plate (the Vibro-Mixer concept). Agitation is frequency controlled to provide very low shear during mixing for fragile cells, and aeration has been improved by introducing air-gas mixtures directly below the vibrating disk.

Process control varies widely with different manufacturers. The most basic systems simply display analog readings for temperature, fluid level, RPM, and pH. Dissolved oxygen concentration readings require a sterilizable oxygen probe, usually sold as an optional feature. In-dwelling, sterilizable probes for glucose, lactic acid, and ammonia have not been developed. These parameters can be monitored manually from daily samples taken to assay product levels. As a result, process control features are very limited for mammalian cell culture applications since several parameters must still be assayed and adjusted by the operator. Semiautomated operation is available for many systems along with data recording features that provide a profile of measurements and responses to changes in RPM. pH, and oxygen. These parameters become critical for longer term, continuous cultures made possible by accessory equipment.

For example, continuous feed-harvest modes of operation are relatively easy to set up with pumps, solenoid valves, and sensors. Some manufacturers now offer these accessories as options. Similarly, continuous perfusion operation of stirred tank bioreactors can be achieved by using ultrafilter membrane cartridges (flat sheet and hollow fiber modules are available). Microporous membrane modules can also be employed to continuously harvest secreted proteins from the bioreactor. At least one manufacturer now offers hollow fiber accessories for their stirred tank bioreactors. These features are important not only for improving the performance of cells in the laboratory but also to model the performance of those cells when scaled up for production. Fortunately, most conventional stirred tank bioreactors can be retrofitted with process control and perfusion culture technologies to greatly improve their performance for sustaining long-term, high-density cultures (Leist et al., 1986). In spite of several systems being offered as "computer-controlled bioreactors," only a few systems are available today which are fully capable of integration with a series of bioreactors under single computer or operator control. This advanced process control system features distributed logic control and redundant microprocessor control modules that can be retrofitted to almost any laboratory- or pilot-scale stirred tank bioreactor system.

Contract production services are available for testing the performance of a client's cell line with laboratory- and pilot-scale stirred tank bioreactors. One American and over twenty European and Japanese companies sell laboratory-scale stirred tank bioreactors.

D. Airlift Bioreactors

Like stirred tank systems, airlift bioreactors for mammalian cell culture evolved from fermentors designed for yeast and aerobic bacterial cultures. The transition involved modifications in vessel shape and draft tubes to improve circulation, more efficient heat exchangers, more sophisticated process controls, and less damaging spargers. The conversions have been swift, but the results have been disappointing because, once again, these bioreactors cannot sustain cell densities greater than those obtained with a conventional roller bottle or spinner flask. The same limitations to the mass transfer of oxygen, nutrients, and metabolites prevent airlift bioreactors from generating higher concentrations and higher purity of product. Also, without accessory technology for perfusion culture operation, the airlift bioreactors are usually limited to batch runs.

For scaleup of hybridoma cells, however, the airlift suspension culture technology has proven efficient and cost effective (Arathoon and Birch, 1986; Birch et al., 1985a,b; Cortessis and Proby, 1987). Several major contract production companies (both inside and outside the United States) have established good track records using modified airlift bioreactor technology to commercially produce clinical trial quantities of therapeutic monoclonal antibodies (e.g., Celltech, Slough, UK). However, more delicate suspension cells or microcarrier cultures have yet to be shown as tolerant to this approach.

A viable solution to the shear problems and mass transfer limitations of conventional airlift bioreactors has been the combination of cell immobilization or encapsulation techniques with a redesigned, funnel-shaped glass bioreactor vessel. A 3 liter bench top system has been developed that more effectively provides airlift agitation and oxygenation to encapsulated suspension or immobilized adherent cells through a stainless steel frit at the base of the tapered vessel. Microprocessor controls regulate medium flow and recirculation rates through the system. High cell densities of encapsulated hybridomas and adherent CHO cells (attached to fiber glass, silica particle, or stainless steel matrices) were achieved, permitting the harvesting of high-purity secreted cell products.

Continuous feed-harvest and perfusion culture operations are possible on airlift bioreactors retrofitted with the approrpiate semipermeable membrane cartridges, pumps, sensors, and controls. Current laboratory-scale airlift bioreactors are designed for batch runs and are not sold with these accessories or process control options. Fewer than five manufacturers of airlift equipment sell bioreactors for the laboratory scale, including LH Fermentation (Haywood, CA) and Chemap (Woodbury, NY).

E. Membrane Perfusion Bioreactors

True perfusion bioreactors maintain cells in a constant environment separated from a circulation medium by a semipermeable membrane (Seaver, 1987; Schreirer, 1988; Adamson et al., 1983). Unlike the conventional stirred or airlift tank bioreactors, a continuous perfusion cell culture system allows the perpetual removal of waste products and the uninterrupted replenishment of oxygen and essential nutrients necessary for optimal production. As a result, high densities of cells can be maintained on one side of the membrane while rapid medium flow rates can be tolerated without concern for damage to the cells. The efficiency in mass transfer accounts for the fact that cells can be grown at one or two orders of magnitude higher densities in perfusion systems than would ever be possible in nonperfusion bioreactors.

Laboratory-scale membrane perfusion systems are available in two basic configurations: hollow fibers and flat sheet membranes (Altshuler et al., 1986; Andersen and Gruenberg, 1987; Brown et al., 1985; Evans et al., 1986; Hirschel, 1986; Hopkinson, 1985; Ku et al., 1981). They are similar in concept but differ widely in the configuration and composition of the membranes.

Small bench top hollow fiber bioreactors are available as separate modules with or without pumps, plumbing, probes, and control systems (e.g., by CD Medical Inc., Miami Lakes, FL, and Endotronics, Coon Rapids, MN). All hollow fiber growth chambers have the same cylindrical configuration with bundles of fibers running through the center. They resemble kidney dialysis cartridges because the technology is an offshoot of that industry. The cells live in the extracapillary space surrounding the fibers. The hollow fiber capillary membranes are then able to deliver a constant influx of freshly oxygenated medium supplemented with fresh nutrients and protein supplements (serum or defined medium components).

Oxygenation of high-density cultures is usually achieved by installing a silicon rubber hollow fiber aerator in line with the medium upstream of the growth chamber. The oxygenated medium flows through the lumen of the hollow fiber and diffuses out to the cells growing in the extracapillary space. Because the diffusion distance for oxygen required for high densities of cells appears to be limited to around 100 μm, conventional hollow fiber growth chambers need to have supplemental convective flow maintained in the extracapillary space. This limitation was not appreciated in the early days of using hollow fiber technology for growing cells. Today, several techniques have been developed to improve the circulation of cells and medium outside the hollow fibers, but these options are usually not offered with conventional bench top hollow fiber bioreactors.

The hollow fibers themselves vary widely in composition and are either "microporous" membranes, referring to nominal pore sizes of around 0.2 μm, or "ultrafilter" membranes with molecular scale porosities available in different cutoff ranges from 10,000 to over 100,000 molecular weight. The fibers may be hydrophobic or hydrophilic, and some are autoclavable, depending on the composition and the manufacturer (e.g., Microgon, Inc., Laguna Hills, CA).

The first commercial hollow fiber units sold for mammalian cell culture by Amicon Corp. were made from cellulose acetate, another carryover from the original medical dialysis technology (Schratter, 1976). These fibers are typically available as anisotopic, ultrafilter membranes that are susceptible to mechanical damage and cannot be autoclaved or steam sterilized. For cell culture purposes, they have some major limitations, particularly since cells grow into the fibrous matrix of the hollow fiber (the ultrafilter qualities of the fiber are contributed by an inner layer of cellulose acetate, hence the term "anisotopic"). Furthermore, it is desirable in most cell culture situations to allow the secreted product to diffuse away from the high-density cells. Ultrafilter membranes do not permit this diffusion, so product must be recovered by harvesting the cells and medium in the extracapillary space. More recent ultrafilter hollow fiber bioreactors are isotopic, with thinner fiber walls, and are now available as part of a bioreactor system with very basic process controls (lacking, for example, dissolved oxygen monitoring and control). Ultrafilter membranes in hollow fiber growth chambers can be used to produce high concentrations of monoclonal antibodies and viruses from suspension cells. This technology should be confined to laboratory-scale applications in which the product can tolerate prolonged exposure to very high densities of cells. If cell viability is not maintained at a very high level in these units, the desired product is continually subjected to degradative enzymes released from necrotic cells.

Higher viabilities and continuous product recovery are two critical reasons that microporous membranes (e.g., 0.2 μm pores) have proven important to hollow fiber growth chambers, particularly in mass culture applications. Typically the membranes are made of hydrophobic polysulfone, polypropylene, or hydrophilic cellulose acetate-cellulose nitrate. These hollow fiber units are usually autoclavable and are available in a variety of sizes and configurations. One manufacturer has introduced a laboratory-scale bioreactor system based on microporous hollow fibers using semiautomated microprocessor controls.

An important breakthrough for hollow fiber technology came with the development of the "tricentric" bioreactor by Setec (Livermore, CA). This fiber-within-a-fiber configuration creates a 200 μm thick annular space in which all cells grow without every being more than 100 μm from an oxygen-nutrient source. Conceivably cells could grow to near tissue densities in this space without ever being more than 10 cell diameters from the semipermeable membrane. Another important feature is that the membranes and the growth chamber module itself are made entirely of polypropylene and thus are amenable to repeated cleaning and autoclaving. This new product is currently undergoing field tests.

Flat membrane sheet bioreactors have also been developed by Millipore Corp. (Bedford, MA) for laboratory-scale production of secreted biologicals from mammalian cells. This small unit consists of alternating layers of microporous (0.65 μm) hydrophilic membranes that create laminar compartments in which cells grow to high densities. Tangential flow circulation of medium above and below each of the 10-cell compartment layer provides continual diffusion of nutrients, removal of waste products, and gas exchange (via silicone tubing) as in other perfusion systems. The bioreactor is not sold with any process control technology and is not available in a larger scale.

Hollow fiber units and flat sheet membrane bioreactors are designed to be scaled up by increasing the number of cartridge modules; problems with gradients and diffusion rates limit the scale of each membrane cartridge module.

In conclusion, the standard hollow fiber membrane bioreactors can provide an efficient and cost-effective mechanism to grow high densities of hybridomas and other suspension cells secreting biologicals. However, they have not been found to be as effective for growing attachment-dependent cells because of limited surface areas, incompatible surface chemistries, and diminished diffusion rates as cells clog the membranes. In fact, the formation of gradients and "dead zones" in the extracapillary space has been the major criticism made against hollow fiber bioreactors, particularly for noncirculating cells.

F. Ceramic Matrix Bioreactor

An entirely different approach to perfusion cell culture is to immobilize cells on a surface and recirculate medium directly over the cells rather than relying on the diffusion of nutrients, metabolites, and gasses across semipermeable membrane barriers. This direct perfusion technique uses the previously described ceramic material to anchor both adherent cells and suspension cells at high densities in a continuous feed-harvest operation (Bognar et al., 1983; Putnam, 1987; Lydersen et al., 1985). The Opticell cell culture system incorporates this extruded ceramic cartridge bioreactor into a bench top process control unit complete with stand-alone microprocessor automation, continuous feed-harvest modes, pH and gas adjustments, and a sterile laminar flow cabinet (Charles River Biotechnical Services, 1987). Continuous calculation of the oxygen consumption rate (OCR) using dual dissolved oxygen probes (oxygen in versus oxygen out) provides a reliable indicator of cell respiration and metabolic activity that is sensitive to changes in culture conditions. Hence, the unit can serve as an informative process development tool for testing new culture conditions or cell lines, as well as performing as a research-scale bench top production unit. When equipped with a single 0.5 m^2 surface area ceramic cartridge growth chamber, the Opticell 5200R research unit can produce over a gram of monoclonal antibody per week from murine hybridoma cells. The system is also available in pilot- and production-scale configurations.

G. Radial Flow Packed Bed Bioreactor

Another emerging perfusion cell culture technology developed by Sepragen (San Leandro, CA) stems from the recent development of radial flow chromatography systems known for their high flow rates and amenability to scaleup. Unlike standard vertical chromatography columns or packed bed cylindrical bioreactors, these columns have a flow path that starts at the top by moving radially outward from an entry port to an outer cylindrical barrier, then proceeding down the outside of a permeable sleeve while penetrating the sleeve to pass radially inward through the packed bed to a smaller diameter inner collection tube, also of a permeable material. The inner and outer sleeves retain the packed bed of immobilized cells while the medium flows continuously through it and is removed through a port at the bottom of the inner collection sleeve. Scaleup is achieved by extending the length of the column along the vertical axis without significantly changing the radius of the column, optimized for different culture methods and cell types. When protein-free medium is circulated through the bioreactor after stabilizing a high-density cell culture, high concentrations of nearly pure product can be harvested routinely from the system.

The concept of radial flow recirculation of culture medium through a packed bed of cells (e.g., cells immobilized on glass beads, microcarriers, collagen matrices, or stainless steel shavings) is appealing because of the high flow rates and rapid mass transfer that should be achieved through very high densities of attached cells (e.g., in excess of 10^8 cells per ml). The concept is equally suited to suspension cells immobilized on biocompatible surfaces (e.g., ceramic matrices) or encapsulated in permeable matrix materials (e.g., sodium alginate, collagen, or gelatin). Coupled with hollow fiber or other membrane-based aeration and dialysis systems (Tharakan and Chau, 1986), the radial flow bioreactor has great potential for efficiently producing secreted products from mammalian cells at the laboratory scale and beyond.

H. Future Trends

Perfusion cell culture systems are clearly growing in popularity and sophistication as researchers strive to maintain high densities of mammalian cells over prolonged periods of time in small-volume bench top bioreactors. Through modifications of conventional tank designs and the introduction of several novel bioreactor concepts, perfusion culture technology has grown beyond the simple ultrafilter hollow fiber dialysis cartridges that set the precedent for this approach over 10 years ago. More reliable and intelligent process control instrumentation has helped automate laboratory-scale culture units, but it has pushed up the costs for these systems as well. The oxygen consumption rate has emerged as a prime physiological parameter for monitoring the respiration and metabolic activity of immobilized cells in response to changes in the culture environment. Future systems will employ additional probes and integrate multiple physiological parameters to further optimize the secretion of desired products from mammalian cells in long-term cultures. The overall trend in the industry is therefore a move away from batch cultures toward more controlled, stable long-term perfusion cultures. The trend naturally calls for improved automation process control technologies, which will eventually lead to an integration of "upstream" cell culture operations with "downstream" processing of the secreted mammalian cell products.

VI. PILOT AND PRODUCTION-SCALE CULTURE SYSTEMS

Several of the commercially available laboratory-scale bioreactors are also sold as pilot- or production-scale systems, which makes them particularly attractive to industrial customers anticipating scaleup needs for valuable mammalian cell products. Each category of cell culture scaleup technology is considered here in general terms, starting with those that have emerged out of laboratory-scale systems, followed by those that have been designed exclusively for large-scale operations. Several large-scale bioreactor technologies are not for sale but are offered to customers as a contract production service instead. Keep in mind that pilot-scale culture systems for experimental quantities of a therapeutic biological may be more than adequate for generating actual production-scale quantities of biologicals used only for diagnostic applications. Recent reviews include those by Glacken et al. (1983), van Brunt (1987b), and von Wedel (1986a,b).

A. Stirred Tank Bioreactors

Modified stirred tank bioreactors (e.g., those of Chemap, Woodbury, NY, and LSL Biolafitte, Princeton, NJ) are being used successfully in biotechnology and pharmaceutical companies for growing a wide variety of mammalian cells, either as suspension cultures or as microcarrier cultures for adherent cells (Reuveny et al., 1986a-c). Conventional stirred tank systems are used primarily for batch cultures as in producing virus vaccines (e.g., polio vaccine from green monkey kidney cells).

Modification of these tank designs with hollow fiber dialysis and production harvesting modules affords the opportunity to maintain stable cultures at higher densities over a period of weeks or months as with dedicated perfusion culture systems. The advantages in terms of decreased labor, increased production yields, and improved purity levels suggest that most firms that have already invested heavily in stainless steel in the form of conventional stirred tank "fermentors" would be anxious to retrofit their equipment for perfusion culture operation. The conversion often also requires the installation of more sophisticated process control instrumentation to properly monitor and control long-term, high-density cultures and the harvesting of secreted products (Fleischaker, 1986). Fortunately, fairly cost-effective process control technologies are becoming available to assist in the retrofitting of conventional deep tank bioreactors for perfusion culture operation.

Contract production using stirred tank equipment is primarily limited to scaleup services based on the polylysine encapsulation technology described earlier. This approach has been most successful for generating large amounts of murine monoclonal antibodies;

it is not recommended for the scaleup production of a highly labile or protease-sensitive biological.

B. Airlift Bioreactors

The largest contract production service in the world for mammalian cell products, largely murine monoclonals at this time, has been established in England using modified large-scale airlift bioreactor systems (Celltech, Slough, UK). By now several kilograms of monoclonal have been produced at this facility alone. The service is expanding to provide hundred gram quantities of other biologicals, such as erythropoietin, from recombinant cell lines in-house and from cell lines provided by contract customers. Smaller, but similar airlift technology contract production services are emerging in the United States and Australia.

At least three European manufacturers of airlift fermentation equipment are now offering complete pilot- (e.g., 250 liter) and production-scale bioreactors "redesigned" for mammalian cell culture applications (i.e., LH Fermentation, Chemap, and Andritz in Graz, Austria). These systems have been well received by pharmaceutical industry plant managers accustomed to simple operating procedures and all-steam sterilizable stainless steel production equipment, in spite of the larger space required, lower cell densities, and overall inefficiency compared with current perfusion bioreactor systems. As indicated before, however, all airlift bioreactor systems can be retrofitted to improve performance (secreted product yields and purity) with hollow fiber or other membrane perfusion cartridge accessories and process controls.

C. Packed Glass Bead Bioreactors

Although currently available only through a contract production service by BioResponse, packed glass bead bioreactor columns have proven effective for long-term perfusion cultures of attachment-dependent cell lines. Like other cartridge-designed growth chambers, the packed bed modules have severe size limitations in terms of scaleup, which are normally overcome by simply increasing the number of units. Alternative packing materials may be more suitable for some cell types, but with the glass bead packing, cell densities of over 10^{10} viable cells in a 1 liter culture chamber can be reached with moderate medium flow rates. The major limitations to scaleup appear to be in attaining sufficient mass transfer of oxygen, nutrients, and metabolites without damaging cells with excessive high flow rates.

The recent introduction of radial flow bioreactor technology as a laboratory-scale unit may open the way for larger scale packed cell bed bioreactor systems in the future due to the greater flow rates and scaleability of the design.

D. Hollow Fiber Bioreactors

Pilot- and production-scale hollow fiber growth chamber modules can be purchased from several manufacturers (e.g., Setec, Amicon, and Microgen), but complete systems with process controls are currently available from only one company (Endotronics). In spite of the success of hollow fiber membrane bioreactor technology, when employed as a contract production service, the equipment has not been easy to sell for in-house industrial scaleup. Part of the hesitation to develop more sophisticated bioreactor systems based on hollow fiber growth chambers appears to be due to difficulties in process control and automated sampling inherent in the current module designs. Furthermore, only the all-polypropylene and, to a limited extent, the polysulfone-cellulose acetate hollow fiber cartridges appear to really tolerate autoclaving or in situ steam sterilization (within a stainless steel jacket). Scaleability may also be a factor, since the only way to expand a system is to increase the number of cartridge modules. Concerns over diffusion gradients, voids, and "dead zones" further hamper acceptance of the hollow fiber bioreactors as long-term production-scale systems in industrial settings. Fortunately, new configurations and combinations of materials are emerging that promise to rectify prior concerns and allow this efficient cell culture technology to regain credibility for at least pilot-scale applications.

E. Ceramic Matrix Bioreactors

The Opticell perfusion cell culture system described previously in the context of ceramic matrix growth supports and laboratory-scale units is also commercially available as pilot- and production-scale units for in-house installation. The pilot-scale unit offers roughly a 10-fold increase in cell culture surface area over the laboratory unit, and the production model is yet another 10-fold larger than the pilot unit. The manufacturer also offers a contract production service and experimental trial runs for clients interested in testing the productivity of their cell lines in this system.

F. Microsphere and Fluidized Bed Bioreactors

The Verax process technology combines the efficient mass transfer capabilities of a fluidized bed bioreactor (particularly critical for oxygenation of high-density mammalian cells) with the protective properties of a matrix immobilization technique using weighted collagen microspheres as a growth support, as described earlier.

The system was designed as a large-scale industrial production cell culture system for in-house pharmaceutical manufacturing facilities. The production-scale package is now marketed in combination with "supplemental" contract production services and the opportunity to culture a customer's cells in a smaller scale bioreactor unit. Recent data on the production of tissue plasminogen activator from recombinant CHO cells (averaging 23 g/day), as well as monoclonal antibodies from murine hybridomas (averaging over 30 g/day), indicate the process is indeed an efficient, albeit costly, perfusion culture technology. The full-scale system may be a bit ahead of its time, but not unreasonable if, for example, monoclonal antibody therapies for cancer continue to require the raw production of 10-20 g antibody per patient.

G. Static Maintenance Reactors

One large-scale perfusion cell culture technology has combined a packed bed cell culture approach with membrane perfusion methods to create a unique cell growth chamber called the Static Maintenance Reactor (Feder and Tolbert, 1983; Tolbert and Feder, 1983; Tolbert et al., 1985; Tolbert and Srigely, 1987). The technology is available only on a contract production service basis from Invitron Corp. (St. Louis, MO) and, generally, is geared for larger orders of monoclonal antibodies and other biologicals secreted from mammalian cells. The culture method relies on the efficient transfer of gasses, nutrients, and metabolites across semipermeable membrane tubing coiled through densely packed cells in a stainless steel growth chamber. The cells are grown almost to tissue densities in a biocompatible, inert packing material such that all the available space is occupied either by the packing or by the cells. There is physically no room for the cells to expand, and hence, the presumption is that the cells then remain indefinitely in a "static" mode, producing the desired product under efficient perfusion conditions. Cell product yields and purity are therefore high, as with any high cell density, high-viability perfusion culture system. By the same token, a stable long-term perfusion culture of this design can remain productive for months under fairly automated operation.

H. Future Trends

The pharmaceutical industry, which is the only realistic customer for manufacturing scale cell culture technologies at this time, has been slow in accepting more efficient and productive perfusion technologies, favoring the conventional stirred tank and airlift bioreactor technologies akin to microbial fermentors. However, as the emerging perfusion culture technologies demonstrate their savings in labor, time, and downstream processing that come with more efficient design and process control, the industry is certain to make the transition to meet the large-scale production needs for biologicals in the next decade. Simultaneously, the cell culture technologies will become increasingly automated, as better in-line biosensors are developed and integrated directly into downstream processing of the biological products (Holtz, 1986, 1987; Hu and Wang, 1986; Schneider, 1987). Meanwhile, the contract production business will hold its ground, providing needed biological products for companies engaged in the transition to in-house production facilities with full GMP compliance.

VII. REGULATORY CONSIDERATIONS FOR CELL CULTURE PRODUCTS

Licensed biologicals produced in mammalian cell culture must conform to a variety of regulatory requirements. These differ in detail somewhat among the various countries, but in general are similar. In addition to country-by-country regulations, the World Health Organization publishes its own recommendations (1981, 1987). These are intended to serve as minimum standards and are especially useful to developing nations. In some countries, such as the United States, rules are not necessarily in one book but are decentralized in multiple but complementary regulations (FDA Drugs and Biologics), guidelines (Food and Drug Administration, 1984a, 1987a), and "points to consider" documents (Food and Drug Administration 1983, 1984b, 1985, 1987b).

Newcomers to the field may be overwhelmed initially by the breadth and complexity of these regulations (Fox, 1987; Ghignone, 1987). Indeed, the U.S. pharmaceutical industry is possibly the most heavily regulated of all product-producing industries. Cell culture biologicals are no exception to this statement. The reasons for this reality are found by examination of the history of these products, the science of these products, and the history of science. These topics are examined elsewhere in greater detail (Petricciani, 1985; Hopps, 1985; Lubiniecki, 1987a) but are briefly discussed here to provide a summary overview.

A. Historical Perspectives

Regulations frequently are promulgated to correct problems known to have occurred. A number of tragic events have shaped the regulations for cell culture biologicals. Among them are outbreaks and sporadic cases of life-threatening illnesses (herpes simplex B virus and Marburg agent) in vaccine workers and scientists employing primary cell cultures (Davidson and Hummeles, 1960; Hennessen, 1968). These incidents and related laboratory studies led to extensive requirements for the testing of cell culture products and the cells producing them (substrates) for the presence of adventitious agents. The occurrence of several hundred cases of polio in recipients of incompletely inactivated poliovirus vaccine led to tighter standards for final product safety testing and for process validation (Nathanson and Langmuir, 1963; Bodian and Horstmann, 1965).

Other regulations are devised to prevent hypothetical risks from becoming real problems. Often these reflect the capability of experimental science at the time to evaluate the risk factor in experimental settings. For example, the scientific growth of cytogenetics and cytogenetic methodology in the 1950s and 1960s led to the implementation of a cell substrate karyology specification for release of each lot of attenuated polio vaccine prepared in human diploid cells in the 1970s [FDA Biologics 630.13c(12)]. Chromosome aberrations were viewed by scientists of that time as a possible marker of abnormal cells with potentially tumorlike properties (Yerganian et al., 1960). By placing a relatively conservative limit on the proportion of substrate cells expressing such defects, it was thought that the risk of tumorigenicity to recipients from ill-defined mechanisms might be minimized (Proceedings, 1963). The scientific basis behind the need for this specification is perhaps weak by the standards of modern knowledge; however, it was the best available idea at the time. In part, it did permit the granting of licenses for products prepared in human diploid cells, which do not carry significant risk of contamination by adventitious agents associated with primary cells. Worldwide experience with diploid substrates for nearly two decades has been consistent with the conclusion that their products are safe from risk of tumorigenicity.

Concern that a putatively tumorigenic property might be inadvertently transferred to recipients of cell culture products has existed for over 30 years. In 1955, the Armed Forces Epidemiological Board rejected the use of continuous cell lines for vaccine preparation because of the resemblance of these cell types to tumor cells (Hilleman, 1979). In the 1960s, fear that human diploid cells contained undetectable leukemia viruses (Committee Report, 1963) delayed the implementation of these cells as vaccine substrates for over 10 years. Lack of appropriate scientific tools to address the question directly was a major factor in prolonging this debate. The same question arose again in the early 1980s when scientific advances made possible the production of therapeutic, prophylactic, and diagnostic proteins through recombinant DNA and hybridoma technologies practiced in continuous cell lines. This time, the scientific tools to address the questions of tumorigenicity risk were highly developed. The details of the resolution of the long-standing debate over the use of continuous cells are detailed elsewhere (Lubiniecki, 1987a;

Petricciani and Hopps, 1985). The effects can be seen in a variety of new regulatory documents issued during the past several years (World Health Organization, 1987; Food and Drug Administration, 1983, 1984b, 1985, 1987b).

The main points of these new documents are that any product of mammalian cell lines (recombinant, hybridomas, or conventional) should possess the following attributes: purity, safety, potency, efficiency, consistency, and stability, just like any other drug. How these attributes are achieved is generally decided on a case-by-case basis by discussion with the national control authorities from each country for each product indication. This system permits a flexible approach tailored to each special situation that can take into account myriad differences among various proteins, target diseases, cell lines, manufacturing processes, and other factors. The tools used to obtain the quality attributes are a variable mixture of cell bank preparation and characterization, scientific investigations of risk factors, process validation, product characterization, and product lot release testing. The subsequent sections describe the principal facets of each tool. The products listed in Table 1 and 2 have recently been licensed or allowed to begin clinical trial under these regulations.

B. Cell Banking

The use of a single pool of cells for all manufacturing and development activities has several advantages over the use of primary cells from animals (Hayflick, 1963). First, the cell bank can be subjected to characterization and extensive testing to demonstrate freedom from adventitious agents and other putative risks. Second, the cell bank is prepared from a homogeneous pool, providing more consistent results than primary cells obtained from different animals. Generally, the system is two tiered, with a master seed bank whose individual vials are used sparingly to prepare hundreds of working bank vials, which are in turn used for actual manufacturing efforts. Vials of banked cells are indefinitely stored in or over liquid nitrogen. The cell bank is prepared and handled under conditions designed to prevent its contamination by adventitious agents and other cell lines (Lubiniecki and May, 1985).

The cell bank is characterized for the presence of adventitious agents, including bacteria, fungi, and mycoplasma, as well as cytopathic, hemagglutinating, and hemadsorbing viruses. These tests are performed by conventional microbial culture methods (Lubiniecki and May, 1985; Jacobs et al., 1981). In addition, immunologically intact animals are inoculated and observed for 28 days to test for pathogens not easily detected by standard microbial methods. Mice or other rodents are also inoculated with banked cells, observed for 28 days, and exsanguinated; the serum is tested for the presence of antibodies to various rodent viruses that might have contaminated the cells (Collins and Parker, 1972). Special in vivo tests are also performed for lymphocytic choriomeningitis virus.

In addition to these tests, thin sections of cells are examined by transmission electron microscopy for evidence of adventitious agents and endogenous viruses. Endogenous viruses are the external manifestiation of viral genetic information whose normal home is the genome of the host cell. The archetypes of this mode of behavior are endogenous retroviruses. All vertebrate cells possess genetic information for endogenous retroviruses; some cells express these genes better than others (Bishop, 1983; Lieber et al., 1973). In addition to electron microscopy, a variety of biological, biochemical, and biophysical methods exist to detect and characterize retroviruses (Lubiniecki and May, 1985). Continuous cell lines are more likely than diploid cells to express endogenous retroviruses of their genes. Almost all hybridomas exhibit retroviruslike particles. In many cases, these particles may represent defective (noninfectious) viruses, since they may lack the detectable biological and biochemical hallmarks of functional retroviruses, such as virion-associated reverse transcriptase, RNA genes, and infectivity (Lubiniecki and May, 1985; Lubiniecki, 1987b).

Two other types of testing are performed. One type is designed to confirm the identity of the cell line. A variety of cytogenetic, immunologic, and isoenzyme electrophoretic methods are available (Seabright, 1971; O'Brien et al., 1980). The other type of testing is designed to assess the tumorigenicity of the cell line in immunosuppressed rodents (Shin, 1979).

In general, tests for adventitious agents produce negative results. Current U.S. regulations preclude utilization of any cell line contaminated by lymphocytic choriomeningitis virus or reovirus. The presence of retroviruses or retroviruslike particles does not prevent the cell line from serving as a substrate, but it requires effective steps to inactivate and/or remove retroviruses from the final product (see following section).

Tumorigenicity tests in immunosuppressed rodents are usually negative for diploid cells and often positive for continuous cell lines. As with the presence of retroviruses, this information is used to guide subsequent development of the product and is discussed in the next section. Although the characterization of the cell substrate is an important part of the regulatory process culminating in product approval, it must be remembered that it is products that are licensed, not cell lines.

C. Scientific Investigation of Putative Risk Factors and Process Validation

An important step in the assurance of product safety is to identify those risk factors that may putatively be present in the product. These may enter the product via the cell substrate or process raw materials. The same type of logic applied to the cell substrate is extended to other potential risk factors, such as tumorigenic DNA sequences, tumorigenic proteins, live tumorigenic cells, tumorigenic viruses, and harmful chemicals. Each putative risk factor must be clearly identified, its chemical and biological nature defined in emirical terms, and its sources listed.

Once this identification process is completed, the concentration or level of each putative risk factor is determined quantitatively. If negative results are obtained, the level is estimated as the limit of sensitivity of the methods employed. Next, acceptable levels of each putative risk factor are established, and when compared with estimated levels, the difference represents the degree to which the manufacturing process must remove the potential risk factor. For example, if (1) a product is accepted as safe when the residual cellular DNA content is 10 pg or less per dose, (2) the harvested cell culture fluid contains 0.1 mg DNA per liter, and (3) 100 ml harvested fluid equivalent is required for one dose, then a clearance factor for cellular nucleic acid of 1 million-fold is required during downstream processing.

Finally, removal or inactivation of each putative risk factor is evaluated under process-relevant conditions to document that the product will be safe. This topic is described in greater detail elsewhere (Lubiniecki, 1987a; van Wezel et al., 1982; Finter and Fantes, 1980). It is sufficient to say that much of the assurance of product safety rests on these studies.

D. Product Characterization and Lot Release

As the molecule of interest develops into a pharmaceutical product, its properties are studied intensely. Some of these characterizations are biophysical, such as determination of its size, shape, charge, structure, and degree of aggregation. Studies are performed to measure its activity in nonliving systems, its integrity, and other facets of its biochemical behavior. Biological and clinical studies are also included to determine its activity in vivo, its pharmacokinetic performance, its efficacy, and its side effects. In addition to the performance of the molecule of interest, characterization studies also focus on the stability of the formulated product in its proposed final container under a variety of storage and environmental conditions.

Once these are reasonably complete, and depending on the product indication, a subset of these many methods will emerge as candidates for lot release assays. This group of assays will be performed on a sample of final product vials each time a lot is filled from a single pooled container. The assays are chosen to assure the purity, potency, identity, and safety of the product for its stated medicinal indication. As such, the choice of specific tests is made on a case-by-case basis.

Tests of potency are usually biological or complex biochemical assays of some type that measure an event highly predicative of efficacy. For example, the efficacy of thrombolytic enzymes like streptokinase, urokinase, and tissue plasminogen activator can be inferred from appropriately designed in vitro laboratory potency assays of clot lysis. Since most cell culture products are glycoproteins, protein content is determined and employed in the calculation of the specific activity of each lot relative to a standard lot of product.

Purity assays are of several general types. Biophysical methods look for impurities based upon chromatographic or electrophoretic separations. Analytical ultracentrifugation and chromatographic procedures are employed to detect molecular aggregates. Biochemical methods are used to detect specific noxious substances; for example, the presence of pyrogenic lipopolysaccarides are frequently detected by coagulation of *Limulus* amebocyte lysate (Hochstein et al., 1973). Residual moisture determination is important for lyophilized products, since inadequate or excessive moisture may affect product stability and hence efficacy. In addition, immunochemical tests are being employed on some recombinant pharmaceuticals because the high purity levels of these products exceed the sensitivity of most biochemical and biophysical methods to detect extraneous substances (Anicetti et al., 1986). Accordingly, sensitive immunoassays have been devised to quantitate contamination by host cell proteins and foreign protein from such raw materials as animal serum.

Identity testing is performed to confirm the presence of the molecule of interest in the final vial. The methods used vary accordingly to the nature of the molecule of interest and its known properties. These tests must distinguish the molecule of interest from any others processed in the same laboratory. For example, identity testing for interferon products is sometimes accomplished by the ability of the product to induce an antiviral response in susceptible cells, which is neutralizable by prior incubation with an appropriate antiserum.

Safety testing is done in several ways. Routine sterility tests are performed to detect bacteria, fungi, and mycoplasma. Additional tests for adventitious agents consist of the inoculation of susceptible cell culture or intact animals, looking for viruses or fastidious microorganisms. The exact nature of this test battery depends upon the nature of the cell line and the manufacturing process. For example, monoclonal antibodies prepared from ascites fluid are tested for adventitious agents more extensively than if the same cell bank was used to make identical monoclonal antibodies in cell culture. A general safety test is usually performed by inoculation of mice and guinea pigs. These are observed for 7 days for any sign of illness or weight loss, reflecting possible noxious substances in the final product.

Some virus vaccines are safety tested in normally susceptible animals to assure their attenuation or inactivation (such as polio vaccines and plasma-derived hepatitis B surface antigen vaccine). This is not done for recombinant protein vaccines, such as yeast-derived hepatitis B surface antigen. The cloning and manufacturing process and characterization studies preclude the presence of infectious hapatitis B virions, its infectious genome, or those of other human viruses that may contaminate the plasma-derived product. Alternatively, cloning a viral surface gene into live vaccinia virus may generate a vaccinia virus with altered tissue tropism and perhaps should be evaluated by these types of in vivo safety tests.

Lot release assays are thoroughly evaluated under a wide range of operating conditions. Some are validated by challenge studies. In all cases, limits or other criteria are set by which assay results for a given lot are considered acceptable. Lots with values outside these ranges are rejected, and the materials are either reworked or destroyed. The details of each assay and its acceptance limits receive careful scrutiny by national control authorities prior to licensing. Some national control authorities, such as the U.S. Center for Biologics Evaluation and Research, also release each lot of final product (and in some cases bulk lot) for some products following completion of its own testing and review of the manufacturer's test results.

Prior to licensing cell culture products in countries like the United States, a manufacturer normally prepares qualification lots. These are usually three to five lots that are representative of the manufacturing process, facilities and equipment, and testing procedures to be employed following licensing. Samples from these lots actually form part of the application. The U.S. Center for Biologics Evaluation and Research issues both a product license for a specific product and an establishment license for the manufacturing facilities of that product.

E. CGMP

In addition to valid licenses from national control authorities of each country where the product is distributed, a number of other requirements exist for manufacturers, collectively known as good manufacturing practices (GMP, or current GMP, CGMP). These are largely the codification of common sense and preventive planning. Regulations are generally promulgated only after control authorities familiarize themselves with the science of the situation and the current practices of the industry, allowing liberal opportunity for comment from industry and other knowledgeable parties. In this way, manufacturing regulations are generally effective in assuring product quality.

For example, cell culture facilities operate most efficiently when they are designed to prevent contamination by microorganisms in the environment, in process raw materials, or from workers. Accordingly, regulations exist to preclude the handling of cell substrates after working with other infectious agents unless outer clothing is completely changed or covered. Certain types of work are excluded from the same area, for example, work with spore-forming organisms. Use of other cell lines is generally not permitted in the manufacturing area unless adequate separation is demonstrated. Personnel unnecessary to manufacturing, supervision, or maintenance efforts are excluded. Equipment must be properly maintained and kept clean and free of sources of contamination. Staff must possess an understanding of the manufacturing operations they perform and their relationship to the regulations. Records and samples of each lot must be maintained to document legibly and permanently what was done in an understandable manner. Records and samples are usually retained for the later of 5 years from manufacture or 6 months past lot expiration. Establishments are inspected for compliance with U.S. CGMP at least every 2 years.

Other regulations specifically address process considerations. Compliance is usually achieved prior to licensing by effective process design. For example, cell culture products may not contain penicillin owing to its propensity to cause allergic reactions. Use of other antibiotics should be minimized; some manufacturers demonstrate their removal during purification of the product. Reduction in the calculated concentration of extraneous proteins (such as animal serum protein) is required by a factor of at least 1 million-fold between cell propagation stages and product collection for virus vaccines. To achieve this, manufacturers design their processes to obtain effective removal of animal serum.

Regulations also usually oblige manufacturers to demonstrate the integrity of cell characteristics at periodic intervals. For example, a cell culture procedure for the preparation of monoclonal antibodies or recombinant proteins should provide harvested material of consistent titer. A loss of titer may indicate the presence of an unwanted mutation or selection of a nonproducing variant cell population. For recombinant cell lines, demonstration of the genetic stability of the inserted genetic information is requested by points to consider documents (Food and Drug Administration, 1984b, 1985).

When the substrate cells generally die during product preparation (for example, most virus vaccines), additional controls for adventitious agents are imposed beyond standard lot release tests for contamination. In most cases, a portion of up to 25% of the cultures in any batch are not infected with vaccine virus seed but rather are observed for several weeks for signs of contamination by adventitious agents.

Vaccine seed materials are also subjected to additional scrutiny in terms of their history, preparation, characterization, documentation, and use. They require extensive study of the biology, genetics, and pathogenesis of virus seed material. These regulations seek to maintain the consistency, safety, and efficacy of virus vaccines.

In conclusion, the application of quality and regulatory considerations to biological products prepared in cell culture is multifaceted, befitting the complexity of human health and modern science. Multiple tools exist to help assure product quality, including the preparations and characterization of cell banks, evaluation of potential risk factors, process validation, product characterization, lot release tests, and adherence to CGMP. Each alone is unable to assure product quality. Together these methods cooperatively provide the assurance of quality.

VIII. SUMMARY AND FUTURE TRENDS

In the mid-1970s, when mammalian cells were first becoming important substrates for the production of human proteins, many scientists believed that this technology would be a short-lived, intermediate-level technology bridging the gap between extraction of the products from body fluids or tissues and their production in genetically engineered microorganisms. In the intervening years, however, it has become clear that this belief was somewhat simplistic. Although production in microbes has proved possible for certain proteins, it is now apparent that many others require mammalian cell substrates in order to yield biologically stable and active molecules.

This realization has shifted the scientific focus to improving the techniques required to use mammalian cells to produce products. Major advances have been made in many areas. For example, genetic engineering of mammalian cells has allowed selected proteins to be produced in heterologous cells with increased product yields. New media have reduced both costs and labor while increasing product safety. Large-scale, computer-aided production systems have been developed to increase total yields and decrease costs.

Despite these advances, however, making products in mammalian cells is not an easy task. Many technological improvements are needed in order to increase yields and reduce costs further. Some of these advances include better genetic engineering techniques to insert and control multiple copies of product genes in order to increase efficiency of production and yields, less expensive production of needed growth factors for use with serum-free media, better techniques for controlling contamination, new probes to monitor cell condition and productivity under large-scale conditions, and more sophisticated computer controlled bioreactor systems, to name just a few.

It is still possible that production of products by mammalian cell culture will turn out to be an intermediate-level technology and that ultimately these products may be successfully manufactured by microorganisms or by using chemical methods. It is clear, however, that this technology will not be a short-term technology. Already its use for generating complex human products has established mammalian cell culture as a major production technique. Moreover, the improvements in this process, which are certain to occur within the next decade, will further increase the ease and cost effectiveness of this technology. This in turn will enhance the usefulness and longevity of using mammalian cell culture to produce important products for human use.

REFERENCES

Adamson, S. R., Fitzpatrick, S. L., Behie, L. A., Gaucher, G. M., and Lesser, B. H. (1983). In vitro production of high titre monoclonal antibody by hybridoma cells in dialysis culture. *Biotechnol. Lett. 5*: 573-578.

Alt, F., Kellems, R., Bertino, J., and Schimke, R. (1978). Selective multiplication of dihydrofolate reductase genes in methotrexate-resistant variants of cultured murine cells. *J. Biol. Chem. 253*: 1357-1370.

Altshuler, G. L., Dziewulski, D. M., Sowek, J. A., and Belfort, G. (1986). Continuous hybridoma growth and monoclonal antibody production in hollow fiber reactors/separators. *Biotechnol. Bioeng. 28*: 646-658.

Ambesi-Impiombato, F., Parks, L., and Coon, H. (1980). Culture of hormone-dependent functional epithelial cells from rat thyroids. *Proc. Natl. Acad. Sci. USA 77*: 3455-3459.

Andersen, B. G., and Gruenberg, M. L. (1987). Optimization techniques for the production of monoclonal antibodies utilizing hollow-fiber technology. In *Commercial Production of Monoclonal Antibodies*, S. Seaver (Ed.). Marcel Dekker, New York, pp. 175-195.

Anicetti, V., Fehskens, E., Reed, B., Chen, A., Moore, P., Geier, M., and Jones, A. (1986). Immunoassay for the detection of *E. coli* proteins in recombinant DNA derived human growth hormone. *J. Immunol. Methods 91*: 213-224.

Arathoon, W. R., and Birch, J. R. (1986). Large scale cell culture in biotechnology. *Science 232*: 1390-1395.

Armstrong, D. (1973). Contamination of tissue culture by bacteria and fungi. In *Contamination of Tissue Culture*, J. Fosh (Ed.). Academic Press, New York, pp. 51-64.

Barnes, D. (1984). Attachment factors in cell culture. In *The Use of Serum-Free and Hormone-Supplemented Media*, J. Mather (Ed.). Plenum Press, New York, pp. 195-237.

Barnes, D. (1987). Serum-free animal cell culture. *BioTechniques 5*(6): 534-542.

Barnes, D., and Sato, G. (1980). Methods for growth of cultured cells in serum-free medium. *Anal. Biochem. 102*: 255-320.

Barnes, D., Sirbasku, D., and Sato, G. (Eds.) (1984). *Cell Culture Methods for Molecular and Cell Biology*, vols. 1-4. Alan R. Liss, New York.

Bettger, W., and Ham, R. (1981). Effects of non-steroidal anti-inflammatory agents and antioxidants on the clonal growth of human diploid fibroblasts. *Prog. Lipid Res. 20*: 265-268.

Bettger, W., and Ham, R. (1982). The nutrient requirements of cultured mammalian cells. *Adv. Nutr. Res. 4*: 249-286.

Birch, J. R., Boraston, R., and Wood, L. (1985a). Bulk production of monoclonal antibodies in fermentors. *Trends Biotechnol. 3*: 162-166.

Birch, J. R., Thompson, P. W., Lambert, K., and Boraston, R. (1985b). The large scale cultivation of hybridoma cells producing monoclonal antibodies. In *Large Scale Mammalian Cell Culture*, J. Feder and W. Tolbert (Eds.). Academic Press, New York.

Bishop, J. (1983). Cellular oncogenes and retroviruses. *Annu. Rev. Biochem. 52*: 301-354.

Bodian, D., and Horstmann, D. (1965). In *Viral and Rickettsial Infections of Man*, 4th ed., F. Horsfall and I. Tamm (Eds.). J. B. Lippincott, Philadelphia, pp. 430-473.

Bognar, E. A., Pugh, G. G., and Lydersen, B. J. (1983). Large scale propagation of BHK-21 cells using the OPTICELL culture system. *J. Tissue Culture Methods 8*: 147-154.

Brown, P. C., Costello, M. A. C., Oakley, R., and Lewis, J. L. (1985). Applications of the mass culturing technique (MCT) in large scale growth of mammalian cells. In *Large Scale Mammalian Cell Culture*, J. Feder and W. Tolbert (Eds.). Academic Press, New York, pp. 59-71.

Brown, P. C., Figueroa, C., Costello, M. A. C., Oakley, R., and Maciukas, S. M. (1988). Protein production from mammalian cells grown on glass beads. In *Animal Cell Biotechnology*, vol. 3, R. Spier and J. Griffiths (Eds.). Academic Press, New York.

Butler, M. (1986). Serum-free media. In *Mammalian Cell Technology*, W. Thilly (Ed.). Butterworths, Boston, pp. 91-108.

Canaani, D., and Berg, P. (1982). Regulated expression of human interferon beta-1 gene after transduction into cultured mouse and rabbit cells. *Proc. Natl. Acad. Sci. USA 79*: 5166-5170.

Charles River Biotechnical Services (1987). Technical bulletins on Opticell cell culture systems. Wilmington, Massachusetts.

Cherry, W. R. and Hull, R. N. (1960). Growth of mammalian cells in agitated fluid medium. *J. Biochem. Microbiol. Tech. Eng. 2*: 267-287.

Collins, M., and Parker, J. (1972). Murine viral contaminants of leukemia viruses and transplantable tumors. *JNCI 49*: 1139-1143.

Committee Report (1963). Continuously cultured cells and viral vaccines. *Science 139*: 15-20.

Coriell, L. (1973). Methods of prevention of bacterial, fungal, and other contaminants. In *Contamination of Tissue Culture*, J. Fosh (Ed.). Academic Press, New York, pp. 29-49.

Cortessis, G. P., and Proby, C. M. (1987). Airlift bioreactors for production of monoclonal antibodies. *Biopharm. Manuf. 1*: 30-33.

Davidson, W., and Hummeles, K. (1960). B virus infections in man. *Ann. N.Y. Acad. Sci. 85*: 970-979.

de Bruyne, N. A., Weiss, L., and Huber, D. (1974). A pneumatic stirrer for suspension cultures of mammalian cells. *In Vitro 10*: 184-187.

Detrisac, C., and Moon, R. (1986). Establishment of a cell strain of murine transitional epithelium in serum free medium. In *Abstracts of the Tissue Culture Association*, Chicago 1986, p. 18A.

Dulbecco, R., and Freeman, G. (1959). Plaque production by polyoma virus. *Virology 8*: 396-397.

Eagle, H. (1955). Nutritional needs of mammalian cells in tissue culture. *Science 122*: 501-504.

Eagle, H. (1959). Amino acid metabolism in mammalian cell cultures. *Science 130*: 432-437.

Eagle, H. (1973). The effect of environmental pH on the growth of normal and malignant cells. *J. Cell. Physiol. 82*: 1-8.

Evans, T. L., Hart, S. M., Nguyen, H. T., Coulter, C., Quentin, T., Miller, R. A., and Fleig, G. E. (1986). Large scale production of monoclonal antibodies using hollow fiber bioreactors with specific molecular weight cutoffs. Poster presentation, Tissue Culture Association Conference, September 22-26. Hershey, Pennsylvania.

Familletti, P. C. (1987). Gel-immobilized cell culture for monoclonal antibody production. *Biopharm. Manuf. 1*: 48-50.

Famílletti, P. C., and Fredericks, J. E. (1988). Techniques for mammalian cell immobilization.

Feder, J., and Tolbert, W. R. (1983). The large-scale cultivation of mammalian cells. *Sci. Am. 248*: 36-43.

Feder, J., and Tolbert, W. R. (1985a). Mass culture of mammalian cells in perfusion systems. *Am. Biotechnol. Lab. 3*: 24-36.

Feder, J., and Tolbert, W. R. (Eds.) (1985b). *Large Scale Mammalian Cell Culture.* Academic Press, New York.

Finter, M., and Fantes, K. (1980). The purity and safety of interferons prepared for clinical use: The use for lymphoblastoid interferons. In *Interferon 1980,* I. Gresser (Ed.). Academic Press, New York, pp. 65-80.

Flam, R. (1982). Importance of the basal nutrient medium in the design of hormonally defined media. In *Growth of Cells in Hormonally Defined Media,* G. Sato, A. Pardee, and D. Sirbaska (Eds.). Cold Spring Harbor Laboratory, Cold Spring Harbor, New York, pp. 39-60.

Fleischaker, R. J., Jr. (1986). Practical matters in instrumentation for mammalian cell cultures. In *Mammalian Cell Culture,* W. Thilly (Ed.). Butterworths, Boston, pp. 199-211.

Fleischaker, R. J., Jr. (1987). Microcarrier cell culture. In *Large Scale Cell Culture Technology*, B. Lydersen (Ed.). Carl Hanser, Munich, pp. 60-79.

Food and Drug Administration, Office of Biologics (1983). Interferon test procedures: Points to consider in the production and testing of interferon intended for investigational use in humans.

Food and Drug Administration, Center for Drugs and Biologics and Center for Devices and Radiological Health (1984a). Guideline on general principles of process validation.

Food and Drug Administration, Division of Blood and Blood Products (1984b). Points to consider in the characterization of cell lines used to produce biologicals.

Food and Drug Administration, Office of Biologics Research and Review (1985). Points to consider in the production and testing of new drugs and biologicals produced by recombinant DNA technology.

Food and Drug Administration, Center for Drugs and Biologics and Center for Devices and Radiological Health (1987a). Guideline on sterile products produced by aseptic processes.

Food and Drug Administration, Office of Biologics Research and Review (1987a). Points to consider in the manufacture of monoclonal antibody products for human use.

Food and Drug Administration. Drugs, General, Code of Federal Regulations. Title 21. Parts 200-299.

Food and Drug Administration. Biologics. Code of Federal Regulations. Title 21. Part 630.13c(2)

Fox, J. L. (1987). The U.S. regulatory patchwork. *Biotechnology* 5(12): 1273-1277.

Ghignone, A. A. (1987). Critical concerns in regulating biotechnology products. *Biopharm. Manuf. 1*: 28-30.

Gillies, S., Morrison, S., Oi, V., and Tonegawa, S. (1983). A tissue specific transcription enhancer element is located in the major intron of a rearranged immunoglobulin heavy chain. *Cell 33*: 717-728.

Glacken, M. W., Fleischaker, R. J., and Sinskey, A. J. (1983). Large-scale production of mammalian cells and their products: Engineering principles and barriers to scale-up. *Ann. N.Y. Acad. Sci. 13*: 355-372.

Goochee, C., Grow, M., Barlow, G., and Abcouwer, S. (1987). The creation of human cell lines through oncogene incorporation for the expression of natural and genetically-engineered products. Abstracts of the American Chemical Society 194th Meeting, Division of Microbial and Biochemical Technology, pp. 26-27.

Ham, R. (1965). Clonal growth of mammalian cells in a chemically defined, synthetic medium. *Proc. Natl. Acad. Sci. USA 53*: 288-293.

Harper, P. A., and Juliano, R. L. (1981). Two distinct mechanisms of fibroblast adhesion. *Nature 290*: 136-139.

Hayflick, L. (1963). A comparison of primary monkey kidney, heteroploid cell lines, and human diploid cell strains for human virus vaccine preparation. *Am. Rev. Respir. Dis. [Suppl.] 88*: 387-393.

Hayflick, L. (1965). The limited in vitro lifetime of human diploid cell strains. *Exp. Cell. Res. 37*: 614.

Hayflick, L., and Moorhead, P. (1961). The serial cultivation of human diploid cell strains. *Exp. Cell Res. 25*: 585.

Hennessen, W. (1968). A hemorrhagic disease transmitted from monkeys to man. *NCI Monograph 29*: 161-171.

Hilleman, M. (1979). Cell line saga—An argument in favor of production of biologics in cancer cells. *Adv. Exp. Biol. Med. 118*: 47-58.

Hirschel, M. (1986). An automated hollow fiber system for the large scale manufacture of mammalian cell secreted products. In *Large Scale Cell Culture Technology*, B. Lydersen (Ed.). Carl Hanser, Munich.

Hockstein, H., Elfin, R., Cooper, J., Seligmann, E., and Wolff, S. (1973). Further development of the *Limulus* amebocyte lysate test. *Bull. Parenteral Drug Assoc. 27*: 139-148.

Holtz, R. B. (1986). A system for computer process control of fermentation. *Am. Biotechnol. Lab. 4*.

Holtz, R. B. (1987). A distributed logic control system for fermentation and downstream processing: Design criteria. *Adv. Instrum. 43*(3): 1443-1453.

Hopkinson, J. (1985). Hollow fiber cell culture systems for economical cell product manufacturing. *Biotechnology 3*: 225-230.

Hopps, H. (1985). Cell substrate issues—a historic perspective. In *Abnormal Cells, New Products and Risks*, J. Petricciani and H. Hopps (Eds.). Tissue Culture Association, Gaithersburg, Maryland, pp. 13-17.

Hoshi, H., Kan, M., Yamane, I., and Minamoto, Y. (1982). Hydrocortisone potentiates cell proliferation and promotes cell spreading on tissue culture substrata of human diploid fibroblasts in a serum-free, hormone-supplemented medium. *Biomed. Res. 3*: 546-552.

Hsiung, N. (1985). Characterization of HBsAg protein produced in animal cells. In *Abnormal Cells, New Products and Risk*, H. Hopps and J. Petricciani (Eds.). Tissue Culture Association, Gaithersburg, Maryland, pp. 101-107.

Hu, W.-S., and Wang, D. (1986). Mammalian cell culture technology: A review from an engineering perspective. In *Mammalian Cell Technology*, W. Thilly (Ed.). Butterworths, Boston, pp. 167-197.

Imamura, T., Crespi, C., Thilly, W., and Brunengraber, H. (1982). Fructose as a carbohydrate source yields stable pH and redox parameters in microcarrier cell culture. *Anal. Biochem. 124*: 353-358.

Iscove, N., and Melchers, F. (1978). Complete replacement of serum by albumin, transferrin, and soybean lipid in cultures of lipopolysaccharide-reactive B lymphocytes. *J. Exp. Med. 147*: 923-933.

Jacobs, J., Magrath, D., Garrett, A., and Schild, G. (1981). Guidelines for the acceptability, management and testing of serially propagated human diploid cells for the production of live virus vaccines for use in man. *J. Biol. Stand. 9*: 331-342.

Jarvis, A. P., Jr., and Gardina, T. A. (1983). Production of biologicals from microencapsulated living cells. *BioTechniques 1*: 22-27.

Juliano, R. L., and Gagalano, E. (1977). Adhesion of CHO cells: Effects of temperature, metabolic inhibitors and proteolytic dissection of cell surface macromolecules. *J. Cell. Physiol. 92*: 209-220.

Kaufman, R., and Sharp, P. (1982). Amplification and expression of sequences cotransfected with a modular dihydrofolate reductase complementary DNA gene. *J. Mol. Biol. 159*: 601-621.

Kaufman, R., Dorner, A., Wasley, L., Murtha, P., Gure, M., and Raney, P. (1987). The potentials and limitations of high level expression of secreted proteins in mammalian cells. Abstracts of the American Chemical Society 194th Meeting, Division of Microbial and Biochemical Technology, p. 27.

Kauvala, H., Carter, W. G., and Hakomori, S. I. (1981). Studies on cell adhesion and recognition: Extent and specificity of cell adhesion triggered by carbohydrate reactive proteins (lectins) and fibronectin. *J. Cell Biol. 88*: 127-137.

Kleinman, H. K., Luckenbill-Edds, L., Cannon, F. W., and Sephel, G. C. (1987). Use of extracellular matrix components for cell culture. *Anal. Biochem. 166*: 1-13.

Kluft, C., van Wezel, A., van der Velden, C., Emeis, J., Verheijen, J., and Wijngaards, G. (1983). Large scale production of extrinsic (tissue-type) plasminogen activator from human melanoma cells. In *Adv. Biochem. Proc. 2*: 97-110.

Kolanko, W., and Symbol, R. (1987). Improved yields of monoclonal antibodies in flexible plastic containers. Proceedings of Biotech USA, 4th Annual Industry Conference and Exhibition, Santa Clara, California.

Ku, K., Kuo, M. J., Delente, J., Wildi, B. S., and Feder, J. (1981). Development of a hollow-fiber system for large scale culture of mammalian cells. *Biotechnol. Bioeng. 23*: 79-95.

Law, M.-F., Lowy, D., Dvoretzky, I., and Howley, P. (1981). Mouse cells transformed by bovine papilloma virus contain only extrachromosomal viral DNA sequences. *Proc. Natl. Acad. Sci. USA 78*: 2727-2731.

Lebherz, W. B. (1987). Batch production of monoclonal antibody by large-scale suspension culture. In *Commercial Production of Monoclonal Antibodies*, S. Seaver (Ed.). Marcel Dekker, New York, pp. 93-118.

Leibowitz, A. (1963). The growth and maintenance of tissue-cell cultures in free gas exchange with the atmosphere. *Am. J. Hyg. 78*: 173-180.

Leist, C., Meyer, H. P., and Feichter, A. (1986). Process control during the suspension culture of a human melanoma cell line in a mechanically stirred loop bioreactor. *J. Biotechnol. 4*: 235-246.

Liang Li, M., Aggeler, J., Farson, D. A., Katier, C., Hassell, J., and Bissell, M. J. (1987). Influence of a reconstituted basement membrane and its components on casein gene expression and secretion in mouse mammary epithelial cells. *Proc. Natl. Acad. Sci. USA 84*: 136-140.

Lieber, M., Benveniste, R., Livingstone, R., and Todaro, G. (1973). Mammalian cells in culture frequently release type C viruses. *Science 182*: 56-59.

Lim, F. (Ed.) (1984). Microencapsulation of living cells and tissued—theory and practice. In *Biomedical Applications of Microencapsulation*. CRC Press, Boca Raton, Florida, pp. 137-154.

Lubiniecki, A. (1987a). Safety considerations for cell culture-derived biologicals. In *Large Scale Cell Culture*, B. Lyderson (Ed.). Carl Hanser, Munich, pp. 231-247.

Lubiniecki, A. (1987b). Pharmaceutical applications of recombinant DNA-modified mammalian cells. *Dev. Ind. Microbiol. 28*: 133-138.

Lubiniecki, A., and May, L. (1986). Cell bank characterization for recombinant-DNA mammalian cell lines. *Dev. Biol. Stand. 60*: 141-146.

Lydersen, B. K. (Ed.) (1987). *Large Scale Cell Culture Technology*. Carl Hanser, Munich.

Lydersen, B. K., Pugh, G. G., Paris, M. S., Sharma, B. P., and Noll, L. A. (1985). Ceramic matrix for large scale animal cell culture. *Biotechnology 3*: 63-67.

Martin, N., Brennan, A., Denomme, L., and Shevitz, J. (1987). High productivity in mammalian cell cultures. *Biotechnology 5*: 838-840.

Mather, J., and Sato, G. (1979). The growth of mouse melanoma cells in hormone-supplemented, serum-free medium. *Exp. Cell Res. 120*: 191-200.

McCormick, D. (1987). Pharmaceutical markets for the 1990's. *Biotechnology 5*: 27.

Mendelsohn, J., Caviles, A., Jr., and Castagnola, J. (1982). Proliferation of normal human lymphocytes in serum free, albumin free medium. In *Growth of Cells in Hormonally Defined Media*, G. Sato, A. Pardee, and D. Sirbasku (Eds.). Cold Spring Harbor Laboratories, Cold Spring Harbor, New York, pp. 677-690.

Morgan, J., Morton, H., and Parker, R. (1950). Nutrition of animal cells in tissue culture. I. Initial studies on a synthetic medium. *Proc. Soc. Exp. Biol. Med. 71*: 1-8.

Morton, H. (1970). A survey of commercially available tissue culture media. *In Vitro 6*: 89.

Nardelli, L., and Panina, G. F. (1976). Large scale roller bottle production. *Dev. Biol. Stand. 37*: 133-138.

Nathanson, N., and Langmuir, A. (1963). The Cutter incident. Poliomyelitis following formaldehyde-inactivated poliovirus vaccination in the United States during the spring of 1955. *Am. J. Hyg. 78*: 16-81.

Nilsson, K. (1987). Methods for immobilizing animal cells. *Tib-Technology 5*: 73-77.

O'Brien, S., Shannon, J., and Gail, M. (1980). A molecular approach to the identification and individualization of human and mammalian cells in culture: Isoenzyme and allozyme genetic signatures. *In Vitro 16*: 119-135.

Orly, J., and Sato, G. (1979). Fibronectin mediates cytokinesis and growth of rat follicular cells in serum-free medium. *Cell 17*: 295-306.
Orr, T. (1988). Verax System scales-up recombinant mammalian cell culture-derived T-PA. *Genet. Eng. News 8*(2): 22.
Pavlakis, G., and Hamer, D. (1983). Regulation of a metallothionein-growth hormone hybrid gene in bovine papilloma virus. *Proc. Natl. Acad. Sci. USA 80*: 397-401.
Petricciani, J. (1985). Background and objectives of the workshop. In *Abnormal Cells, New Products and Risks*, J. Petricciani and H. Hopps (Eds.). Tissue Culture Association, Gaithersburg, Maryland, pp. 9-12.
Petricciani, J., and Hopps, H. (Eds.) (1985). *Abnormal Cells, New Products and Risk.* Tissue Culture Association, Gaithersburg, Maryland.
Pharmacia/LKB Biotechnology, Inc. (1984). *Cytodex Reference List and Technical Bulletins*, Piscataway, New Jersey.
Posillico, E. G. (1986). Microencapsulation technology for large scale antibody production. *Biotechnology 4*: 114-117.
Posillico, E. G., Kallelis, M. S., and George, J. M. (1987). Large-scale production and purification of monoclonal antibodies using cellular microencapsulation. In *Commercial Production of Monoclonal Antibodies*, S. S. Seaver (Ed.). Marcel Dekker, New York, pp. 139-158.
Proceedings of the Symposium on Characterization and Uses of Human Cell Strains (1963). Optija, Yugoslavia, pp. 709-738.
Propst, C. (1987). Recent developments in the use of mammalian cell culture for the production of pharmaceuticals and other important products. In *Dev. Ind. Microbiol. 28*: 125-132.
Putnam, J. E. (1987). Monoclonal antibody production in a ceramic matrix. In *Commercial Production of Monoclonal Antibodies*, S. S. Seaver (Ed.). Marcel Dekker, New York, pp. 119-138.
Ratafia, M. (1987). Mammalian cell culture: Worldwide activities and markets. *Biotechnology 5*: 692-694.
Reitzer, L., Wice, B., and Kennell, D. (1979). Evidence that glutamine, not sugar, is the major energy source for cultured HeLa cells. *J. Biol. Chem. 254*: 2669-2676.
Reuveny, S., Velez, D., MacMillan, J. D., and Miller, L. (1986a). Factors affecting cell growth and monoclonal antibody production in stirred reactors. *J. Immunol. Methods 86*: 53-59.
Reuveny, S., Velez, D., Miller, L., and MacMillan, J. D. (1986b). Comparison of cell propagation methods for their effect on monoclonal antibody yield in fermentors. *J. Immunol. Methods 86*: 61-69.
Reuveny, S., Zheng, Z.-B., and Eppstein, L. (1986c). Evaluation of a cell culture fermentor. *Am. Biotech. Lab. 4*(1): 28-36.

Rundstadler, P. W., Jr., and Young, M. W. (1987). Large-scale, fluidized bed bioreactor systems. Veraz technical memo 218, presented at Fronteirs in Bioprocessing, Boulder, Colorado, June 1987.

Rupp, R. (1985). Use of cellular microencapsulation in large scale production of monoclonal antibodies. In *Large Scale Mammalian Cell Culture*, W. Tolbert and J. Feder (Eds.). Academic Press, New York, pp. 97-123.

Sato, G., Pardee, A., and Sirbaska, D. (1982). Growth of cells in hormonally defined media, Books A & B. Cold Spring Harbor Laboratory, Cold Spring Harbor, New York.

Schneider, I. (1987). Biosensors and computers help optimize bioprocessing system production. *Genet. Eng. News 7*(5): 16, 20.

Schratter, P. (1976). Cell culture with synthetic capillaries. Technical Bulletin from Amicon Division, W. R. Grace & Co., Danvers, Massachusetts.

Schreirer, W. (1988). High density growth of animal cells within cell retention fermentors equipped with membranes. In *Animal Cell Biotechnology*, vol. 3, R. Spier and J. Griffiths (Eds.). Academic Press, New York.

Schreirer, W., Nilsson, K., Merten, O. W., Katinger, K. W. D., and Mosbach, K. (1982). Entrapment of animal cells for the production of biomolecules such as a monoclonal antibodies. *Dev. Biol. Stand. 55*: 155.

Seabright, M. (1971). Rapid banding technique for human chromosomes. *Lancet 2*: 971-972.

Seaver, S. S. (Ed.) (1987). Culture method affects antibody secretion of hybridoma cells. In *Commercial Production of Monoclonal Antibodies*. Marcel Dekker, New York, pp. 49-71.

Shimke, R. (1984). Gene amplification in cultured animal cells. *Cell 37*: 705-713.

Shin, S. (1979). Use of nude mice for tumorigenicity testing and mass propagation. *Methods Enzymol. 58*: 370-379.

Spier, R. E., and Griffiths, J. B. (1985). *Animal Cell Biotechnology*, vol. 1. Academic Press, New York, p. 247.

Spier, R. E., and Griffiths, J. B. (1988). *Animal Cell Biotechnology*, vol. 3. Academic Press, New York.

Swartz, R. (1985). Alternatives for the production of mammalian cell products: A survey. *Genet. Eng. News 5*(7): 16-21.

Tharakan, J. P., and Chau, P. C. (1986). The radial flow hollow fiber bioreactor for the large scale culture of mammalian cells. *Biotechnol. Bioeng. 18*: 329-342.

Thilly, W. (Ed.) (1986). *Mammalian Cell Technology*. Butterworths, Boston.

Thomas, J. (1986). Nutrients, oxygen, and pH. In *Mammalian Cell Technology*, W. Thilly (Ed.). Butterworths, Boston, pp. 109-130.

Tolbert, W. R. and Feder, J. (1983). Large-scale cell culture technology. *Annu. Rep. Ferment. Proc. 6*: 35-74.

Tolbert, W. R., and Srigely, W. R. (1987). Manufacture of pharmacologically active proteins by mammalian cell culture. *Biopharm. Manuf. 1*: 42-48.

Tolbert, W. R., Lewis, C., White, P. J., and Feder, J. (1985). Perfusion culture systems for production of mammalian cell biomolecules. In *Large Scale Mammalian Cell Culture*, J. Feder and W. Tolbert (Eds.). Academic Press, New York, pp. 97-123.

United Kingdom Patent Application (1983). GB 2119804 A.

Vaheri, A., and Mosher, D. (1978). High molecular weight, cell surface-associated glycoprotein (fibronectin) lost in malignant transformation. *Biochim. Biophys. Acta 516*: 1-25.

van Brunt, J. (1986). Immobilized mammalian cells: The gentle way to productivity. *Biotechnology 4*: 505-510.

van Brunt, J. (1987a). Solving the shortage of human cell lines. *Biotechnology 5*: 1117.

van Brunt, J. (1987b). A closer look at fermentors and bioreactors. *Biotechnology 5*(11): 1133-1138.

van Wezel, A. L. (1967). Growth of cell strains and primary cells on microcarriers in homogeneous culture. *Nature 216*: 64.

van Wezel, A., van der Marel, P., van Bevern, C., Verma, I., Salk, P., and Salk, J. (1982). Detection and elimination of cellular nucleic acids in biologicals produced on continuous cell lines. *Dev. Biol. Stand. 50*: 59-69.

Varani, J., Dane, M., Beals, T. F., and Wass, J. A. (1983). Growth of three established cell lines on glass microcarriers. *Biotechnol. Bioeng. 25*(5): 1359-1372.

Varani, J., Dane, M., Rediske, J., Beals, T. F., and Hillegas, W. J. (1985). Substrate-dependent differences in growth and biological properties of fibroblasts and epitheleal cells grown in microcarrier culture. *J. Biol. Stand. 13:* 67-76.

Varani, J., Hasday, J. D., Sitriu, R. G., Brabaker, P. G., and Hillegas, W. J. (1986). Proteolytic enzymes and arachidonic acid metabolites produced by MRC-5 cells on various microcarrier substrates. *In Vitro Cell. Dev. Biol. 22*: 575-582.

Varani, J., Fliegel, S., Hinman, D., Helmreich, D., and Hillegas, W. J. Substrate-dependent differences in production of extracellular matrix molecules by squamous carcinoma cells. Submitted for publication.

von Wedel, R. J. (1986a). Lab scale fermenters can solve in-house mammalian cell culture needs. *Genet. Eng. News 6*(7): 16, 18.

von Wedel, R. J. (1986b). Integrated cell culture systems appear ready to emerge in marketplace. *Genet. Eng. News 6*(10): 52-53.

von Wedel, R. J. (1987). Mass culture of mouse and human hybridomas in hollow-fiber culture. In *Commercial Production of Monoclonal Antibodies*, S. Seaver (Ed.). Marcel Dekker, New York, pp. 159-173.

Weymouth, L., and Barsoum, J. (1986). Genetic engineering in mammalian cells. In *Mammalian Cell Technology*, W. Thilly (Ed). Butterworths, Boston, pp. 9-62.

Whiteside, J. P., and Spier, R. E. (1985). Factors affecting the productivity of glass sphere propagators. *Dev. Biol. Stand. 60*: 305-311.

Wolfe, R. A., Braatz, J. A., Miller, D. A., and Heifetz, A. H. (1988). A new serum-free medium for monoclonal antibody production. *BioTechniques 6*(1): 62-67.

World Health Organization (1981). WHO expert committee on biological standardization. Technical report series No. 658.

World Health Organization (1987). Acceptability of cell substrates for production of biologicals. Technical report series No. 747.

Yamada, K., and Olden, K. (1978). Fibronectin-adhesive glycoproteins of cell surface and blood. *Nature 275*: 179-183.

Yamane, I. (1978). Development and application of serum-free culture medium for primary culture. In *Nutritional Requirements for Cultured Cells*, H. Katsuta (Ed.). University Park Press, Baltimore, pp. 1-21.

Yerganian, G., Kato, R., Leonard, M., Gagnon, H., and Grodzino, L. (1960). Sex chromosomes in malignancy, transplantability of growths and aberrant sex determination. In *Cell Physiology of Neoplasia*. Austin University-Texas Press, pp. 49-93.

Young, M. W., and Dean, R. C., Jr. (1987). Optimization of mammalian cell bioreactors. *Biotechnology 5*: 835-840.

Zinn, K., Mellon, P., Ptashne, M., and Maniatis, T. (1982). Regulated expression of an extrachromosomal human beta interferon gene in mouse cells. *Proc. Natl. Acad. Sci. USA 79*: 4897-4901.

Zwerner, R. K., Cox, R. M., Lynn, J. E., and Acton, R. T. (1981). Five year perspective of the large-scale growth of mammalian cells in suspension culture. *Biotechnol. Bioeng. 23*: 2717-2735.

6

Growth of Yeast

DAVID R. BERRY

University of Strathclyde, Glasgow, Scotland

I. INTRODUCTION

Although brewing, distilling, wine making, and the baking of bread, all involving the growth of the yeast *Saccharomyces cerevisiae*, have been carried out for several thousand years, it is only over the past century that the role of yeast in these processes has been appreciated and investigated. Each of these traditional processes involves the growth of yeast in anaerobic conditions during which carbohydrate substrates are converted into ethanol and carbon dioxide by the process of fermentation. Ethanol is the key product in the production of alcoholic beverages, but it is the carbon dioxide that is important in baking and is responsible for raising the bread. Growth occurs in these processes, but the yield of yeast biomass is not important, although sufficient growth occurs so that subsequent fermentations may be inoculated with the surplus yeast from previous fermentations. A piece of dough may be transferred from one batch of baking to the next; however, it was traditional for surplus yeast from brewing to be used for baking. The recycling of yeast from one production vessel to the next is still widely practiced in the brewing industry. However, most bakers, distillers, and wine producers now require freshly grown, specially developed yeast strains to inoculate each fermentation. These strains provide a high yield of yeast from the substrate supplied and minimize ethanol production.

It was Louis Pasteur who first observed that the growth of yeast in aerobic conditions gave a higher yield of yeast cells at the expense of ethanol production. This effect, which has been referred to as the Pasteur effect and has been extensively studied since that time, is central to our understanding of yeast growth and is discussed later. The development of specialized processes to obtain high yields of yeast led to two key developments in the field on biotechnology: the introduction of air into submerged fermentation, which was achieved by blowing air through a perforated tube in the base of the fermentor, and the use of an incremental feed system rather than a batch system for supplying the growth substrate (Rosen, 1987). The use of incremental feeds was introduced when it was realized that even in aerobic conditions, the yield of yeast was reduced and ethanol fermentation occurred in the presence of high levels of readily metabolizable sugars. This phenomenon, which involves the suppression of aerobic respiration, has been variously referred to as the Crabtree effect, catabolite repression, or glucose repression.

In spite of the dramatic advances in biotechnology in recent years, industries that involve the growth of yeast still constitute the largest fermentation industries. The world market for alcoholic beverages has been estimated at 25 billion pounds, and baker's yeast at 1 million tonnes per annum (Dunnill and Rudd, 1984) constitutes a major biomass product. Both these fermentations are still the subject of active research and have evolved to very large scale and sophisticated processes. This chapter is concerned with these processes as well as with the growth of yeast for newer processes, such as fuel alcohol production or the production of genetically engineered products.

II. NUTRITIONAL REQUIREMENTS

Although the nutritional requirements of yeast are not very demanding, they do vary with the mode of growth and they must be controlled precisely if maximum growth rates and optimum yields are to be obtained. Since this topic has been discussed in detail in several recent reviews (Suomalainen and Oura, 1971; Jones et al., 1981; Jones and Greenfield, 1984; Berry and Brown, 1987), only the general requirements of yeast nutrition are described here (Table 1). All yeasts clearly need a carbon source, which also normally acts as an energy source, and a supply of those inorganic elements that occur in yeast. Yeasts also have a limited requirement for growth factors, the actual amount varying with the species of yeast and the growth conditions. Most of this section is concerned with the baker's and brewer's yeast *S. cerevisiae*; however, reference is made to other yeasts when appropriate.

The elemental composition of dried baker's yeast has been determined as 47% C, 6.3% H, 33% O, 8% N, 1.2% P, and 4.5% salts. This is equivalent to a gross formula of $C_8H_{12}O_4NP_{0.1}$. In ideal conditions, it has been reported that 245 g dry weight of yeast can be

Table 1 Medium Composition for Growth of *Saccharomyces cerevisiae* in Batch Culture (Amounts per Liter)

Glucose	50 g
Mineral Salts	
$FeSO_4(NH_4)_2 \cdot 6H_2O$	35 mg
KCl	120 mg
NaCl	60 mg
$CaCl_2 \cdot 6H_2O$	135 mg
$MgCl_2 \cdot 6H_2O$	520 mg
$MnSO_4 \cdot 4H_2O$	5 mg
$CuSO_4 \cdot 5H_2O$	0.5 mg
KH_2PO_4	2-3 g
$(NH_4)_2SO_4$	3.0 g
Trace Elements	
$ZnSO_4 \cdot 7H_2O$	2.3 μg
$CoSO_4 \cdot 7H_2O$	2.3 μg
$Na_2MoO_4 \cdot 2H_2O$	3.3 μg
H_3BO_4	7.3 μg
KI	1.7 μg
$NiSO_4 \cdot 6H_2O$	2.5 μg
Vitamins	
Aneurine hydrochloride	5.0 mg
Pyridoxine hydrochloride	6.25 mg
Nicotinic acid	5.0 mg
D-biotin	3.0 mg
Ca pantothenate	6.25 mg
Myoinositol	125 mg

obtained from 1 kg of molasses containing 480 g sucrose (Rosen, 1987), which represents a substrate yield of almost 50% on a weight basis. Such yields, however, can only be achieved in aerobic conditions, and much lower biomass yields are obtained in fermentative conditions when most of the carbohydrate is converted to ethanol.

A. Carbon Nutrition

Although some yeasts can utilize hydrocarbons as their main carbon source (see Sec. VIII), carbohydrates represent the most common carbon substrates for most yeasts, including *S. cerevisiae* (Barnett, 1976; Stewart and Russell, 1987). Different strains of *S. cerevisiae* can grow aerobically and anaerobically on a range of sugars such as glucose, galactose, mannose, fructose, sucrose, maltose, melibiose (*Saccharomyces uvarum*), and starch (formerly *Saccharomyces diastaticus*). However, other carbohydrates, such as glycerol and xylose, can only support aerobic growth and cannot be fermented to ethanol (see Sec. IV.B).

B. Nitrogen Nutrition

S. cerevisiae can grow on ammonium salts or urea as the sole nitrogen source, but unlike some yeasts, it cannot utilize nitrate. It can utilize amino acids, purines, and pyrimidines as the nitrogen source, and higher growth rates are normally achieved when mixtures of amino acids, such as casein hydrolysate, are supplied. Few yeasts have been reported that are able to grow on protein as the sole nitrogen source; however, small peptides are assimilated by brewing yeast. When supplied with complex media, ammonium ions are taken up in preference to organic nitrogen and the uptake of amino acids is repressed in the presence of ammonia.

C. Other Inorganic Nutrients

Phosphate is assimilated as the dihydrogen orthophosphate ion, and any phosphate esters present in the medium are broken down by an acid phosphatase present in the yeast cell wall. Sulfur can be assimilated as sulfate, sulfite, or thiosulfate ions and as methionine, but other sulfur amino acids are not assimilated.

D. Metals

The metal and growth factor requirements of *S. cerevisiae* are given in Table 1. Potassium is required as an effector for several glycolytic enzymes but also plays a key role in the maintenance of cation balance in the cell (Maiorella et al., 1984). The requirement for potassium is not constant, and anaerobically growing yeast has been reported to require twice as much potassium as aerobically growing yeast (Jones and Greenfield, 1984). Magnesium is required as a cofactor or effector for over 100 enzymes and also for the stability of nucleic acids and cell membranes. However, calcium does not appear to be essential for yeast growth although it can be assimilated and can influence the budding and flocculence processes as well as having a stimulatory effect on growth. Iron, zinc, and manganese are also important as enzyme cofactors. The requirement for iron, which is present in heme enzymes, such as cytochromes, is greater in aerobically grown yeast, whereas the requirement for zinc, which is required for alcohol dehydrogenase activity, is greater in anaerobic conditions.

E. Growth Factors

The requirement of yeast for growth factors varies with species, strain, and growth conditions. *S. cerevisiae* is reported to have a requirement for inositol, pantothenate, pyridoxine, thiamine, nicotinic acid, and biotin, although inositol and nicotinic acid can be synthesized by the yeast. A few strains of *Saccharomyces* have also been reported to have a requirement for folic acid, paraaminobenzoic acid, and riboflavin, although these can be synthesized by most yeasts.

When yeast is grown in strictly anaerobic conditions, it is unable to synthesize unsaturated fatty acids and sterols, which must be added to the medium. Although the requirement for oxygen or these growth factors is very low, oxygen is frequently the limiting nutrient in large-scale alcohol fermentations (Aires and Kirsop, 1977).

Although yeast is not a difficult organism to grow, great care must be taken in medium composition if yeast is to be grown efficiently on a large scale. A medium composition that has proved satisfactory for one set of growth conditions may not be correct if the growth rate, aeration, or carbohydrate feed is altered (Fiechter et al., 1981). In addition, interactions occur between different inorganic nutrients, which must be evaluated if maximum growth rates are to be achieved (Nagamune et al., 1981).

III. MODES OF YEAST GROWTH

S. cerevisiae is an unusual fungus in that it is able to grow equally well in both aerobic and anaerobic conditions. In anaerobic conditions it can grow with a specific growth rate of up to 0.5 hr^{-1}; however, the yield of biomass produced is low, and most of the substrate supplied is metabolized to ethanol and carbon dioxide by the classic process of ethanol fermentation. During this process, other minor products also appear, such as glycerol, acetate, and a wide variety of higher alcohols, acids, esters, and carbonyl compounds that contribute to the flavor of alcoholic beverages (Engan, 1981; Berry and Watson, 1987). The relative amounts of these compounds can be modified by varying the growth conditions, and the process can be manipulated to give glycerol as the major fermentation product (Sec. IV.C).

The control of growth in aerobic conditions is more complicated. If yeast is inoculated into an aerobic batch fermentation containing a high level of readily assimilable sugar, then the sugar is initially metabolized to ethanol in a manner similar to that occurring in anaerobic conditions. However, once the sugar has been assimilated the yeast continues to grow after a short lag phase using the ethanol produced during the first phase of the fermentation as the carbon substrate. This suppression of mitochondrial function and aerobic respiration is very important in the control of yeast growth. Fiechter has classified yeasts into glucose-sensitive species, such as *Saccharomyces* and *Schizosaccharomyces* (see Fig. 1), and glucose-insensitive yeasts, such as *Rhodotorula, Trichosporon, Pichia, Torulopsis, Hansenula,* and some *Candida* species (Fiechter et al., 1981). The glucose-sensitive yeasts were then divided into those that can grow on the ethanol produced and those, like *Schizosaccharomyces pombe*, that cannot.

If high yields of yeast biomass are required, then it is essential that it is grown in conditions in which catabolite repression does not occur so that the carbon substrate is directed into biomass, not diverted into ethanol. This can be achieved by using a glucose-insensitive yeast, such as the fodder yeast *Candida utilis*, or if the process requires *S. cerevisiae*, by keeping the level of sugars low or using a nonrepressing carbon substrate, such as acetate, galactose, or glycerol. Since economics normally require that cheap sugars like glucose and sucrose are used, it is necessary to keep the level of sugar low, below 100 mg/liter (van Dijken and Scheffers, 1984). This can be achieved either in exponentially batch-fed culture or in continuous culture.

In continuous culture, *S. cerevisiae* grows aerobically at low dilution rates, giving a high yield of biomass and a respiratory quotient (RQ) of around 1. However if the dilution rate, and hence

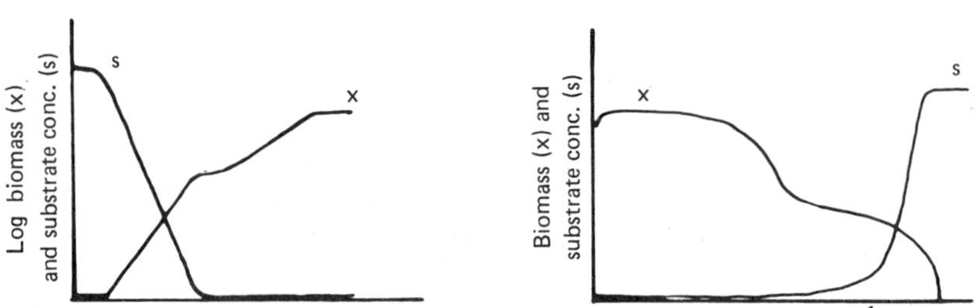

Figure 1 Kinetics of growth and substrate utilization of a glucose-sensitive yeast, such as *Saccharomyces cerevisiae*, grown in (a) batch and (b) continuous culture.

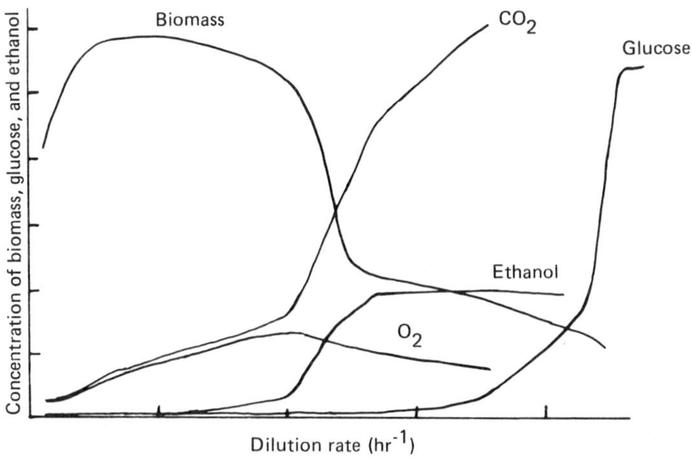

Figure 2 Growth of a glucose-sensitive yeast in continuous culture illustrating the transition from aerobic respiration to ethanol fermentation and its effect on oxygen consumption and carbon dioxide evolution.

the specific growth rate, is increased above values of around 0.2 then the RQ begins to increase as a result of increased CO_2 production and the level of glucose and ethanol in the medium begins to increase (Fig. 2). The level of biomass in the fermentor decreases, corresponding to a decrease in substrate yield. These changes indicate a switch to fermentative growth at high dilution rates and have been interpreted in terms of the rate of sugar feed exceeding the ability of the yeast cell to metabolize it aerobically, resulting in excess substrate being metabolized fermentatively. However, the dilution rate at which this transition occurs is influenced by the concentration of sugar in the inlet medium (Fiechter et al., 1981).

IV. PHYSIOLOGY OF YEAST GROWTH

A. Anaerobic Metabolism

The metabolism of the carbon substrate during yeast growth must satisfy the cells' requirements for energy in the form of ATP, reducing power as NADPH, and intermediates. These key metabolic intermediates are generated by the key pathways of intermediary metabolism, the Embden-Meyerhof-Parnas pathway, the hexose monophosphate pathway, the tricarboxylic acid cycle (TCA), and the glyoxylate cycle. However, the details of how they operate and their relative importance varies with different growth conditions.

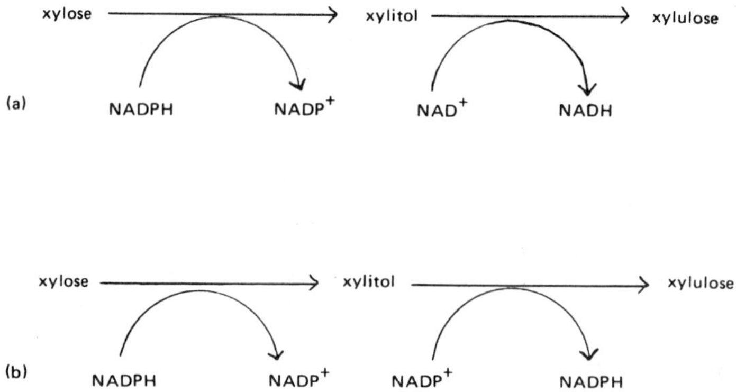

Figure 3 Metabolic pathways leading to metabolism of xylose in *Candida utilis* (a), which is incapable of fermenting xylose, and in two xylose-fermenting yeasts, *Pichia stipitis* and *Pachysolen tannophilus* (b).

Although it has been reported that yeast can survive at an energy charge of 0.15, the normal level is between 0.7 and 1.0 (Ball and Atkinson, 1975; Franco et al., 1984). In anaerobic conditions, the ATP generated by substrate-level phosphorylation reactions in the EMP pathway is the sole source of ATP so rapid growth in anaerobic conditions, involving a high demand for ATP for biosynthesis, requires a high rate of throughput in the EMP pathway. This leads to an accumulation of reduced NADH and of pyruvate. In the classic ethanol fermentation, pyruvate is decarboxylated to acetaldehyde by the action of pyruvate decarboxylase, and the acetaldehyde produced is reduced to ethanol using the NADH as a source of reducing power, thus releasing NAD for reuse in the EMP pathway. The trace levels of oxygen required to permit growth in anaerobic conditions, 0.7 mm Hg oxygen tension (Cysewski and Wilke, 1976), are insufficient to permit any production of ATP by oxidative phosphorylation. In the early stages of a simple batch fermentation the rate of ethanol production increases exponentially in parallel with the increase in biomass. However, when growth ceases due to a depletion of nitrogen or oxygen, then ethanol production continues in a linear manner. The fate of the ATP generated by glycolysis during this linear period of the fermentation is not clear. However, it may be required for maintenance functions or for the biosynthesis of storage carbohydrates.

The biosynthesis of cellular materials in anaerobic conditions involves the formation of intermediates that are normally synthesized by the tricarboxylic acid cycle or the glyoxylate cycle. Normal mitochondrial development is repressed in anaerobic conditions in yeast; however most of the enzymes are present at a low level (Wales et al., 1980), and some of the pathways operate at a sufficient level to provide intermediates.

B. Redox Balance

NADH cannot be used as a source of reducing power for biosynthetic reactions. These require NADPH, which is generated by the hexose monophosphate pathway. Maintaining redox balance plays a key role in the control of yeast metabolism. NADPH metabolism has been studied in detail in *C. utilis* (Bruinenberg et al., 1983a,b). It was calculated that 930.5 mmol NADPH was required to produce 100 g yeast when growing on glucose and supplied with ammonia as the sole nitrogen source. This is equivalent to the metabolism of 2% of the available glucose via the hexose monophosphate pathway. This value was increased to 20% when the nitrogen was supplied as nitrate as a result of the increased demand for NADPH to reduce nitrate to ammonium.

Redox balance is of particular importance when yeasts are grown anaerobically on pentose sugars. Xylose, a pentose sugar that is abundant in hemicelluloses, has to be converted to xylulose before it can be metabolized by yeast. In aerobic conditions many yeasts, such as *C. utilis*, can assimilate and metabolize xylose. However, most yeasts cannot grow on xylose in anaerobic conditions, although they can on xylulose. In aerobic conditions, the xylose is converted to xylulose in a two-step reaction. The xylose is first reduced to xylitol in a reaction involving NADPH. The xylitol is then oxidized to xylulose in a reaction that requires NAD and generates NADH (Fig. 3). *C. utilis*, however, does not produce a transhydrogenase enzyme, which would permit the NADH to be used to produce the NADPH required in the first step, so the process comes to a halt. In *Pichia stipitis* and *Pachysolen tannophilus*, two yeasts capable of fermenting xylose at a rapid rate, the xylose reduction step involves an enzyme that can use either NADH or NADPH (Bruinenberg et al., 1984; van Dijken and Scheffers, 1984), so the reducing power generated in the oxidation of the xylulose can be utilized to reduce xylose.

C. Glycerol Formation

Glycerol is frequently found as a minor product in ethanol fermentations. It is generally considered that it is produced by the cell as a mechanism of retaining redox balance if the NADH produced during glycolysis cannot be used to reduce acetaldehyde to ethanol. This situation can arise if glycolytic intermediates have been diverted into other cellular products, such as acetate or carbon skeletons for amino acids, or if the acetaldehyde to ethanol step is blocked, as it is in the classic glycerol fermentation (Peppler, 1979). The NADH produced under these conditions is reoxidized by reducing 2-dihydroxyacetone phosphate to glycerol phosphate, which is readily converted into glycerol by the action of glycerophosphatase.

In addition to being involved in the control of redox balance, it is now becoming clear that glycerol production can also occur in response to osmotic stress. Glycerol production increases at the expense of ethanol production when brewer's yeast is grown in media of high osmotic strength (Panchal and Stewart, 1980). However, it can also be induced in aerobically grown yeast by the addition, for example, of NaCl at approximately 0.5 M concentration (Maiorella et al., 1984). This is associated with a loss of biomass yield and reduced ethanol yields in anaerobic conditions. This osmotic effect may be important in the operation of fermentations, such as the biostill process in which liquid recycle is practiced, and the level of inorganic salts can increase each time the stillage is recycled (Maiorella, 1986). Osmophilic yeasts, that is, those capable of growing in an environment that has a low water activity, accumulate large concentrations of glycerol in the cell. The basic enzymology of glycerol metabolism in osmophilic yeasts, such as *Saccharomyces rouxii*, does not appear to differ from that of *S. cerevisiae*; however, *S. rouxii* appears to have the ability to retain a higher percentage of the glycerol produced in conditions of low water activity within the cell (Edgely and Brown, 1978, 1983).

D. Ethanol Toxicity

Although ethanol is a major product of *S. cerevisiae* when it is grown anaerobically, it is also toxic to yeasts. Extensive studies have been carried out on the nature of ethanol tolerance, and attempts have been made to develop strains with improved ethanol tolerance for use in the production of fuel alcohol (Tubb, 1984). Ethanol concentrations of around 10% appear to be sufficient to inhibit growth and ethanol production in most yeast strains. However, some saké and sherry yeasts can continue to ferment up to 160 g/liter. The degree of toxicity is influenced by many factors, however, such as nutritional status of the medium (Casey et al., 1983), osmotic strength (Panchal and Stewart, 1980), and temperature (Leao and van Uden, 1982). One of the difficulties that has hampered studies of the site of action of ethanol toxicity has been the difficulty in measuring the intracellular concentration of ethanol. Several workers have reported that during ethanol fermentations the intracellular ethanol concentration was greater than the extracellular concentration. However, recent work using radioactive isotopes suggests that this is not the case (Guijarro and Lagunas, 1984). The primary site of ethanol action during alcohol inhibition has not been precisely defined. Although ethanol has been shown to be inhibitory to several glycolytic enzymes (Millar et al., 1982), it is generally considered that the main site of action of ethanol at concentrations that occur in fermentations is at the cell membrane. However, evidence has recently been presented that many of the effects of high concentrations of ethanol can be attributed to the effect of ethanol on the water activity of the growth medium (Jones and Greenfield, 1986). There can be little doubt that the effects of ethanol on the yeast cell are complex (Ingram and Bukkte, 1984). Perhaps because of this, considerable difficulty has been experienced in selecting for yeast strains with improved ethanol tolerance. However, a strain that fermented glucose at higher rates in the presence of 10% ethanol was isolated by Brown and Oliver (1982) using a selection technique that involved the use of continuous culture.

E. Aerobic Metabolism

The physiology of aerobic growth is much more complex since respiratory metabolism is influenced not just by the availability of oxygen but also by the level and nature of the carbohydrate substrate in the medium and to some extent by the history of the yeast cell. It is best studied in continuous culture since it is difficult to achieve conditions in which ethanol fermentation is completely suppressed, especially on a small scale, in batch culture. In continuous culture at low dilution rates, mitochondria are fully developed and fully active. The RQ of the fermentation in these conditions is 1, indicating that respiration is highly aerobic, ATP generation is by oxidative phosphorylation, and there is no ethanol fermentation. If the dilution rate is increased, this physiological state is not maintained. The RQ of the fermentation increases, the residual glucose level increases, and ethanol is produced. The increase in RQ can be attributed to an increase in CO_2 production associated with the onset of ethanol fermentation. In these conditions, however, mitochondrial function is not fully repressed since oxygen uptake continues at a high level. Several workers have reported that washout occurs when the dilution rate approaches 0.3; however, Barford and Hall (1979) obtained high rates of oxygen uptake in cultures grown at a dilution rate of 0.6. These results were obtained after a period of adaptation of the yeast. The RQ of yeast is a valuable guide to its metabolic state. Whereas values above 1 indicate the onset of fermentation, values below 1 can be obtained when acetate or ethanol are being metabolized. A value of 0.2 has been reported for growth on ethanol (Rickard and Hogan, 1978).

Changes in the physiological state of the yeast cell at increased growth rates have also been demonstrated by studying the levels of intermediates of the glycolytic pathways and the tricarboxylic and glyoxylate cycles. The levels of a range of intermediates were measured at seven different dilution rates in yeast grown on three different media. In these experiments, the onset of ethanol fermentation occurred at D = 0.16 and was associated with an increase in the level of several glycolytic intermediates. However, the presence of TCA cycle intermediates at higher dilution rates indicated continued mitochondrial function (Franco et al., 1984). The results also demonstrated that the physiological state of the yeast cell is continuously changing and that it is an oversimplification to consider yeast as in a glucose repressed or nonrepressed state. This view is supported by studies on the levels of TCA and glyoxylate cycle enzymes in yeast grown in different conditions. Under aerobic conditions, the levels of most TCA enzymes were reduced by 10% glucose but were not completely repressed. The levels were further reduced in anaerobic conditions, but only one enzyme, 2-oxoglutarate dehydrogenase, was completely absent (Wales et al., 1980). The role of oxygen in the control of mitochondrial development is not clear; however, it cannot be replaced by unsaturated fatty acids and sterols so it appears to involve more than a requirement for these types of compound.

The role of carbon metabolism in the control of yeast growth has been reviewed by Kappeli (1986), who reported that in addition to changes in metabolic state associated with changes in growth condition, cell cycle-associated changes may also occur. Spontaneous synchronization of growth can occur during transition from batch to continuous culture in a carbon-limited chemostat. In these circumstances endogenous glycogen and trehalose were metabolized to ethanol during the initial phases of the cell cycle and the level of gluconeogenic enzymes increased (Kappeli, 1986; De Lorencez and Kappeli, 1987). In asynchronous cultures no such accumulation of ethanol was observed, presumably because all stages of the cell cycle are represented in the yeast population. However, a persistent synchronization of yeast growth has been reported using a fermentation system, a compact loop reactor, in which the mixing time was very short, below 1 sec, and the dilution rate was controlled to within 1%. In this system the synchronization of the yeast metabolism could be monitored by measuring oxygen, ethanol, and carbon dioxide continuously in the exhaust gas (Sonnleitner et al., 1987).

Studies on *C. utilis*, a glucose-insensitive yeast, indicate that high levels of glucose only have a small effect on the levels of mitochondrial enzymes (Fiechter et al., 1981).

The molecular basis of glucose repression of mitochondrial and other enzymes in yeast is not well understood. It does not appear to involve cAMP, as does catabolite repression in *Escherichia coli*, and as yet the key regulatory molecule has not been identified.

Several studies have implicated the initial steps in the uptake and metabolism of glucose, and recent studies on the genetics of catabolite repression have pointed to the enzyme hexokinase as having an important role. Strains of *S. cerevisiae* that lack hexokinase PII do not exhibit catabolite repression; however, transformation of the gene for hexokinase PII into such strains restores both hexokinase activity and catabolite repression (Frohlich et al., 1984). Other mutations that cause derepression of several hydrolytic enzymes, such as invertase and maltase, as well as mitochondrial enzymes, also exhibit elevated levels of hexokinase (Ciriacy, 1978; Bailey and Woodward, 1984). One of these strains was derepressed irrespective of whether it was grown on glucose or maltose, suggesting that the defect in the mechanism of catabolite repression cannot be attributed solely to a defect in the glucose transport system (Ciriacy, 1978). The transition from anaerobic growth to aerobic growth is associated with an increase in cell yield from around 10% to over 50%. This has been considered the effect of the much higher levels of ATP generated per molecule of glucose in aerobic conditions; however, it is now accepted that this mechanism is backed by a variety of biochemical control mechanisms (Sols et al., 1971; Fraenkel, 1982).

F. Storage Products

Up to 40% of the dry weight of the yeast cell can be storage materials. These can be either glycogen or trehalose in *S. cerevisiae* or more usually a mixture of both. In sporulating yeast, glycogen has been considered the storage material for ascus development whereas trehalose was the reserve material for the germinating ascospores (Tingle et al., 1973; Berry, 1983). In vegetative yeast, both glycogen and trehalose are usually present, with glycogen present in greater quantities. However, the ratio of trehalose to glycogen is higher in aerobically grown rather than anaerobically grown yeast. At the cellular level, it has been reported that the level of storage carbohydrates increases during the unbudded phase of the cell cycle and that these reserves are then mobilized during the budding phase of the cell cycle. The implications of this are discussed by Barford (1987).

When yeast is grown in batch fermentation, the level of both storage carbohydrates increases during the early stages of the fermentation when the level of sugar in the medium is high and growth is nitrogen limited (Patel and Ingledew, 1973). The stored carbohydrate is then mobilized when the exogenous sugars have been assimilated. In both beer and whiskey fermentation, ethanol production continues after wort sugars have been metabolized at the expense of stored carbohydrate (Lavery and Berry, 1983). The level of glycogen has been reported to be important in pitching yeast, a high level of glycogen being essential for giving vigorous growth and a rapid rate of attenuation in lager fermentations (Murray et al., 1984). Some of the glycogen is metabolized during the storage of the pitching yeast. However, it is important that a high level remains at pitching. When yeast is inoculated into fresh wort, the level of glycogen decreases rapidly (Quain et al., 1981). This decrease has been associated with a period of rapid lipid synthesis by the yeast.

In aerobic continuous culture, the level of storage carbohydrates is highest at low dilution rates (Kuenzi and Fiechter, 1972).

V. TECHNOLOGY OF ETHANOL PRODUCTION

The choice of bioreactor and fermentation system for ethanol fermentations is dependent upon the nature of the final product. The requirements of a process to produce ethanol in large volumes for fuel alcohol or chemical feedstock are very different from those required in the traditional processes for the production of alcoholic beverages. In the beverage industry the requirement to maintain the characteristic flavor of the product takes precedence over substrate and volumetric yields, although these are by no means ignored in the modern brewing and distilling industries. The maintenance of a constant flavor requires that the levels of a large number of organoleptic compounds produced in the fermentation must remain constant between batches over many years. In view of this it can be argued that the "problems facing the fermentation technologist in the beverage industry are much more complex than when fuel ethanol is the objective" (Margaritis and Merchant, 1987). The production of ethanol is the largest fermentation process, and most of this is in the form of alcoholic beverages. These are normally carried out as batch processes in a variety of fermentation vessels. Some of the traditional processes, such as the Yorkshire Stone Square and the Burton Union (Pollock, 1981), are still used in traditional breweries. However, these have largely given way to large rectangular open tanks characteristic of traditional lager fermentations and more recently to large (20 m or more in height) stainless steel cylindrical vessels that have either a flat sloping base, the Asahi tank, or a conical base, the cylindricoconical tank, to facilitate the harvesting of the spent yeast (Lindsay and Larson, 1975). Improved sterility is achieved in such vessels because they are enclosed, jacket cooled, and readily cleaned. These vessel contents are mixed by the carbon dioxide being released during the fermentation, and in vessels with a conical base the release of carbon dioxide sets up convection currents that lead to improved cooling and mass transfer characteristics.

A. Use of Continuous Culture

Several attempts have been made to produce beer using continuous culture since its use leads to reduced capital, reduced downtime when the fermentor is not in operation, improved volumetric efficiency, and reduced production costs. These processes have normally used tower fermentors (Smith and Greenshields, 1973; Kirsop, 1982). Although continuous culture is used to produce beer in one factory in New Zealand, the technique has not been developed by most companies. The problems associated with use of continuous culture in the brewing industry have been discussed by Portno (1979). Some of the difficulties are of a microbiological nature, such as problems of contamination and strain stability and the difficulties of reproducing the subtle metabolic changes that occur in batch culture and are critical to the development of the flavor of alcoholic beverages. Other difficulties result from industrial practice. The operation of a continuous culture requires a continuous input of engineering and scientific expertise, which may not be available on a round-the-clock basis. In addition, a continuous system only offers the greatest advantage in long uninterrupted production runs. However, many modern breweries produce a variety of beers, sometimes under license, so flexibility in production facilities is desirable.

B. Fuel Alcohol and Immobilized Cell Systems

Although the batch system for yeast growth and ethanol production can be used for industrial alcohol production, it has a low volumetric efficiency. Although this can be improved by inoculating the fermentation with a high yeast concentration using yeast that has been harvested by centrifugation from previous fermentations (the Melle Boinot process), the highest productivities and the fastest production rates are obtained using continuous processes (Maiorella, 1986). Productivities of 7 g/liter per hr, three times the level that can be obtained in batch fermentations, have been obtained using a single continuous stirred tank reactor (Cysewski and Wilke, 1976). However, this process is itself limited by the inhibitory effects of ethanol in the fermentor and by the cell densities that can be achieved, and much of the recent research work has been concerned with obtaining higher ethanol productivities. The problem of ethanol toxicity can be minimized either by using a two-stage or more process, so that the cells are only exposed to the highest concentrations of ethanol in the final stage of the process (Ghose and Tyagi, 1979), or by the continuous removal of ethanol throughout the process. Removal of ethanol at temperatures that do not inhibit the activities of the yeast can be achieved by carrying out the fermentation at reduced pressures whereby the ethanol boils off as it is formed (Cysewski and Wilke,

1977). Using such techniques, sugar feed concentrations of up to 50% have been fermented and productivities of 82 g/liter per hr have been obtained. High cell densities can be maintained in the fermentor using a variety of cell recycle techniques. The simplest of these involves exploiting the flocculent character of many brewing yeasts. Flocculent yeast can be concentrated by allowing it to settle out from the fermentation medium in conditions that are free of turbulence. Flocculence has been exploited to achieve cell recycle in stirred tank reactors by passing the wash into an external gravity settler in which the yeast is concentrated before being returned to the fermentor. The use of a tower fermentor, however, offers a simpler method for recycling flocculent yeast. Special tower fermentors have been designed in which the top of the fermentor is adapted to permit the separation of the carbon dioxide, leaving a still volume in which yeast settling can occur. The tower fermentor becomes full of densely packed flocs of yeast, and the fermentation proceeds as the fresh medium is pumped in at the base of the fermentor and passes up the tower (Greenshields and Smith, (1974). This process, however, suffers from poor mixing, breakup of the yeast flocs by the carbon dioxide released during the fermentation, and loss of yeast by washout, especially at high dilution rates. Many of these problems can be minimized by using a variety of immobilization techniques to maintain a high biomass level in the fermentor. Margaritis has distinguished three categories of immobilization technique: (1) containment of yeast cells in membrane reactors, (2) attachment of yeast cells to solid supports, and (3) entrapment of yeast cells in polymeric gels (Margaritis and Merchant, 1984).

Space does not permit a detailed description of all the different systems that have been employed to immobilize yeast cells; however, the subject has been extensively reviewed by Linko and Linko (1984) and Margaritis and Merchant (1984, 1987). The production of ethanol by yeast that has been entrapped in calcium alginate beads has been extensively studied in recent years. In a process that has recently been described (Oda, 1983; Nagashima, 1984), up to 60% of the reactor volume was filled with Ca-alginate beads containing 250 g dry weight yeast cells per liter of gel. The fermentation was carried out in a 4 m^3 two-stage fluidized bed reactor that gave a volumetric ethanol productivity of 50 liter per hr. Other gels that have been used for yeast immobilization include carrageenan, gelatin, agar, and polyacrylamide. Using entrapped cell systems, high productivities can be achieved as a result of the high cell densities and maintained for long periods of time. Agar-entrapped yeast cells have been kept actively fermenting molasses for up to 30 weeks. However, the use of entrapped cells does have limitations. The most important of these is that the support matrix presents a diffusion barrier between the medium and the yeast cells, which reduces the rate of transfer of nutrients into the cell and the rate of products, some of which may be toxic, out of the cell. The accumulation of carbon dioxide in the matrix can present a particular problem since the formation of bubbles can disrupt the gel matrix.

The technique of surface adsorption of cells has also been used to immobilize yeast. Cells immobilized in this manner can be used in ethanol fermentation processes that exhibit high biomass densities, increased ethanol productivities, and improved stability compared with free cell fermentation systems (Margaritis and Merchant, 1987). Yeast binds to support materials, such as wood chips, porous brick particles, stainless steel mesh, and polyester foam, by electrostatic forces. However, it can also be bound covalently to gelatin-coated surfaces using glutaraldehyde as the cross-linking agent.

VI. TECHNOLOGY OF AEROBIC GROWTH

The selection of fermentation techniques for the aerobic growth of yeast is influenced by several factors, such as the physiology of the yeast strain or species, the nature of the product, the level of sterility required, the volume of production, and the product value. The production of baker's yeast is the largest and probably most intensively studied process for yeast production, with an annual rate of production of 1574×10^6 kg/year (1980 figures, Peppler, 1983). However, the production of animal feed, usually using *C. utilis* as the producer organism, is carried out on a large scale in several countries, especially in Eastern Europe. Smaller quantities of yeast are also used for the production of enzymes and fine chemicals, which may be produced in special fermentations or obtained as by-products from other processes. It is not the purpose of this review to enter into a detailed discussion of the design of individual fermentors used for producing yeast or to discuss in detail individual processes. An attempt is made, however, to indicate how the nature of the product influences the choice of process.

The critical feature of the physiology of *S. cerevisiae* is that its aerobic metabolism is subject to glucose repression, so if high yields of biomass are to be achieved then the level of sugar in the fermentor must remain low. This can be achieved either by the use of continuous culture or by using a batch-fed fermentation system in which the sugar feed rate follows the biomass level closely. Continuous culture is the favored process for high-volume products, such as animal feed production, and is particularly suited to the growth of *C. utilis*, which does not exhibit glucose repression, so it can be grown at a high yield at all dilution rates below its μmax.

Continuous culture operation, however, is not without problems. Contamination by faster growing organisms can limit the value of the process, so fermentors designed for continuous use must be built and operated to a high standard. Special fermentors have been designed for large-scale continuous processes. One of these, designed for the production of biomass using *Candida parafinica*, consists of four bubble towers, each with a volume of 1.26×10^6 liters (Zanetti, 1984). The use of airlift fermentors, which may be fitted with a draft tube, offers the advantages of no moving parts, low shear, and low energy requirements as well as reduced capital cost (Barford, 1987).

Although baker's yeast can be described as a low-price, high-volume product, it is normally produced in fermentors that do not exceed 200 m^3 (200,000 liters). It is a product required to be clean but not necessarily free of contamination and is produced in different forms, such as compressed and dried yeast, which require different fermentation regimes (Reed, 1982; Burrows, 1979; Rosen, 1987). These requirements are more readily satisfied by the use of batch-fed production techniques, which are used throughout the world for baker's yeast production. The history of growth of the yeast appears to be important in determining the final metabolic state of the yeast as well as the final growth conditions. The use of serial batch cultures allows the inoculum stages and the early stages of the growth of the yeast to be varied independently of the final growth and ripening phases. This subtlety of control cannot be achieved in continuous culture, unless several vessels are used, which makes the process complex and difficult to control.

Recent improvements in the techniques available for the analysis of effluent gases using mass spectroscopy and for the analysis of cellular intermediates using nuclear magnetic resonances (NMR) should facilitate our understanding of the physiology of yeast growth and lead to improved control systems (Sonnleitner et al., 1987; Campbell-Burk and Shulman, 1987).

The selection of production vessel is influenced by a requirement for a high level of aeration and good mixing at a low cost to prevent ethanol production. Traditionally, this has been achieved by the use of static air spargers without mechanical agitation, stirred tank reactors, or proprietary brands of aeration system, such as the Vogelbusch system (Barford, 1987). The appearance of ethanol in the air outlet is a sensitive indicator of any deviation from aerobic growth, which leads to a loss of yield and quality. Processes have therefore been developed in which the molasses feed is controlled by an alcohol sensor fitted in the air outlet line. However, the growth of baker's yeast is a complex process in which the medium composition and feed rate are influenced by the need for a high substrate yield, good volumetric productivities, and the need to produce a yeast in the correct physiological state for a baking yeast. This involves not only a high fermentation capacity in the dough but also a good keeping quality in either the compressed or dried state. One of the problems facing the commercial producer is that such conditions as high nitrogen levels, which favor a high fermentative activity, tend to produce a yeast with a poor keeping quality; compromises in the selection of fermentation parameters must therefore be made. The processes used for the production of baker's yeast are also widely used for the production of yeast for the isolation of other yeast products, such as enzymes (e.g., invertase and lactase), yeast autolysates, and nucleotides. However, when special growth conditions or special species of yeast are required to produce fine chemicals, yeast is grown on a small scale (Peppler, 1979).

The production of high-value pharmaceutical products by genetically engineered yeast is still in its infancy. Some of the unique problems associated with these processes are discussed in Section VII. However, some general considerations are appropriate here. The production of pharmaceutical products requires a high level of sterility at all stages of the production process, including the fermentation stage. A high level of control is also required over the quality and composition of the substrates used such that defined media are preferred when possible. These requirements make the process more expensive and dictate that the fermentation equipment and process should be designed to permit the high standards of control to be maintained. At present these high-value products are produced on a small scale, 10-100 liters, more akin to a laboratory fermentation than a factory process.

VII. GROWTH OF GENETICALLY ENGINEERED YEAST

It has been possible to manipulate the genotype of *S. cerevisiae* for many years using classic genetic techniques and more recently protoplast fusion (Mortimer and Hawthorne, 1975; Jones et al., 1987). However, these techniques permit only the rearrangement of genes that are already present in the *S. cerevisiae* gene pool. It was the development of cloning techniques for yeast (Botstein and Davis, 1982; Panchal, 1987) that permitted the introduction of genes from other species, including plants and animals as well as other fungi.

A large use of yeast as the producer organism for a wide range of genetically engineered products in the near future seems probable since *S. cerevisiae* offers many advantages in such processes. In addition to being a simple eukaryotic organism that has a well-developed and intensively investigated genetic system, it is a food-grade organism that has a high level of consumer acceptability and can be readily grown on a large scale in industry. Some of the proteins that may be produced in yeast are shown in Table 2.

Table 2 Selected Examples of Heterologous Proteins Produced in *Saccharomyces cerevisiae*

	Reference
Viral	
Hepatitis B surface antigen	Hitzeman et al., 1983a
Influenza viral hemagglutinin	Jabber et al., 1985
Herpes simplex IgD particle	Valenzuela et al., 1985
Human	
γ_1, γ_2, γ_3 Interferons	Hitzeman et al., 1983b
Insulin	Stepien et al., 1983
α-Antitrypsin	Cabezon et al., 1984
Tissue plasminogen activator	Kingsman et al., 1985
Epidermal growth factor	Urdea et al., 1983
Animal	
Mouse interleukin-2	Miyajima et al., 1985
Calf chymosin	Mellor et al., 1983
Calf prochymosin	Goff et al., 1984
Plant	
α-Amylase	Rothstein et al., 1984
Thaumatin	Edens et al., 1984
Bacterial and other fungal	
Cellulase (*Clostridium*	Skipper et al., 1985
β-Glucanase (*Bacillus subtilis*)	Hinchliffe and Box, 1984
β-Lactamase (*Escherichia coli*)	Chevalier and Aigle, 1979
Glucoamylase (*Aspergillus awamori*)	Innes et al., 1985

Since the first reports of transformation in yeast (Hinnen et al., 1978), a wide variety of techniques have been developed to introduce nonyeast genes into *S. cerevisiae* (Botstein and Davis, 1982) and several other yeasts (Gunge, 1983). These techniques have been used both to develop new products (Hitzeman et al., 1983b) and to alter the properties of yeasts used in existing processes (Hinchliff and Box, 1984). The desire to achieve a high level of production, and in some cases excretion of heterologous proteins, has led to the development of a range of vectors or plasmids that can be used for genetic engineering in yeast.

These have been classified into eight groups by Bostian (Parent et al., 1985) based primarily on their mode of replication and on their expression characteristics. It is important to have some understanding of the nature of the plasmid being used, since the stability of the cloned gene and its level of expression in a fermentation are very dependent upon the characteristics of the plasmid used. Autonomously replicating plasmids that contain segments of the 2μ plasmid of *S. cerevisiae* (YEp) are probably used most widely at the present time since they give high frequencies of recombination and can give good stability in the correct selective conditions. Yeast replicating plasmids (YRp), which contain ARS segments, transform yeasts at a high frequency and can have a high copy number but are mitotically and meiotically unstable (Parent et al., 1985). The stability of this type of plasmid can be increased by introducing a centromere to produce a yeast centromeric plasmid (YCp). However, this leads to a reduction in copy number. Linear plasmids that function like small eukaryotic chromosomes can also be produced by the introduction of telomeres. Yeast integrating plasmids (YIp) have also been constructed that give a low frequency of transformation but cause the cloned gene to be integrated into the chromosome such that the stability of the gene is not dependent upon the autonomous replication of the plasmid. The integration of genes in this manner is probably most valuable in situations in which a high copy number and a high level of gene expression are not required. Similar results can be achieved using transposons, although use of transposons can give integration at several different sites in the genome. The rate of production of a plasmid-borne genetically engineered product is controlled not only by the level of gene expression, which is a product of the genetic construction of the plasmid, but also by the copy number of the plasmid and by the stability of the plasmid. Any satisfactory study of the growth of such genetically engineered organisms should involve monitoring the frequency of plasmid bearing cells (Xp) and cured cells (Xc), plasmid copy number (Np), and plasmid stability, as well as studying cell and product yield. Some studies of plasmid stability in yeast using both batch (Futchers and Cox, 1984) and continuous culture (Walmsley et al., 1983) have been published. Walmsley presented equations for predicting the kinetics of appearance of plasmid free, in their case leu$^-$, cells in continuous culture in conditions in which the plasmid could be considered at a selective advantage (Fig. 4a), selective disadvantage (Fig. 4b), or selectively neutral (Fig. 4c). Other models for predicting the stability and expression of yeast plasmids have been presented (Hjortso and Bailey, 1984; Srienc et al., 1986). However, few detailed studies of the growth of genetically engineered yeast are available. If the yeasts are to be grown

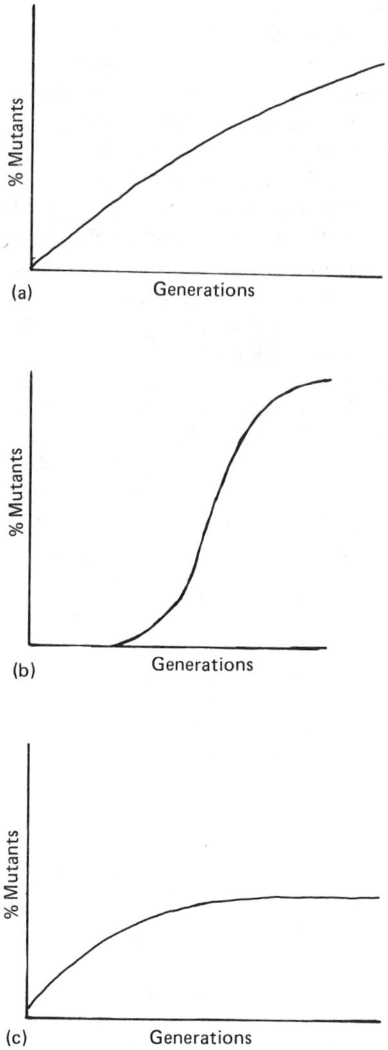

Figure 4 Idealized curves predicting the pattern of appearance of leu^- cells arising from plasmid loss, in yeast grown in a nonselective continuous culture (see text). (From Walmsley et al., 1983.)

on synthetic medium, as may be so for the production of high-value pharmaceutical products, then the use of auxotrophic selective markers to achieve plasmid stability may be acceptable. A yeast strain that required tryptophan and a plasmid containing the appropriate tryptophan gene were used in the production of hepatitis B surface antigen in a 380 liter scale fermentation (Carty et al., 1984). The use of a production strain carrying a double mutant for uracil and uracil transport deficiency has been reported in a process to produce human α_1-antitrypsin (Marquet et al., 1986). If polyploid yeasts are essential to the process, then the use of auxotrophic selection techniques is not possible and other selection procedures, such as the use of resistance markers, like copper (Fogel and Welch, 1982) or G418 (Webster and Dickson, 1983), must be adopted or the genes should be integrated. Equally, auxotrophic markers cannot be used if the yeast is to be grown on complex substrates; however, a similar result can be obtained if the plasmid contains a key metabolic enzyme, such as triose phosphate isomerase (Klotz, 1983).

It has been clearly demonstrated in bacterial systems that the stability of the plasmid is also influenced by the level of expression of the plasmid-borne genes. This has led several groups to investigate the use of conditional expression systems that allow the yeast to be grown up in conditions that do not favor the expression of the cloned genes, thus favoring plasmid stability, then transferred to conditions that promote a high level of expression of the cloned genes when growth has been completed. Conditional expression may be controlled by substrate-level induction by using, for example, the galactokinase promoter or by temperature shift (Klotz, 1983; Primrose et al., 1983; Caulcott and Rhodes, 1986). The use of such conditional expression systems favor batch-fed systems of propagation, although two-stage continuous processes may also be used (Siegel and Ryu, 1985). The stability of a plasmid is also influenced by the nature of the host cell and by the growth conditions, however. It is a common experience of geneticists that the level of expression of a vector varies if different recipient cells are used. However, quantitative data are less readily available. It has been demonstrated that the copy number of a 2μ-based plasmid, pJDB219, differed in two haploid strains of *S. cerevisiae* and was lower in both haploids than in the diploid derived from them. The higher copy number in the diploid led to a higher plasmid stability (Mead et al., 1986). However, Takagi investigated the production of β-galactosidase in haploid, diploid, and tetraploid cells using three different types of plasmid, a YEp, a YRp, and a YCp, and found that although the stability of the plasmid increased with ploidy, the level of expression on a cellular basis did not. Since there were fewer cells per milliliter in the polyploid culture, the volumetric yield of enzyme decreased significantly (Takagi et al., 1985). The effect of growth rate on plasmid stability has been investigated using continuous culture. A 2μ-based plasmid, pJDB248, was shown to be more stable when the yeast was grown at low dilution (Kleinman et al., 1986). A high plasmid stability was obtained in continuous culture using hosts that had a ura^{3+} mutant and fur^{1} mutation, which impedes uracil uptake, and a *ura*$^{3+}$-bearing plasmid (Marquet et al., 1986).

However, plasmid stability may not be the most important characteristic in controlling product formation. It has been reported that although the plasmid for producing hepatitis B virus antigen was more stable when the yeast was grown on defined selective medium, yeast growth and antigen production were much higher in nonselective complex medium (Lee et al., 1986).

One reason that has frequently been advanced for producing proteins using yeast rather than bacterial vector systems is that they have the ability to glycosylate proteins. Most of the periplasmic and extracellular proteins produced by *S. cerevisiae* are glycoproteins. Although it has been reported that some glycosylated forms of heterologous proteins, such as antitrypsin and mouse immunoglobulin, have been produced in yeast, the natural glycoproteins of yeast contain mainly mannan, whereas mammalian glycoproteins have a more varied carbohydrate composition. Variation in the degree of glycosylation can also occur, leading to a product with a variable molecular weight that may be difficult to extract, so it seems unlikely that the direct glycosylation of heterologous proteins by yeast will be an important feature of yeast fermentations (Kingsman et al., 1985).

Because of the problems of hyperglycosylation by *S. cerevisiae*, it is possible that other yeasts will find favor for the production of heterologous proteins. A cloning system for *Pichia pastoria* has been developed, and in *Pichia*, using an invertase secretion system, glycosylation was restricted to core glycosylation (van Brunt, 1986). A further limitation of *S. cerevisiae*, its inability to excise intervening sequences from transcripts of higher eukaryotic genes, may be overcome by the use of another yeast, *S. pombe*, which contains introns in its genome and has been shown to be capable of accurately excising such intervening sequences (Dawes, 1986).

VIII. GROWTH ON NONCARBOHYDRATE SUBSTRATES

Although the commercial applications are as yet limited, no chapter on yeast growth would be complete without some reference to the growth of yeast on noncarbohydrate substrates, such as methanol and hydrocarbons. Most of the work carried out in this area was stimulated by the desire to produce large quantities of single-cell protein for animal feed using either methanol, produced directly from methane, or n-alkanes (Litchfield, 1979; Cartledge, 1987). However, there is also interest in the production of organic acids, especially citric acid from hydrocarbons (Marchal et al., 1977), and fats (Ratledge, 1982).

Such yeasts as *Candida tropicalis* and *Saccharomycopsis lipolytica* can grow on n-alkanes in the range C4-C18. These alkanes pass into the cell in an unaltered form, where they are then oxidized to the corresponding alkanol. The alkanols are then oxidized first to the alkanals and then to the corresponding fatty acids, which are then metabolized by the normal β oxidation of their acyl-CoA derivatives. The initial oxidation of the alkane involves a mono-oxygenase enzyme that contains a cytochrome P_{450} component and requires NADP. This initial reaction involves the generation of hydrogen peroxide, which is degraded by catalase (Kawamoto et al., 1977). Several plants for the production of single-cell protein from hydrocarbons were designed and constructed in the early 1970s; however, the subsequent increase in oil prices led to all the plants outside Comecon countries being shut down before 1977 (Cartledge, 1987). Production was carried out in either Waldhof fermentors or airlift fermentors; the cells were then separated off and washed with polar and nonpolar solvents to give a final product that contained 60% protein and 10% lipid (Shacklady, 1975; Litchfield, 1979).

Growth on methanol has been intensively studied in such yeasts as *Hansenula polymorpha* and *Candida boidinii*. The methanol is oxidized to formaldehyde by a nonspecific alcohol oxygenase. This can be oxidized directly to carbon dioxide; however, growth on methanol requires the fixation of formaldehyde into dihydroxyacetone phosphate via a reaction that involves xylulose monophosphate as one of the initial intermediates (Waites and Quayle, 1980). The production of yeast single-cell protein on a pilot plant scale has been reported by the Mitsubishi Chemical Company (Kuraishi et al., 1979). In this process production was carried out in the airlift fermentor fitted with a perforated plate and a draft tube to obtain improved oxygen transfer characteristics. Optical growth occurs at methanol concentrations of between 1 and 3%. Above these concentrations growth can be inhibited by the accumulation of toxic products of methanol oxidation, such as formaldehyde and formate.

Several processes for the production of citric acid from n-alkanes using *S. lipolytica* have been described. These produce around 1.5 g citric acid per gram of alkane used with a citrate-isocitrate ratio of up to 93:3 (Akiyama et al., 1972). Since in these fermentations the biosynthesis of citric acid occurs in the stationary phase, production is carried out in batch fermentations.

Several yeasts produce a high level of lipid when grown in low nitrogen conditions. This may occur in yeasts grown on carbohydrates; however, it has been reported that growth on hydrocarbons gives a higher percentage of the biomass as lipid. Growth on hydrocarbon also offers the advantage that the chain length of fatty acids produced can be predetermined by the substrate (Ratledge and Boulton, 1986).

IX. POSTSCRIPT

The growth of yeast has been studied by industrialists and scientists for many centuries, and a wealth of knowledge has been accumulated. It is not possible in a single chapter to do justice to all aspects of this fascinating subject since the production of fuel alcohol, alcoholic beverages, baker's yeast, biomass from carbohydrates and hydrocarbons, food products, fine chemicals, and pharmaceutical products are all specialist subjects in their own right requiring special expertise in the fields of biochemical engineering, genetics, and biochemistry. Although it is relatively easy to grow yeast in the laboratory, the large-scale cultivation of yeast for special purposes in situations in which yields and productivity are critical requires a much more precise understanding of the process and a high level of control. In this chapter an attempt has been made to introduce some of the distinctive features of different yeast processes and indicate some of the critical parameters that influence these processes.

REFERENCES

Akiyama, S., Suzuki, T., Sumino, Y., Nako, Y., and Fukuda, H. (1972). Production of citric acid from n-paraffins by fluoroacetate sensitive mutant strains of *Candida lipolytica*. *Agr. Biol. Chem. Jpn. 36*: 339-341.

Aries, V., and Kirsop, B. H. (1977). Sterol synthesis in relation to growth and fermentation by brewing yeasts inoculated at different concentrations. *J. Inst. Brew. 83*: 220-223.

Bailey, R. B., and Woodward, A. (1984). Isolation and characterisation of a pleiotropic glucose repression resistant mutant of *Saccharomyces cerevisiae*. *Mol. Gen. Genet. 193*: 507-512.

Ball, W. J., and Atkinson, D. E. (1975). Adenylate charge in *Saccharomyces cerevisiae* during starvation. *J. Bacteriol. 121*: 975-982.

Barford, J. P. (1987). The technology of aerobic growth. In *Yeast Biotechnology*, D. R. Berry, I. Russell, and G. G. Stewart (Eds.). George Allen & Unwin, London, pp. 200-230.

Barford, J. P., and Hall, R. J. (1979). An examination of the Crabtree effect in *Saccharomyces cerevisiae*, the role of respiratory adaptation. *J. Gen. Microbiol. 114*: 267-275.

Barnett, J. A. (1976). Utilisation of sugars by yeasts. *Adv. Carbohydrate Chem. Biochem. 32*: 125-234.

Berry, D. R. (1983). Ascospore formation in yeast. In *Fungal Differentiation*, J. E. Smith (Ed.). Marcel Dekker, New York, pp. 147-174.

Berry, D. R., and Brown, C. (1987). Physiology of yeast growth. In *Yeast Biotechnology*, D. R. Berry, I. Russell, and G. G. Stewart (Eds.). George Allen & Unwin, London.

Berry, D. R., and Watson, D. C. (1987). Production of organoleptic compounds. In *Yeast Biotechnology*, D. R. Berry, I. Russell, and G. G. Stewart (Eds.). George Allen & Unwin, London , pp. 345-368.

Botstein, D., and Davis, R. W. (1982). Principles and practices of recombinant DNA research with yeast. In *Molecular Biology of the Yeast Saccharomyces*, J. N. Strathern, E. W. Jones, and J. R. Broach (Eds.). Cold Spring Harbor Monograph Series, New York, 607-636.

Brown, S. W., and Oliver, S. G. (1982). Isolation of ethanol tolerant mutants by continuous selection. *Eur. J. Appl. Microbiol. Biotechnol. 16*: 119-122.

Bruinenberg, P. M., van Dijken, J. P., and Scheffers, W. A. (1983a). A theoretical analysis of NADPH production and consumption in yeast. *J. Gen. Microbiol. 126*: 953-964.

Bruinenberg, P. M., van Dijken, J. P., and Scheffers, W. A. (1983b). An enzymic analysis of NADPH production and consumption in *Candida utilis*. *J. Gen. Microbiol. 129*: 265-271.

Bruinenberg, P. M. De Bot, P. H. M., van Dijken, J. P., and Scheffers, W. A. (1984). NADH linked aldose reductase: The key to alcoholic fermentation of xylose by yeast. *Appl. Microbiol. Biotechnol. 19*: 256-260.

Burrows, S. (1979). Bakers yeast. In *Economic Microbiology*, vol. 4, *Microbial Biomass*, A. H. Rose (Ed.). Academic Press, London, pp. 31-64.

Cabezon, T., DeWilde, M., Herion, P., Lorain, R., and Bollen, A. (1984). Expression of human α antitrypsin cDNA in the yeast *Saccharomyces cerevisiae*. *Proc. Natl. Acad. Sci. USA 81*: 6594-6598.

Campbell-Burke, S. L., and Shulman, R. G. (1987). High resolution NMR studies of *Saccharomyces cerevisiae*. *Annu. Rev. Microbiol. 41*: 595-616.

Cartledge, T. G. (1987). Substrate utilisation by yeast; non-carbohydrate substrates. In *Yeast Biotechnology*, D. R. Berry, I. Russell, and G. G. Stewart (Eds.). George Allen and Unwin, London, pp. 311-342.

Carty, C. E., Kouach, F. Z., McAteer, W. J., and Maigetter, R. Z. (1984). Hepatitis B surface antigen production during fermentation by a recombinant yeast. Abstracts Annual Meeting Society for General Microbiology, vol. 194.

Casey, G. R., Magnus, C. A., and Ingledew, W. M. (1983). High gravity brewing: Nutrient enhanced production of high concentrations of ethanol by brewing yeast. *Biotechnol. Lett.* 5(6): 429-435.

Caulcott, C. A., and Rhodes, M. (1986). Temperature induced synthesis of recombinant proteins. *Trends Biotechnol. 4*: 142-146.

Chevalier, M. R., and Aigle, M. (1979). Qualitative detection of penicillinase produced by yeast strains carrying chimeric yeast-coli plasmids. *FEBS Lett. 108*: 179-180.

Ciriacy, M. (1978). A yeast mutant with glucose resistant formation of mitochondrial enzymes. *Mol. Gen. Genet. 129*: 329-335.

Cysewski, G. R., and Wilke, C. R. (1976). Utilisation of cellulosic materials through enzymatic hydrolysis. I. Fermentation of hydrolysate to ethanol and single cell protein. *Biotechnol. Bioeng. 18*: 1297-1313.

Cysewski, G. R., and Wilke, C. R. (1977). Rapid ethanol fermentations using vacuum and cell recycle. *Biotechnol. Bioeng. 19*: 1125-1143.

Dawes, I. W. (1986). A more extensive role for *Schizosaccharomyces pombe* in biotechnology. *Trends Biotechnol. 4*(Feb.): 30-31.

De Lorencez, M. A., and Kappeli, O. (1987). Regulation of gluconeogenic enzymes during the cell cycle of *Saccharomyces cerevisiae* growing in a chemostat. *J. Gen. Microbiol. 133*: 2517-2522.

Dunnill, P., and Rudd, M. (1984). *Biotechnology and British Industry*. A report to the Biotechnology Directorate of the Science and Engineering Research Council, Swindon, U.K.

Edens, L., Born, I., Lederboer, A. M., Marjolain, Y. T., Visset, C., and Verrips, C. T. (1984). Synthesis and processing of the plant protein thaumatin in yeast. *Cell 37*: 629-633.

Edgley, M., and Brown, A. D. (1978). Response of xerotolerant and non-xerotolerant yeasts to water stress. *J. Gen. Microbiol. 104*: 343-345.

Edgley, M., and Brown, A. D. (1983). Yeast water relations: Physiological changes induced by solute stress in *Saccharomyces cerevisiae* and *Saccharomyces rouxii*. *J. Gen. Microbiol. 129*: 3453-3463.

Engan, S. (1981). Beer composition: Volatile substances. In *Brewing Sciences*, vol. 2, J. R. A. Pollock (Ed.). Academic Press, London, pp. 93-165.

Fiechter, A., Fuhrmann, G. F., and Kappeli, O. (1981). Regulation of glucose metabolism in growing yeast cells. *Adv. Microb. Physiol. 22*: 123-183.

Fogel, S., and Welch, J. W. (1982). Tandem gene amplification mediates copper resistance in yeast. *Proc. Natl. Acad. Sci. USA 79*: 5432-5436.

Fraenkel, D. G. (1982). Carbohydrate metabolism. In *Molecular Biology of the Yeast* Saccharomyces, J. N. Strathern, E. W. Jones, and J. R. Broach (Eds.). Cold Spring Harbor Monographs, New York, pp. 38-57.

Franco, C. M. M., Smith, J. E., and Berry, D. R. (1984). Effect of nitrogen and phosphate on the levels of intermediates in baker's yeast grown in continuous culture. *J. Gen. Microbiol. 130*: 2465-2472.

Frohlich, K. U., Entian, K. D., and Mecke, D. (1984). Cloning and restriction analysis of the hexokinase gene of the yeast *Saccharomyces cerevisiae*. *Mol. Gen. Genet. 194*: 144-148.

Futchers, A. B., and Cox, B. S. (1984). Copy number and stability of 2 μm circle based artificial plasmids of *Saccharomyces cerevisiae*. *J. Bacteriol. 157*: 283-290.

Ghose, T. K., and Tyagi, R. D. (1979). Rapid ethanol fermentation of cellulose hydrolysate. II. Product and substrate inhibition and optimisation of fermenter design. *Biotechnol. Bioeng. 21*: 1401-1420.

Goff, C. G., Mair, D. T., Kohno, T., Gravius, T. C., Smish, R. A., Yamaski, E., and Taunton-Rigby, J. (1984). Expression of calf prochymosin in *Saccharomyces cerevisiae*. *Gene 27*: 35-46.

Greenshields, R. N., and Smith, E. C. (1974). Tubular reactors in fermentation. *Proc. Biochem. 9*(4): 11-17.

Guijarro, J. M., and Lagunas, R. (1984). *Saccharomyces cerevisiae* does not accumulate ethanol against a concentration gradient. *J. Bacteriol. 160*: 874-878.

Gunge, N. (1983). Yeast-DNA plasmids. *Annu. Rev. Microbiol. 37*: 253-276.

Hinchliffe, E., and Box, W. (1984). Expression of the cloned endo-1,3-1,4-β-glucanase gene of *Bacillus subtilis* in *Saccharomyces cerevisiae*. *Curr. Genet. 8*: 471-476.

Hinnen, A., Hicks, J. B., and Fink, G. R. (1978). Transformation of yeast. *Proc. Natl. Acad. Sci. USA 75*: 1129-1933.

Hitzeman, R. A., Chen, C. Y., Hagie, F. E., Patzer, E. G., Lin, C. C., Estell, P. A., Miller, J. V., Yaffe, A., Kleid, D. G., Levinson, A. D., and Opperman, H. (1983a). Expression of hepatitis B virus surface antigens in yeast. *Nucleic Acids Res. 11*: 2745-2763.

Hitzeman, R. A., Leung, D. W., Perry, L. T., Kohr, W. J., Levine, H. L., and Goeddel, D. W. (1983b). Secretion of human interferons by yeast. *Science 21*: 620-625.

Hjortso, M. A., and Bailey, J. E. (1984). Plasmid stability in budding yeast populations: Steady state growth with selection pressure. *Biotechnol. Bioeng. 26*: 528-536.

Ingram, L. A., and Bukkte, T. M. (1984). Effect of alcohol on microorganisms. *Adv. Microbiol. Physiol. 25*: 253-303.

Innes, M. A., Holland, M. J., McCabe, P. C., Cole, G. E., Wittman, V. P., Tal, R., Watt, K. W. K., Gelford, D. H., Holland, J. P., and Meade, J. H. (1985). Expression glycosylation and secretion of an *Aspergillus* glucoamylase by *Saccharomyces cerevisiae*. *Science 228*: 21-26.

Jabber, M. A., Sisasubramanian, N., and Nayak, D. P. (1985). Influenza viral haemagglutinin is expressed and glycosylated in the yeast *Saccharomyces cerevisiae*. *Proc. Natl. Acad. Sci. USA 82*: 2019-2023.

Jones, R. M., Russell, I., and Stewart, G. G. (1987). Classical genetic and protoplast fusion techniques in yeast. In *Yeast Biotechnology*, D. R. Berry, I. Russell, and G. G. Stewart (Eds.). George Allen & Unwin, London, pp. 55-79.

Jones, R. P., and Greenfield, P. F. (1984). A review of yeast ionic nutrition, Part 1: Growth and fermentation requirements. *Proc. Biochem. April*: 48-60.

Jones, R. P., and Greenfield, P. F. (1986). Role of water activity in ethanol fermentations. *Biotechnol. Bioeng. 28*: 29-40.

Jones, R. P., Pamment, N., and Greenfield, P. F. (1981). Alcohol fermentation by yeast, the effect of environment and other variables. *Proc. Biochem. April/May:* 42.

Kappeli, O. (1986). Regulation of carbon metabolism in *Saccharomyces cerevisiae* and related yeasts. *Adv. Microb. Physiol. 28*: 170-209.

Kawamoto, S., Tanaka, A., Yamamura, Y., Teranishi, Y., Fukui, S., and Osumi, M. (1977). Microbody of n-alkane grown yeast, enzyme localization in the isolated microbody. *Arch. Microbiol. 112*: 1-8.

Kingsman, S. M., Kingsman, A., Dobson, M. J., Mellor, J., and Roberts, N. A. (1985). Heterologous gene expression in *Saccharomyces cerevisiae*. In *Biotechnology & Genetic Engineering*, vol. 3, G. E. Russell (Ed.). Intercept, Newcastle-upon-Tyne, pp. 377-416.

Kirsop, B. H. (1982). Developments in beer fermentation. In *Topics in Enzyme and Fermentation Biotechnology*, vol. 3, A. Wiseman (Ed.). Ellis Horwood, Chichester, pp. 80-131.

Kleinman, M. J., Gingold, E. B., and Stanbury, P. F. (1986). The stability of the yeast plasmid pJDB248 depends upon the growth rate of the culture. *Biotechnol. Lett. 8*: 225-230.

Klotz, L. C. (1983). Overproduction of proteins in recombinant organisms. *Ann. N.Y. Acad. Sci. 413:* 1-11.

Kuenzi, M. T., and Fiechter, A. (1972). Regulation of carbohydrate composition of *Saccharomyces cerevisiae* under growth limitations. *Arch. Mikrobiol. 84*: 254-265.

Kuraishi, M., Terao, I., Ohkauchi, H., Matsuda, N., and Nagai, P. (1979). Single cell protein-process development with methanol as substrate. *Doechema Monographs, Proc. XII Int. Cong. Microbiol. 83*: 11-24.

Lavery, M., and Berry, D. R. (1983). Amylolytic activities during Scotch Whisky fermentation. In *Current Developments in Malting, Brewing and Distilling*, F. G. Priest and I. Campbell (Eds.). Institute of Brewing, London, pp. 293-296.

Leao, C., and van Uden, N. (1982). Effect of ethanol and other alkanols on the kinetics and the activation parameters of thermal death in *Saccharomyces cerevisiae*. *Biotechnol. Bioeng. 24*: 1581-1590.

Lee, G. M., Song, K. B., Rhee, S. I., and Han, M. H. (1986). Plasmid maintenance and growth of recombinant *Saccharomyces cerevisiae* producing hepatitis B virus surface antigen. *Biotechnol. Lett. 8*: 385-390.

Lemontt, J. L., Wei, C. M., and Dackowski, W. R. (1985). Expression of active human uterine tissue plasminogen activator in yeast. *DNA 4*: 419-428.

Lindsay, J. H., and Larson, J. W. (1975). Large volume tanks—A review. *Master Brew. Assoc. Am. Technol. Q. 12:* 264-272.

Linko, P., and Linko, Y. Y. (1984). Industrial applications of immobilised cells. *CRC Crit. Rev. Biotechnol. 11*: 289-338.

Litchfield, J. H. (1979). Production of single cell protein for use in food or feed. In *Microbial Technology, Microbial Processes*, vol. 1, 2nd ed., H. J. Pepper and H. Perlman (Eds.), pp. 93-105.

Maiorella, B. L. (1986). Ethanol. In *Comprehensive Biotechnology*, vol. 3, M. Moo Young (Ed.). Pergamon Press, Toronto.

Maiorella, B. L., Blanch, H. W., and Wilke, C. R. (1984). Feed component inhibition of ethanolic fermentations by *Saccharomyces cerevisiae*. *Biotechnol. Bioeng. 26*: 1155-1166.

Marchal, R., Odile, C., and Metche, M. (1977). Production of citric acid from n-paraffins by *Saccharomycopsis lipolytica*: Kinetics and balance of the fermentation. *Eur. J. Appl. Microbiol. 4*: 111-123.

Margaritis, A., and Merchant, F. J. A. (1984). Advances in ethanol production using immobilised cell systems. *CRC Crit. Rev. Biotechnol. 1*: 339-393.

Margaritis, A., and Merchant, F. J. A. (1987). Technology of anaerobic yeast growth. In *Yeast Biotechnology*, D. R. Berry, I. Russell, and G. G. Stewart (Eds.). George Allen & Unwin, London, pp. 231-276.

Marquet, M., Alouani, S., and Brown, S. W. (1986). Plasmid stability during continuous culture in a *Saccharomyces* double mutant transformed by a plasmid carrying a eukaryotic gene. *Biotechnol. Lett. 8*: 535-540.

Mead, D. J., Gardner, D. C. J., and Oliver, S. G. (1986). Enhanced stability of a 2μ-based recombinant plasmid in diploid yeast. *Biotechnol. Lett. 8*: 391-396.

Mellor, J., Dobson, M. J., Roberts, N. A., Tuite, M. F., Erntage, J. S., White, S., Lowe, P. A., Patel, T., Kingsman, A., and Kingsman, S. M. (1983). Efficient synthesis of enzymically active calf chymosin in *Saccharomyces*. *Gene 24*: 1-14.

Miller, D. G., Griffiths-Smith, K., Algar, E., and Scopes, R. K. (1982). Activity and stability of glycolytic enzymes in the presence of ethanol. *Biotechnol. Lett. 4*(9): 601-607.

Miyajima, A., Bond, M. W., Otsu, K., Aria, K., and Aria, N. (1985). Secretion of mature mouse interlukin-2 by *Saccharomyces cerevisiae*: Use of a general secretion vector containing promotor and leader sequences of the mating pheromone α-factor. *Gene 37*: 115-161.

Mortimer, R. K., and Hawthorne, D. C. (1975). Genetic mapping in yeast. In *Methods in Cell Biology*, vol. 11, D. M. Prescott (Ed.). Academic Press, New York, pp. 221-234.

Murray, C. R., Barich, T., and Taylor, D. (1984). The effect of yeast storage conditions on subsequent fermentations. *Master Brewers Assoc. Am. Technol. 21*: 189-194.

Nagamune, T., Indo, I., and Inone, I. (1981). Effects of medium composition on yeast physiological activities for ethanol production. In *Advances in Biotechnology*, vol. 2, *Fuels, Chemicals, Food and Waste Treatment*, M. Moo Young (Ed.). Pergamon Press, Toronto, pp. 219-224.

Nagashima, M., Azuma, S., Noguchi, K., Inuzuki, J., and Sanejima, H. (1984). Continuous ethanol production using immobilised yeast cells. *Biotechnol. Bioeng. 26*: 992-997.

Oda, G., Samejima, H., and Yamada, T. (1983). Continuous alcohol fermentation technologies using immobilised yeast cells. Paper presented to *Biotechnology 83*.

Panchal, C. J. (1987). Yeast plasmids and transformations. In *Yeast Biotechnology*, D. R. Berry, I. Russell, and G. G. Stewart (Eds.). George Allen & Unwin, London, pp. 80-103.

Panchal, C. J., and Stewart, G. G. (1980). The effect of osmotic pressure on the production and excretion of ethanol and glycerol by a brewing yeast strain. *J. Inst. Brewing 86*: 207-211.

Parent, S. A., Fenimore, C. M., and Bostian, K. A. (1985). Vector systems for the expression, analysis and cloning of DNA sequences in *Saccharomyces cerevisiae*. *Yeast 1*: 83-138.

Patel, G. B., and Ingledew, W. M. (1973). Internal carbohydrates of *S. carlsbergensis* during commercial lager brewing. *J. Inst. Brew. 79*: 392-396.

Peppler, H. J. (1979). Production of yeast and yeast products. In *Microbial Technology*, vol. 2, H. J. Peppler and D. Perlman (Eds.). Academic Press, New York, pp. 157-187.

Peppler, H. J. (1983). Ventures in yeast utilisation. *Annu. Rep. Ferment. Proc. 6*: 237-251.

Pollock, J. R. A. (1981). *Brewing Science*. Academic Press, London.

Portno, A. D. (1979). Continuous brewing in the fermentation industry—the future outlook. In *Eur. Brew. Conv. Fermentation and Storage Symposium Monograph V Zoeterwoude*, November 1978, pp. 145-154.

Primrose, S. B., Derbyshire, P., Jones, I. M., Nugent, M., and Tacon, W. (1983). Hereditary instability of recombinant DNA molecules. In *Bioactive Microbial Products*, vol. 2, *Development and Production*, L. J. Nesbit and D. J. Winstanley (Eds.). Academic Press, London, pp. 63-77.

Quain, D. R., Thurston, P. A., and Tubb, R. S. (1981). The structure and storage carbohydrates of *Saccharomyces cerevisiae*: Changes during the fermentation of wort and a role for glycogen catabolism in lipid biosynthesis. *J. Inst. Brew. 87*: 108-111.

Ratledge, C. (1982). Microbial oils and fats: An assessment of their commercial potential. In *Progress in Industrial Microbiology*, vol. 16, M. J. Bull (Ed.). Elsevier, Amsterdam, pp. 119-206.

Ratledge, C., and Boulton, C. A. (1986). Fats and oils. In *Comprehensive Biotechnology*, vol. 3, M. Moo Young (Ed.), pp. 983-1003.

Reed, G. (1982). Production of bakers' yeast. In *Industrial Microbiology*, G. Reed (Ed.). AVI, Westport, Connecticut, pp. 593-633.

Rickard, P. A. D., and Hogan, C. B. J. (1978). The effect of the glyoxylic acid cycle on the respiratory quotient of *Saccharomyces* sp. *Biotechnol. Bioeng. 20*: 1111-1115.

Rosen, K. (1987). Production of bakers' yeast. In *Yeast Biotechnology*, D. R. Berry, I. Russell, and G. G. Stewart (Eds.). George Allen & Unwin, London, pp. 471-500.

Rothstein, S. J., Lazarus, C. N., Smith, W. E., Baulcombe, D. C., and Gatenby, A. A. (1984). Secretion of wheat α-amylase expressed in yeast. *Nature 12*: 663-665.

Shacklady, C. A. (1975). Yeast growth on hydrocarbons. In *International Biological Programme*, vol. 4, *Food Protein Sources*. Cambridge University Press, Cambridge.

Siegel, R., and Ryu, D. D. Y. (1985). Kinetic study of instability of recombinant plasmid pPLc231trp A1 in *E. coli* using two stage continuous culture system. *Biotechnol. Bioeng. 27*: 28-33.

Skipper, N., Sutherland, M., Davis, R. W., Kilburn, D., Miller, R. C., Warren, A., and Wong, R. (1985). Secretion of a bacterial cellulase by yeast. *Science 230*: 958-960.

Smith, E. L., and Greenshields, R. N. (1973). Tower fermentation systems: Their application in continuous brewing. *Biotechnol. Bioeng. Symp. 4*: 519-533.

Sols, A., Gancedo, C., and Delafuente, G. (1971). Energy yielding metabolism in yeast. In *The Yeasts*, vol. 2, A. H. Rose and J. S. Harrison (Eds.). Academic Press, New York, pp. 271-367.

Sonnleitner, B., Straessle, C., and Fiechter, A. (1987). A structured and segregated model: Explanation and description of the spontaneous synchronisation of *Saccharomyces cerevisiae* at low dilution rates. In *Proceedings of the 4th European Congress of Biotechnology*, vol. 3, O. M. Neijssel, R. R. van der Meer, and K. C. A. M. Luyben (Eds.). Elsevier, Amsterdam, pp. 76-79.

Srienc, F., Campbell, J. L., and Bailey, J. E. (1986). Analysis of unstable recombinant *Saccharomyces cerevisiae* population growth in selective medium. *Biotechnol. Bioeng. 28*: 996-1006.

Stepien, P. P., Brousseau, R., Wu, R., Narang, S., and Thomas, D. (1983). Synthesis of human insulin gene. VI. Expression of the synthetic pro-insulin gene in yeast. *Gene 24*: 289-297.

Stewart, G. G., and Russell, I. (1987). Biochemical and genetic control of sugar and carbohydrate metabolism in yeast. In *Yeast Biotechnology*, D. R. Berry, I. Russell, and G. G. Stewart (Eds.). George Allen & Unwin, London, pp. 277-310.

Suomalainen, H., and Oura, H. (1971). Yeast nutrition and solute uptake. In *The Yeasts*, vol. 2, A. H. Rose and J. S. Harrison (Eds.). Academic Press, London, pp. 3-74.

Takagi, A., Chua, E. N., Boonchird, C., Harashima, S., and Oshima, Y. (1985). Constant copy number of plasmids in *Saccharomyces cerevisiae* hosts with different ploidies. *Appl. Microbiol. Biotechnol. 23*: 123-129.

Tingle, M., Klar, A. J. S., Henry, S. A., and Halvorson, H. O. (1973). Ascospore formation in yeast. *Soc. Gen. Microbiol. Symp. 23*: 209-243.

Tubb, R. S. (1984). Genetics of ethanol producing microorganisms. *CRC Crit. Rev. Biotechnol. 1*(3): 241-261.

Urdea, M. S., Merryweather, J. P., Mullenbach, G. T., Coit, D., Haberlain, U., Valenzuela, P., and Barr, P. J. (1983). Chemical synthesis of a gene for human epidermal growth factor urogastrone and its expression in yeast. *Proc. Natl. Acad. Sci. USA 80*: 7461-7465.

Valenzuela, P., Coit, D., Medina-Selby, M. A., Kuo, C. H., Nest, G. V., Burke, R. L., Bull, P., Urdea, M. S., and Graves, P. V. (1985). Antigen engineering in yeast: Synthesis and assembly of hybrid hepatitis B surface antigen-herpes simplex lgD particles. *Biotechnology 3*: 323-326.

Van Brunt, J. V. (1986). Fungi the perfect hosts. *Biotechnology 4*(12): 1057-1062.

Van Dijken, J. P., and Scheffers, W. A. (1984). Studies on alcohol fermentation in yeast. *Prog. Ind. Microbiol. 20*: 497-506.

Waites, M. J., and Quayle, J. R. (1980). Dihydroxyacetone: A product of xylulose 5-phosphate dependent fixation of formaldehyde by methanol grown *Candida boidinii*. *J. Gen. Microbiol. 118*: 321-327.

Wales, D. S. T., Cartledge, T. G., and Lloyd, D. (1980). Effect of glucose repression and anaerobiosis on the activities and subcellular distribution of tricarboxylic acid cycle and associated enzymes in *Saccharomyces carlsbergensis*. *J. Gen. Microbiol. 116*: 93-98.

Walmsley, R. M., Gardner, D. C. J., and Oliver, S. G. (1983). Stability of a cloned gene in yeast grown in chemostat culture. *Mol. Gen. Genet. 192*: 361-365.

Webster, T. D., and Dickson, R. C. (1983). Direct selection of *Saccharomyces cerevisiae* resistant to the antibiotic G418, following transformation with a DNA vector carrying kanamycin-resistance of Tn903. *Gene 26*: 243-252.

Zanetti, R. (1984). Breathing new life into single-cell protein. *Chem. Eng. 91*: 18-21.

Index

D

E